NATURAL ORGANIC MATTER IN WATER

NATURAL ORGANIC MATTER IN WATER

Characterization and Treatment Methods

MIKA SILLANPÄÄ
Laboratory of Green Chemistry
Lappeenranta University of Technology
Sammonkatu 12
FI-50130 Mikkeli
Finland

IWA PUBLISHING

ELSEVIER

Amsterdam • Boston • Heidelberg • London
New York • Oxford • Paris • San Diego
San Francisco • Singapore • Sydney • Tokyo
Butterworth-Heinemann is an imprint of Elsevier

BH

Butterworth-Heinemann is an imprint of Elsevier
The Boulevard, Langford Lane, Kidlington, Oxford, OX5 1GB, UK
225 Wyman Street, Waltham, MA 02451, USA

Notices
Knowledge and best practice in this field are constantly changing. As new research and experience broaden our understanding, changes in research methods, professional practices, or medical treatment may become necessary.

Practitioners and researchers must always rely on their own experience and knowledge in evaluating and using any information, methods, compounds, or experiments described herein. In using such information or methods they should be mindful of their own safety and the safety of others, including parties for whom they have a professional responsibility.

To the fullest extent of the law, neither the Publisher nor the authors, contributors, or editors, assume any liability for any injury and/or damage to persons or property as a matter of products liability, negligence or otherwise, or from any use or operation of any methods, products, instructions, or ideas contained in the material herein.

Library of Congress Cataloging-in-Publication Data
Sillanpää, Mika E.T.
Natural organic matter in water : characterization and treatment methods / Mika Sillanpää. – First edition.
pages cm
Includes bibliographical references and index.
ISBN 978-0-12-801503-2
1. Organic water pollutants. 2. Water–Purification–Organic compounds removal. 3. Drinking water–Purification. I. Title.
TD427.O7S53 2014
363.6'1–dc23

2014033442

British Library Cataloguing in Publication Data
A catalogue record for this book is available from the British Library

ISBN: 978-0-12-801503-2

For information on all Butterworth-Heinemann publications visit our web site at http://store.elsevier.com

CONTENTS

Contributors *ix*
List of Abbreviations *xi*

1. General Introduction **1**
Mika Sillanpää
References 12

2. Characterization of NOM **17**
Mika Sillanpää, Anu Matilainen, Tanja Lahtinen
2.1 Introduction 18
2.2 General parameters 21
2.3 Isolation and concentration 23
2.4 Fractionation 23
2.5 Chromatographic methods 26
2.6 Spectroscopic methods 32
2.7 Biological testing 40
2.8 Other characterization methods 41
2.9 Conclusions 42
References 44

3. NOM Removal by Coagulation **55**
Mika Sillanpää, Anu Matilainen
3.1 Introduction 56
3.2 Aluminum-based coagulants 61
3.3 Ferric-based coagulants 63
3.4 Inorganic polymer flocculants 63
3.5 Organic polyelectrolytes 66
3.6 Composite coagulants 69
3.7 Novel coagulants 71
3.8 Conclusions 73
References 75

4. NOM Removal by Electrochemical Methods **81**

Mika Sillanpää, Heikki Särkkä, Mikko Vepsäläinen

4.1 Introduction 82
4.2 Principles of EC and EO technologies 83
4.3 EC and EO technologies in NOM removal 92
4.4 Conclusions 103
References 105

5. Membranes **113**

Mika Sillanpää, Sari Metsämuuronen, Mika Mänttäri

5.1 Introduction 114
5.2 Microfiltration 116
5.3 Ultrafiltration 117
5.4 Nanofiltration 132
5.5 Reverse osmosis 143
5.6 Membrane fouling 145
5.7 Conclusions 149
References 151

6. NOM Removal by Advanced Oxidation Processes **159**

Mika Sillanpää, Anu Matilainen

6.1 Introduction 160
6.2 AOPs in NOM removal 162
6.3 UV light based applications 163
6.4 Fenton processes 170
6.5 Heterogeneous photocatalysis and catalytic oxidation 180
6.6 Ultrasound irradiation and E-beam irradiation 189
6.7 Conclusions 200
References 203

7. NOM Removal by Adsorption **213**

Mika Sillanpää, Amit Bhatnagar

7.1 Introduction 214
7.2 Adsorption 214
7.3 Adsorbents and their characteristics 216
7.4 NOM removal from water by adsorption 218
7.5 Conclusions 233
References 233

8. Ion Exchange **239**

Mika Sillanpää, Irina Levchuk

8.1 Introduction 240
8.2 Removal of NOM from water by ion exchange 242
8.3 Conclusion 249
References 270

9. Integrated Methods **275**

Mika Sillanpää, Sari Metsämuuronen, Anu Matilainen, Mika Mänttäri

9.1 Introduction 275
9.2 Coupling coagulation with other processes 275
9.3 Coupling membrane technology with other processes 279
9.4 Conclusions 294
References 294

Bibliography *303*
List of Journal Titles with Abbreviations *351*
Index *355*

CONTRIBUTORS

Amit Bhatnagar
University of Eastern Finland, Department of Environmental Science, Faculty of Science and Forestry, P.O. Box 1627, 70211 Kuopio, Finland

Tanja Lahtinen
University of Jyväskylä, Department of Chemistry, Organic Chemistry, Survontie 9 B, 40500 Jyväskylä, Finland

Irina Levchuk
Lappeenranta University of Technology, LUT Faculty of Technology, LUT Chemtech, Laboratory of Green Chemistry, Sammonkatu 12, 50130 Mikkeli, Finland

Mika Mänttäri
Lappeenranta University of Technology, LUT Faculty of Technology, LUT Chemtech, Skinnarilankatu 34, 53850 Lappeenranta, Finland

Anu Matilainen
Finnish Safety and Chemicals Agency, Kalevantie 2, 33100 Tampere, Finland

Sari Metsämuuronen
Lappeenranta University of Technology, LUT Faculty of Technology, LUT Chemtech, Skinnarilankatu 34, 53850 Lappeenranta, Finland

Heikki Särkkä
Lappeenranta University of Technology, LUT Faculty of Technology, LUT Chemtech, Laboratory of Green Chemistry, Sammonkatu 12, 50130 Mikkeli, Finland

Mika Sillanpää
Lappeenranta University of Technology, LUT Faculty of Technology, LUT Chemtech, Laboratory of Green Chemistry, Sammonkatu 12, 50130 Mikkeli, Finland

Mikko Vepsäläinen
CSIRO Mineral Resources Flagship, Box 312, Clayton South, VIC 3169, Australia

LIST OF ABBREVIATIONS

AC	Activated carbon
AFM	Atomic force microscopy
ALD	Atomic layer deposition
AOC	Assimilable organic carbon
AOPs	Advanced oxidation processes
APAMs	Anionic polyacrylamides
APANFs	Aminated polyacrylonitrile fibers
ATR	Attenuated total reflection
BAC	Biologically activated carbon
BDD	Boron-doped diamond
BDOC	Biodegradable dissolved organic carbon
BIEC	Doubled EC
CNT	Carbon nanotube
COD	Chemical oxygen demand
CPAMs	Cationic polyacrylamides
CSZ	Chitosan/zeolite composite
CUR	Carbon usage rate
DBP	Disinfection by-product
DBPFP	Disinfection by-product formation potential
DOC	Dissolved organic carbon
DOM	Dissolved natural organic matter
DSA	Dimensional stable anode
EBCT	Empty bed contact times
EC	Electrocoagulation
EEM	Excitation-emission matrix
EF	Electroflotation
EO	Electro-oxidation
eMBR	Electro-MBR
EMT-BIEC	Electromagnetic treatment–doubled EC
EPS	Extracellular polymeric substance
ESI	Electrospray ionization
FA	Fulvic acids
FIX	Fluidized ion exchange
FIFFF	Flow field-flow fractionation
FTICR-MS	Fourier transform ion cyclotron resonance mass spectrometry
FTIR	Fourier transform infrared
FP	Formation potential
GAC	Granular activated carbon
GC	Gas chromatography
HA	Humic acids
HAAs	Haloacetic acids
HDFe	Iron-impregnated activated carbon
HDN	Ammonia-treated activated carbon
HDTMA	Hexadecyl trimethyl ammonium
HFCVD	Hot filament chemical vapor deposition
HMM	High molar mass
HMW	High molecular weight

HMBC	Heteronuclear multiple bond correlation
HPAC	Composite polyaluminum chloride
HPLC	High performance liquid chromatography
HPSEC	High performance size exclusion chromatography
HR-MAS	High resolution magic-angle spinning
ICP AES	Inductively coupled plasma atomic emission spectroscopy
IE	Ion exchange
IOCS	Iron oxide coated sand
IOP	Iron oxide particle
IPFs	Inorganic polymer flocculants
IR	Infrared
LC-MS	Liquid chromatography–mass spectrometry
LC-OCD	Liquid chromatography–organic carbon detection
LDH	Layered double hydroxide
LMM	Low molar mass
LMW	Low molecular weight
Mag-PCMAs	Magnetic permanently confined micelle arrays
MBR	Membrane bioreactor
MCBR	Membrane coagulation bioreactor
MF	Microfiltration
MIEX®	Magnetic ion exchange resin
MM	Molar mass
MWCNT	Multiwalled carbon nanotube
MWD	Molecular weight distribution
NEU	(hydrophilic) Neutral
NF	Nanofiltration
NMR	Nuclear magnetic resonance
NOM	Natural organic matter
OCD	Organic carbon detection
ORP	Oxidation reduction potential
PAA	Polyacrylamide
PAC	Powdered activated carbon
PACl	Polyaluminum chloride
PAF-SiC	Polyaluminum ferric silicate chloride
PAFC	Polymer-polyaluminum ferric chloride
PARAFAC	Parallel factor
PAS	Polyaluminum sulfate
PASiC	Polyaluminum silicate chloride
PCA	Principal component analysis
PDADMAC	Polydiallyldimethyl ammonium chloride
PES	Polyethersulfone
PFC	Polyferric chloride
PFS	Polyferric sulfate
PFSiS	Polyferric silicate sulfate
PICl	Polymeric iron chloride
PSF	Polysilicate ferric
PSS	Polystyrene sulfonate
PVDF	Polyvinylidene fluoride
Py-GC-MS	Pyrolysis gas chromatography–mass spectrometry
RO	Reverse osmosis
ROM	Residual organic matter
ROS	Reactive oxygen species

RSM	Response surface method
RPHPLC	Reversed-phase high-performance liquid chromatography
SAM	Self-assembled monolayer
SEC	Size exclusion chromatography
SEM	Scanning electron microscopy
SHA	Slightly hydrophobic acid
SPEF	Solar photoelectro-Fenton
SUVA	Specific UV absorbance
TEM	Transmission electron microscopy
THMFP	Trihalomethane formation potential
THM	Trihalomethane
TOC	Total organic carbon
UF	Ultrafiltration
US	Ultrasonic
UV–Vis	Ultraviolet and visible
VHA	Very hydrophobic acid
VUV	Vacuum ultraviolet

CHAPTER 1

General Introduction

Mika Sillanpää
Lappeenranta University of Technology, LUT Faculty of Technology, LUT Chemtech, Laboratory of Green Chemistry, Sammonkatu 12, 50130 Mikkeli, Finland

Abstract

Natural organic matter (NOM) is a complex matrix of organic materials and a key component in aquatic environments. As a result of the interactions between the hydrologic cycle and the biosphere and geosphere, the water sources of drinking water generally contain NOM. The amount, character, and properties of NOM vary considerably according to the origins of the waters and depend on the biogeochemical cycles of their surrounding environments. Also, the interrelation between NOM and climate change has attracted a great deal of attention in recent research. NOM has a significant impact on many aspects of water treatment, including the performance of unit processes, necessity for and application of water treatment chemicals, and the biological stability of the water. As a result, NOM affects potable water quality as a carrier of metals and hydrophobic organic chemicals and by contributing to undesirable color, taste, and odor problems. Moreover, NOM has been found to be the major contributor to disinfection by-product (DBP) formation. Changes in NOM quantity and quality have a significant influence on the selection, design, and operation of water treatment processes. These changes also cause operational difficulties in water utilities. High seasonal variability and the trend toward elevated levels of NOM concentration pose challenges to water treatment facilities in terms of operational optimization and proper process control. To improve and optimize these processes, it is vital to characterize and quantify NOM at various stages during the purification and treatment process. It is also essential to be able to understand and predict the reactivity of NOM or its fractions during different phases of the treatment. Once the composition and quantity of NOM in the water source has been examined, suitable methods for efficient NOM removal can be applied. No single process alone can be used to treat NOM due to its high variability. The most common and economically feasible process available is coagulation and flocculation followed by sedimentation/flotation and filtration. Other treatment options for NOM removal include magnetic ion exchange resin (MIEX®) techniques, activated carbon filtration, membrane filtration methods, and advanced oxidation processes.

Keywords: Characterization; Climate change; Coagulation; Drinking water; Natural organic matter (NOM); Water analysis; Water treatment.

ABBREVIATIONS

AOP Advanced oxidation process
DBP Disinfection by-product
FA Fulvic acids
FTICR-MS Fourier transform ion cyclotron resonance mass spectrometry
GAC Granulated activated carbon
HA Humic acids
HAAs Haloacetic acids
HMM High molecular mass
HMW High molecular weight
LMM Low molecular mass
LMW Low molecular weight
MIEX® Magnetic ion exchange resin
NF Nanofiltration
NMR Nuclear magnetic resonance
NOM Natural organic matter
Py-GC-MS Pyrolysis gas chromatography-mass spectrometry
SEC Size exclusion chromatography
SUVA Specific UV absorbance
THM Trihalomethane
TOC Total organic carbon
UV–Vis Ultraviolet and visible

Natural organic matter (NOM) is ubiquitous in waters, sediments, and soils. Aquatic NOM is derived both from the breakdown of terrestrial plants and as the by-product of bacteria, algae, and aquatic plants. NOM is defined as a complex matrix of organic materials present in all natural waters.

A wide range of terminology is used to describe NOM in the environment. These terms are listed and comprehensively discussed in recent reviews by Filella (2009) and Uyguner-Demirel and Bekbolet (2011). NOM is a key component in aquatic environments and is present in all fresh waters, particularly surface waters. As a result of the interactions between the hydrologic cycle and the biosphere and geosphere, the water sources of drinking water generally contain NOM. It consists of a range of compounds with a wide variety of chemical charges, from highly charged to uncharged. These also vary widely according to chemical composition and molecular size, molecular weight (the molecular masses of humic substances range from several hundreds to tens of thousands), and structure. NOM compounds are complex mixtures possessing unique combinations of various functional groups, including esteric, phenolic, quinine, carboxylic, hydroxyl, amino, and nitroso, which are usually negatively charged at neutral pH (Gjessing, 1976). Humic substances, which are the major constituents of NOM in waters, are amorphous, dark colored, and acidic in nature. Structurally, they consist of substituted aromatic rings linked by aliphatic chains (Uyguner-Demirel and Bekbolet 2011). NOM also influences the acidity, mobility, and toxicity of metals and organic pollutants and weathering (Winterdahl, 2013).

Water systems often have multiple sources of NOM, thus different organic carbon fractions (Rigobello et al., 2011). The amount, character, and properties of NOM vary considerably according to the origins of the waters, and depend on the biogeochemical cycles of their surrounding environments (Fabris et al., 2008). The factors that determine the composition of NOM are location dependent and include the source of organic matter, the water chemistry, temperature, pH, and biological processes (Leenheer and Croué, 2003). Thus, the character of NOM varies with source and season (Sharp et al., 2006a,b; Fabris et al., 2008; Rigobello et al., 2011; Nkambule et al., 2012). Moreover, the range of organic components in NOM can vary seasonally at the same location (Sharp et al., 2006a,b; Smith and Kamal, 2009), for example, due to rainfall, snowmelt runoff, floods or droughts. Floods and droughts are the main indications of the impact of climate change on water availability and quality. It has been suggested that these changes may explain the increase in the total amount of NOM (Delpla et al., 2009; Evans et al., 2005), which has been observed in several parts of the world during the past 20 years (Eikebrokk et al., 2004; Evans et al., 2005; Korth et al., 2004; De Wit et al., 2007; Monteith et al., 2007; Worral and Burt, 2007, 2009; Couture et al., 2012; Jarvis et al., 2012; Winterdahl, 2013; Gough et al., 2014).

NOM is an integral part of the carbon cycle, which may have indirect climatic effects (Winterdahl, 2013). The interrelation between NOM and climate change has attracted a great deal of attention in recent research (Maurice et al., 2002; Hejzlar et al., 2003; Granskog et al., 2007; Park et al., 2007; Clements et al., 2008; Soh et al., 2008; Porcal et al., 2009; Sulzberger and Durisch-Kaiser, 2009; Bunting et al., 2010; Chen et al., 2012; Pautler et al., 2012; Brezonik and Arnold, 2012; Diem et al., 2013; Ritson et al., 2014). Not only the quantity of NOM, but also its quality, have been observed to change when other important characteristics of NOM, such as specific UV absorbance (SUVA), are taken into account (Eikebrokk et al., 2004). Winterdahl (2013) has emphasized that intraannual NOM variability is often larger than year-to-year changes by several orders of magnitude. Unravelling the controls on intraannual NOM dynamics is thus essential to understanding long-term changes, since climate change could alter NOM dynamics in ways not reflected in interannual trends. Several potential factors, including an increase in air and surface water temperature, rainfall intensity, and atmospheric CO_2 and/or a decrease in acid deposition, have been proposed to explain the increased amount of NOM, but there is yet no scientific consensus on this issue (Delpla et al., 2009).

Tang et al. (2014) demonstrated that NOM has complex effects on the environmental behavior of and removal of heavy metal ions by nanomaterials. The impacts were reported to be controversial, depending both on the type of nanomaterials and metal ions and on geochemical conditions. They also demonstrated that the presence of NOM can modify the mechanisms controlling metal ion removal and transportation by nanomaterials in heterogeneous aquatic environments (Tang et al., 2014). Another emerging field of research is the effect of NOM on nanoparticle aggregation (Louie et al., 2013). Collin et al. (2014) recently concluded that NOM decreased the toxicity and bioaccumulation of nanoparticles. The impact of photooxidation on the optical, electrochemical, and photochemical properties of NOM has recently been investigated systematically (Sharpless et al., 2014). This sheds light on the relationships between NOM aromaticity, redox state, and photoreactivity.

NOM found in natural waters consists of both hydrophobic and hydrophilic components, of which the largest fraction is generally hydrophobic acids, making up approximately 50% of the total organic carbon (TOC) in water (Thurman, 1985). These hydrophobic acids can be described as humic substances comprising (1) humic acids (HA), which are soluble in alkali, but insoluble in acid, (2) fulvic acids (FA), which are soluble in both alkali and acid, and (3) humins, which are insoluble in both alkali and acid. FAs constitute a major fraction of these humic substances, which, while structurally comparable, vary in molecular size and functional group content (Table 1.1). Hydrophobic NOM is rich in aromatic carbon, phenolic structures, and conjugated double bonds, while hydrophilic NOM contains more aliphatic carbon and nitrogenous compounds, such as carbohydrates, sugars, and amino acids. The physical and chemical fractionation of aquatic NOM at specific pH can be used to classify organic solutes into these broadly defined hydrophobic and hydrophilic fractions (Chow et al., 2004; Leenher, 2004; Sharp et al., 2006a,c). While these fractions are more operationally than structurally defined, organic compounds can be assigned to a particular fraction according to their chain length and functional groups (Swietlik et al., 2004; Buchanan et al., 2005). The hypothetical molecular structure of HA is presented in Figure 1.1.

Table 1.1 Common properties of humic acid (HA) and fulvic acid (FA) (Snoeyink and Jenkins, 1980; Xing, 2010; Chamoli, 2013)

Property	HA	FA
Elemental composition (% by weight)		
Carbon	50–60	40–50
Hydrogen	4–6	4–6
Oxygen	30–35	44–50
Nitrogen	2–4	<1–3
Sulfur	1–2	0–2
Solubility in strong acid	Not soluble	Soluble
Apparent molecular weight range (atomic mass units)	Few 100 to several million	180–10,000
Functional group distribution (% of oxygen is indicated in functional groups)		
Carboxyl (–COOH)	14–45	58–65
Phenol (–Ph)	10–38	9–19
Alcohol (–R–OH)	13–15	11–16
Carbonyl (–C=O)	4–23	4–11
Methoxyl (–O–CH_3)	1–5	1–2

Aquatic HAs are larger than aquatic FAs, but FAs have more carboxylic functional groups and oxygen, hence less carbon on a mass basis than HAs (Chamoli, 2013). FAs also are more soluble in water, because they have more polar groups per unit mass than HAs.

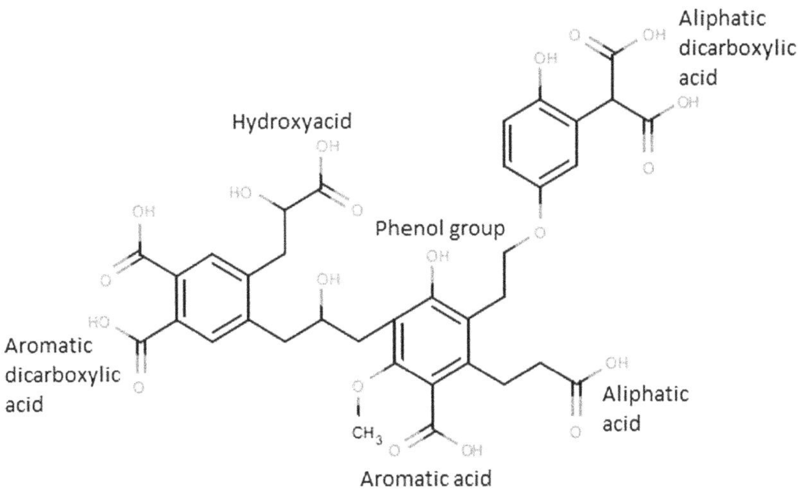

Figure 1.1 Hypothetical molecular structure of humic acid (HA). *Adapted from Duan and Gregory (2003).*

NOM has a significant impact on many aspects of water treatment, including the performance of unit processes, necessity for and application of water treatment chemicals, and the biological stability of the water. As a result, NOM affects potable water quality as a carrier of metals and hydrophobic organic chemicals and by contributing to undesirable color, taste, and odor problems. In addition, NOM necessitates the majority of the coagulant and disinfectant used in water treatment. It tends to interfere with the performance of unit operations, such as biofilm growth on media, causing rapid filter clogging and fast saturation of activated carbon beds. NOM is also responsible for the fouling of membranes. NOM contributes to corrosion, is a source of nutrients for heterotrophic bacteria, and acts as a substrate for bacterial growth in distribution systems (Jacangelo et al., 1995). Moreover, NOM has been found to be the major contributor to disinfection by-product (DBP) formation (Trang et al., 2012). NOM also forms stable complexes with metal ions.

Thus, the removal of NOM from water is an emerging issue, and a robust and efficient treatment technology is needed to address it. The number of publications related to NOM research in indexed journals has been escalating, indicating the scientific community's burgeoning interest in NOM research (Figure 1.2).

Interest in DBPs of water treatment has grown significantly in the past few decades. More than 700 compounds of DBPs have been confirmed;

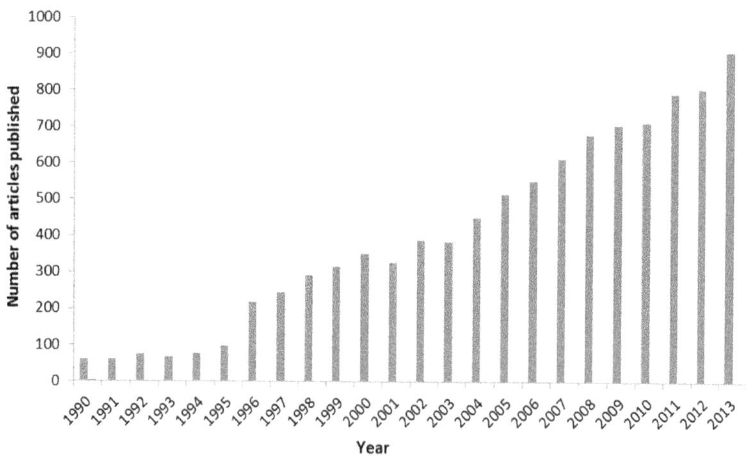

Figure 1.2 Record of the number of publications in indexed journals containing the keyword "natural organic matter" in the abstract between 1990 and 2013. *Scopus, January 27, 2014.*

among them, trihalomethanes (THMs) and haloacetic acids (HAAs) are the two groups found most frequently and in the highest concentrations in drinking waters worldwide (Krasner et al., 2006). Chlorine, ozone, chlorine dioxide, and chloramine are the most common disinfectants used today, and each produces its own suite of chemical DBPs in drinking water (Krasner et al., 2006). Although chlorinated DBPs have been the subject of the greatest concern, brominated DBPs have been considered even more hazardous than their chlorinated counterparts (Singer, 2006), while the formation of iodinated and nitrogen-containing DBPs has been studied with increasing intensity (Krasner et al., 2006; Hua and Reckhow, 2007; Zhao et al., 2008). DBPs have been associated with adverse health effects such as bladder cancer, spontaneous abortions, and birth defects (Singer, 2006; Gough et al., 2014). Thus, their occurrence in drinking water has been regulated in most countries. NOM has generally been considered the main precursor to DBPs, especially hydrophobic and high molecular mass (HMM) NOM, with its high aromatic carbon content (Hua and Reckhow, 2007; Bond et al., 2009; Liu et al., 2007). It has also been observed that hydrophilic and low molecular mass (LMM) NOM play a significant role in DBP formation (Hua and Reckhow, 2007; Bond et al., 2009). On the one hand, bromine and iodine appear more reactive with hydrophilic and LMM fractions of NOM in the formation of THMs and HAAs. On the other hand, chlorine has been shown to react more readily with HMM and hydrophobic NOM compounds (Hua and Reckhow, 2007).

Thus, water treatment should be optimized to remove both hydrophobic and hydrophilic organic compounds, mitigating the formation of DBPs. More efficient, but economical, NOM removal methods are needed to meet stricter drinking water treatment regulations and overcome problems with water quality.

Changes in NOM quantity and quality have a significant influence on the selection, design, and operation of water treatment processes. These changes also cause operational difficulties in water utilities. High seasonal variability and the trend toward elevated levels of NOM concentration pose challenges to water treatment facilities in terms of operational optimization and proper process control. A sustained increase in NOM in raw water will lower the efficiency of water treatment processes and increase the demand for water purification. Changes in the properties of NOM also influence the treatment significantly (Eikebrokk et al., 2004; Sharp et al., 2006a,c). Water treatment facilities thus need to invest in additional NOM removal methods where existing water purification processes become insufficient.

To improve and optimize these processes, it is vital to characterize and quantify NOM at various stages during the purification and treatment process. It is also essential to be able to understand and predict the reactivity of NOM or its fractions during different phases of the treatment. Methods used in the characterization of NOM include resin adsorption, size exclusion chromatography, nuclear magnetic resonance (NMR) spectroscopy, and fluorescence spectroscopy. The amount of NOM in water has been predicted with parameters including ultraviolet and visible (UV–Vis), TOC, and SUVA. More precise methods for determining NOM structures have been developed recently: pyrolysis gas chromatography mass spectrometry, multidimensional NMR techniques, and Fourier transform ion cyclotron resonance mass spectrometry.

Once the composition and quantity of NOM in the water source has been examined, suitable methods for efficient NOM removal can be applied (Nkambule et al., 2012). No single process alone can be used to treat NOM due to its high variability. The development of rapid NOM characterization methods enables the selection of the proper treatment for the water concerned (Nkambule et al., 2012). Thus, NOM characterization and its removal during water treatment are closely connected. Effective selection criteria for precursor removal processes arise from increased knowledge of the nature of the precursor in individual water, facilitating the choice of appropriate technologies for precursor treatment (Bond et al., 2011). One reported variable is the nature of the reactive precursors present. Optimized coagulation treatment may be sufficient in hydrophobic waters, but if the postcoagulation residual remains reactive in DBP formation, technologies such as ion exchange for carboxylic acid precursors and/or granular activated carbon for hydrophobic precursors and/or nanofiltration for hydrophilic precursors are recommended (Bond et al., 2011).

The most common and economically feasible process available is coagulation and flocculation followed by sedimentation/flotation and filtration. Most of the NOM can be removed by coagulation, although the hydrophilic, low molecular weight (LMW) fractions of NOM are apparently removed less efficiently than the hydrophobic, high molecular weight (HMW) compounds (Jacangelo et al., 1995; Matilainen et al., 2010; Sharp et al., 2006a,c). This preference may be due to the more aromatic character, therefore more hydrophobic nature, of the latter (Sharp et al., 2006a,c). Moreover, the hydrophobic fraction generally has a higher specific colloidal charge; more highly charged fractions are more amenable to removal (Sharp et al., 2006a; Bose and Reckhow, 2007). Thus, LMW and hydrophilic NOM dominate the residual organic matter after coagulation (Zhao et al., 2009; Liu et al., 2007).

Other treatment options for NOM removal include magnetic ion exchange resin (MIEX®) techniques, activated carbon filtration, membrane filtration methods, and advanced oxidation processes (AOPs) (Jacangelo et al., 1995; Singer and Bilyk 2002; Matilainen et al., 2006a,b; Zularisam et al., 2006; Toor and Mohseni 2007; Matilainen and Sillanpää 2010; Parsons and Byrne, 2004; Pera-Titus et al., 2004; Suty et al., 2004; Agustina et al., 2005; Malato et al., 2007; Comninellis et al., 2008; Klavarioti et al., 2009; Malato et al., 2009). Although the MIEX® technique has been reported to remove even hydrophilic NOM, none of the above-named alternative treatment methods successfully remove all the NOM fractions present in raw water. Bioprocesses typically entail the development of a biofilm on a sand or activated carbon filter during water treatment (Bond et al., 2011). Biotreatment is reported to have a significant impact on precursor removal where reactive precursors are readily biodegradable, which is more likely in waters with high amounts of biologically derived NOM (Bond et al., 2011). The pros and cons of some treatment methods of NOM removal are presented in Table 1.2.

Table 1.2 Advantages and disadvantages of treatment methods for natural organic matter (NOM) removal (Jarvis et al., 2008; Bond et al., 2011; Shestakova and Sillanpää, 2013)

Treatment method	Advantages	Disadvantages
Adsorption	High NOM removal efficiency is achievable. Systems are available for various flow rates and concentrations of pollutants. Easy to implement. Removes hydrophobic NOM fraction.	Requires multiple regenerations and partial replacement of adsorbent. In some cases, adsorbent is not regenerated and needs to be disposed of, causing secondary pollution of the environment. Pretreatment of influent from suspended solids is required. Effectiveness depends on the temperature and pH.

Continued

Table 1.2 Advantages and disadvantages of treatment methods for natural organic matter (NOM) removal (Jarvis et al., 2008; Bond et al., 2011; Shestakova and Sillanpää, 2013)—cont'd

Treatment method	Advantages	Disadvantages
Advanced oxidation processes (AOP)	Lower chemical consumption in some AOPs. Complete mineralization of pollutant is achievable. Reactions are often rapid. Unselective oxidants. Harmful to any microorganisms that may be present in the water. Can often be installed in existing water treatment plants. Oxidation of Disinfection by-products (DBPs).	Requires powdered photocatalyst separation from treated water. Poorer performance for NOM removal than ferric salt coagulation. Hydrogen peroxide residual is toxic. Effectiveness depends on the pH value. Requires additional reagents. Dependent on the pH value. Radical scavenging. UV irradiation is a high-energy intensive process. Recalcitrant NOM is poorly removed. Ozone is a toxic gas, thus careful safety monitoring is required. Water is corrosive due to high oxidation power. Short duration of ozone exposure.
Biological	Removes some major NOM fractions that are biologically degradable	Requires a relatively large area. Activated sludge from aerobic treatment is a waste that requires disposal. Requires additional nutrients. Process is sensitive to conditions and concentration variations.

Continued

Table 1.2 Advantages and disadvantages of treatment methods for natural organic matter (NOM) removal (Jarvis et al., 2008; Bond et al., 2011; Shestakova and Sillanpää, 2013)—cont'd

Treatment method	Advantages	Disadvantages
Coagulation	Cost efficient. Conventional method with high levels of NOM removal. Suitable for large molecules.	Sludge production.
Electrochemical methods	Electrocoagulation produces compact flocs. Innovative, inexpensive, and effective. Little or no chemical needed to facilitate NOM removal. Less coagulant is needed in electrocoagulation; consequently, less sludge is formed.	Energy costs may limit practical use. Electrode materials can be expensive for electro-chemical oxidation. Passivation of electrodes due to the presence of oxides and precipitation layers on the electrode surface, dissolving of the electrodes, and low conductivity of the surface waters. Formation of DBPs.
Ion exchange	Proven technology. Potentially highly efficient. Very low DBP formation. Efficient in treating the transphilic fraction of NOM.	Additional treatment stage required.
Membrane technology	Large, strong flocs formed. Nanofiltration is efficient in the removal of low charged amino acids and carbohydrates. Potential to remove low NOM fraction. Can be integrated into other processes.	Energy-intensive process. Additional treatment stage required.

This book has two objectives. First, it investigates the techniques for NOM characterization, including conventional bulk parameters and the most advanced instrumental analytical tools. Second, it assesses the water treatment methods relevant to NOM removal, such as coagulation, electrochemical methods, membrane technology, AOPs, adsorption, and ion exchange.

REFERENCES

Agustina, T., Ang, H., Vareek, V., 2005. A review of synergistic effect of photocatalysis and ozonation on wastewater treatment. J. Photochem. Photobiol. C. 6, 264–273.

Bond, T., Goslan, E., Jefferson, B., Roddick, F., Fan, L., Parsons, S., 2009. Chemical and biological oxidation of NOM surrogates and effect on HAA formation. Water Res. 43, 2615–2622.

Bond, T., Goslan, E.H., Parsons, S.A., Jefferson, B., 2011. Treatment of disinfection by-product precursors. Environ. Technol. 32, 1–21.

Bose, P., Reckhow, D., 2007. The effect of ozonation on natural organic matter removal by alum coagulation. Water Res. 41, 1516–1524.

Brezonik, P., Arnold, W., 2012. Water chemistry: fifty years of change and progress. Environ. Sci. Technol. 46, 5650–5657.

Buchanan, W., Roddick, F., Porter, N., Drikas, M., 2005. Fractionation of UV and VUV pretreated natural organic matter from drinking water. Environ. Sci. Technol. 39, 4647–4654.

Bunting, L., Leavitt, P., Weidman, P., Vinebrooke, R., 2010. Regulation of the nitrogen biogeochemistry of mountain lakes by subsidies of terrestrial dissolved organic matter and the implications for climate studies. Limnol. Oceanogr. 55 (1), 333–345.

Chamoli, U., 2013. Disinfection and Self-sensitized Degradation of Natural Organic Matter (NOM) by TiO_2 Photocatalysis with Visible Light (M.Sc. thesis). University of Calgary, Calgary, Canada, 129 p.

Chen, X., Chuaia, X., Yanga, L., Zhao, H., 2012. Climatic warming and overgrazing induced the high concentration of organic matter in Lake Hulun, a large shallow eutrophic steppe lake in northern China. Sci. Total Environ. 431, 332–338.

Chow, C., Fabris, R., Drikas, M., 2004. A rapid fractionation technique to characterize natural organic matter for the optimization of water treatment processes. J. Water Supply Res. Technol. Aqua 53, 85–92.

Clements, W., Brooks, M., Kashian, D., Zuellig, R., 2008. Changes in dissolved organic material determine exposure of stream benthic communities to UV-B radiation and heavy metals: implications for climate change. Global Change Biol. 14, 2201–2214.

Collin, B., Oostveen, E., Tsyusko, O.V., Unrine, J.M., 2014. Influence of natural organic matter and surface charge on the toxicity and bioaccumulation of functionalized ceria nanoparticles in *Caenorhabditis elegans*. Environ. Sci. Technol. 48, 1280–1289.

Comninellis, C., Kapalka, A., Malato, S., Parsons, S., Poulios, I., Mantzavinos, D., 2008. Advanced oxidation processes for water treatment: advances and trends for R&D. J. Chem. Technol. Biotechnol. 83, 769–776.

Couture, S., Houle, D., Gagnon, C., 2012. Increases of dissolved organic carbon in temperate and boreal lakes in Quebec, Canada. Environ. Sci. Pollut. Res. 19, 361–371.

De Wit, H.A., Mulder, J., Hindar, A., Hole, L., 2007. Long-term increase in dissolved organic carbon in streamwaters in Norway is response to reduced acid deposition. Environ. Sci. Technol. 41, 7706–7713.

Delpla, I., Jung, A.-V., Baures, E., Clement, M., Thomas, O., 2009. Impacts of climate change on surface water quality in relation to drinking water production. Environ. Int. 35, 1225–1233.
Diem, S., von Rohra, M., Heringa, J., Kohlera, H.-P., Schirmera, M., von Gunten, U., 2013. NOM degradation during river infiltration: effects of the climate variables temperature and discharge. Water Res. 47, 6585–6595.
Duan, J., Gregory, J., 2003. Coagulation by hydrolysing metal salts. Adv. Colloid Interface Sci. 100–102, 475–502.
Eikebrokk, B., Vogt, R., Liltved, H., 2004. NOM increase in Northern European source waters: discussion of possible causes and impacts on coagulation/contact filtration processes. Water Sci. Technol. Water Supply 4 (4), 47–54.
Gough, R., Holliman, P., Willis, N., Freeman, C., 2014. Dissolved organic carbon and trihalomethane precursor removal at a UK upland water treatment works. Sci. Total Environ. 468–469, 228–239.
Evans, C., Monteith, D., Cooper, M., 2005. Long-term increases in surface water dissolved organic carbon: observations, possible causes and environmental impacts. Environ. Pollut. 137, 55–71.
Fabris, R., Chow, C., Drikas, M., Eikebrokk, B., 2008. Comparison of NOM character in selected Australian and Norwegian drinking waters. Water Res. 42, 4188–4196.
Filella, M., 2009. Freshwaters: which NOM matters? Environ. Chem. Lett. 7, 21–35.
Gjessing, E., 1976. Physical and Chemical Characteristics of Aquatic Humus. Ann Arbor Science Publishers, Michigan.
Granskog, M., Macdonald, R., Mundy, C.-J., Barber, D., 2007. Distribution, characteristics and potential impacts of chromophoric dissolved organic matter (CDOM) in Hudson Strait and Hudson Bay, Canada. Cont. Shelf Res. 27, 2032–2050.
Hejzlar, J., Dubrovskýb, M., Buchtelec, J., Růžičkac, M., 2003. The apparent and potential effects of climate change on the inferred concentration of dissolved organic matter in a temperate stream (the Malše River, South Bohemia). Sci. Total Environ. 310, 143–152.
Hua, G., Reckhow, D., 2007. Characterization of disinfection byproduct precursors based on hydrophobicity and molecular size. Environ. Sci. Technol. 41, 3309–3315.
Jacangelo, J., DeMarco, J., Owen, D., Randtke, S., 1995. Selected processes for removing NOM: an overview. J. Am. Water Works Assoc. 87 (1), 64–77.
Jarvis, P., Banks, J., Molinder, R., Stephenson, T., Parsons, S., Jefferson, B., 2008. Processes for enhanced NOM removal: beyond Fe and Al coagulation. Water Sci. Technol. Water Supply 8, 709–716.
Jarvis, P., Sharp, E., Pidou, M., Molinder, R., Parsons, S., Jefferson, B., 2012. Comparison of coagulation performance and floc properties using a novel zirconium coagulant against traditional ferric and alum coagulants. Water Res. 46, 4179–4187.
Klavarioti, M., Mantzavinos, D., Kassinos, D., 2009. Removal of residual pharmaceuticals from aqueous systems by advanced oxidation processes. Environ. Int. 35, 402–417.
Korth, A., Fiebiger, C., Bornmann, K., Schmidt, W., 2004. NOM increase in drinking water reservoirs – relevance for drinking water production. Water Sci. Technol. Water Supply 4 (4), 55–60.
Krasner, S., Weinberg, H., Richardson, S., Pastor, S., Chinn, R., Sclimenti, M., Onstad, G., Thruston Jr., A., 2006. Occurrence of a new generation of disinfection by-products. Environ. Sci. Technol. 40, 7175–7185.
Leenheer, J., Croué, J.-P., 2003. Peer reviewed: characterizing aquatic dissolved organic matter. Environ. Sci. Technol. 37, 18A–26A.
Leenher, J., 2004. Comprehensive assessment of precursors, diagenesis, and reactivity to water treatment of dissolved and colloidal organic matter. Water Sci. Technol. Water Supply 4 (4), 1–9.
Liu, S., Lim, M., Chiang, K., Amal, R., Fabris, R., Chow, C., Drikas, M., 2007. A study on the removal of humic acid using advanced oxidation process. Sep. Sci. Technol. 42, 1391–1404.

Louie, S., Tilton, R., Lowry, G., 2013. Effects of molecular weight distribution and chemical properties of natural organic matter on gold nanoparticle aggregation. Environ. Sci. Technol. 47, 4245–4254.

Malato, S., Blanco, J., Alarcón, D., Maldonado, M., Fernández-Ibáñez, P., Gernjak, W., 2007. Photocatalytic decontamination and disinfection of water with solar collectors. Catal. Today 122, 137–149.

Malato, S., Fernández-Ibáñez, P., Maldonado, M., Blanco, J., Gernjak, W., 2009. Decontamination and disinfection of water by solar photocatalysis: recent overview and trends. Catal. Today 147, 1–59.

Matilainen, A., Iivari, P., Sallanko, J., Heiska, E., Tuhkanen, T., 2006a. The role of ozonation and activated carbon filtration in the natural organic matter removal from drinking water. Environ. Technol. 27, 1171–1180.

Matilainen, A., Sillanpää, M., 2010. Removal of natural organic matter from drinking water by advanced oxidation processes: a review. Chemosphere 80, 351–365.

Matilainen, A., Vepsäläinen, M., Sillanpää, M., 2010. Natural organic matter removal by coagulation during drinking water treatment: a review. Adv. Colloid Interface Sci. 159, 189–197.

Matilainen, A., Vieno, N., Tuhkanen, T., 2006b. Efficiency of the activated carbon filtration in the natural organic matter removal. Environ. Int. 32, 324–331.

Maurice, P., Cabaniss, S., Drummond, J., Ito, E., 2002. Hydrogeochemical controls on the variations in chemical characteristics of natural organic matter at a small freshwater wetland. Chem. Geol. 187, 59–77.

Monteith, D.T., Stoddard, J.L., Evans, C.D., de Wit, H.A., Forsius, M., Høgåsen, T., Wilander, A., Skjelkvåle, B.L., Jeffries, D.S., Vuorenmaa, J., Keller, B., Kopácek, J., Vesely, J., 2007. Dissolved organic carbon trends resulting from changes in atmospheric deposition chemistry. Nature 450, 537–540.

Nkambule, T., Krause, R., Haarhoff, J., Mamba, B., 2012. A three step approach for removing organic matter from South African water sources and treatment plants. Phys. Chem. Earth 50–52 (2012), 132–139.

Park, J.-H., Lee, J.-H., Kang, S.-Y., Kim, S.-Y., 2007. Hydroclimatic controls on dissolved organic matter (DOM) characteristics and implications for trace metal transport in Hwangryong River Watershed, Korea, during a summer monsoon period. Hydrol. Processes 21, 3025–3034.

Parsons, S., Byrne, A., 2004. Water treatment applications. In: Parsons, S. (Ed.), Advanced Oxidation Processes for Water and Wastewater Treatment. IWA Publishing, London, pp. 329–346.

Pautler, B., Woods, G., Dubnick, A., Simpson, A., Sharp, M., Fitzimons, S., Simpson, M., 2012. Molecular characterization of dissolved organic matter in glacial ice: coupling natural abundance1H NMR and fluorescence spectroscopy. Environ. Sci. Technol. 46, 3753–3761.

Pera-Titus, M., Garcia-Molina, V., Baños, M., Giménes, J., Esplugas, S., 2004. Degradation of chlorophenols by means of advanced oxidation processes: a general review. Appl. Catal. B Environ. 47, 219–256.

Porcal, P., Koprivnjak, J.-F., Molot, L., Dillon, P., 2009. Humic substances—part 7: the biogeochemistry of dissolved organic carbon and its interactions with climate change. Environ. Sci. Pollut. Res. 16, 714–726.

Rigobello, E., Dantas, A., Di Bernardo, L., Vieira, E., 2011. Influence of the apparent molecular size of aquatic humic substances on colour removal by coagulation and filtration. Environ. Technol. 32, 1767–1777.

Ritson, J.P., Grahamb, N.J.D., Templetonb, M.R., Clark, J.M., Gough, R., Freemand, C., 2014. The impact of climate change on the treatability of dissolved organic matter (DOM) in upland water supplies: a UK perspective. Sci. Total Environ. 473–474, 714–730.

Sharp, E., Jarvis, P., Parsons, S., Jefferson, B., 2006a. Impact of fractional character on the coagulation of NOM. Colloids Surf. A Physicochem. Eng. Aspects 286, 104–111.

Sharp, E., Parsons, S., Jefferson, B., 2006b. Seasonal variations in natural organic matter and its impact on coagulation in water treatment. Sci. Total Environ. 363, 183–194.

Sharpless, C.M., Aeschbacher, M., Page, S.E., Wenk, J., Sander, M., McNeill, K., 2014. Photooxidation-induced changes in optical, electrochemical and photochemical properties of humic substances. Environ. Sci. Technol. 48, 2688–2696.
Shestakova, M., Sillanpää, M., 2013. Removal of dichloromethane from ground and wastewater: a review. Chemosphere 93, 1258–1267.
Singer, P., 2006. DBPs in drinking water: additional scientific and policy considerations for public health protection. J. Am. Water Works Assoc. 98 (10), 73–79.
Singer, P., Bilyk, K., 2002. Enhanced coagulation using a magnetic ion exchange resin. Water Res. 36, 4009–4022.
Smith, E., Kamal, Y., 2009. Optimizing treatment for reduction of disinfection by-product (DBP) formation. Water Sci. Technol. Water Supply 9, 191–198.
Snoeyink, V.L., Jenkins, D., 1980. Water Chemistry. John Wiley & Sons, New York.
Soh, Y., Roddick, F., van Leeuwen, J., 2008. The future of water in Australia: the potential effects of climate change and ozone depletion on Australian water quality, quantity and treatability. Environmentalist 28, 158–165.
Sulzberger, B., Durisch-Kaiser, E., 2009. Chemical characterization of dissolved organic matter (DOM): a prerequisite for understanding UV-induced changes of DOM absorption properties and bioavailability. Aquat. Sci. 71, 104–126.
Suty, H., De Traversay, C., Cost, M., 2004. Applications of advanced oxidation processes: present and future. Water Sci. Technol. 49 (4), 227–233.
Swietlik, J., Dabrowska, A., Raczyk-Stanislawiak, U., Nawrocki, J., 2004. Reactivity of natural organic matter fractions with chlorine dioxide and ozone. Water Res. 38, 547–558.
Tang, W.-W., Zeng, G.-M., Gong, J.-L., Liang, J., Xu, P., Zhang, C., Huang, B.-B., 2014. Impact of humic/fulvic acid on the removal of heavy metals from aqueous solutions using nanomaterials: a review. Sci. Total Environ. 468–469, 1014–1027.
Thurman, E., 1985. Organic Geochemistry of Natural Waters. Martinus Nijhoff/Dr W. Junk Publ., Dordrecht, The Netherlands.
Toor, R., Mohseni, M., 2007. UV-H_2O_2 based AOP and its integration with biological activated carbon treatment for DBP reduction in drinking water. Chemosphere 66, 2087–2095.
Trang, V., Phuong, L., Dan, N., Thanh, B., Visvanathan, C., 2012. Assessment on the trihalomethanes formation potential of Tan Hiep Water Treatment Plant. J. Water Sustainability 2, 43–53.
Uyguner-Demirel, C., Bekbolet, M., 2011. Significance of analytical parameters for the understanding of natural organic matter in relation to photocatalytic oxidation. Chemosphere 84, 1009–1031.
Winterdahl, M., 2013. Intra-annual Variability of Natural Organic Matter in Boreal Streams Patterns and Controls, Faculty of Natural Resources and Agricultural Sciences, Department of Aquatic Sciences and Assessment, Uppsala (Doctoral thesis). Swedish University of Agricultural Sciences, Uppsala, Sweden.
Worrall, F., Burt, T., 2007. Trends in DOC concentrations in Great Britain. J. Hydrol. 346, 81–92.
Worrall, F., Burt, T.P., 2009. Changes in DOC treatability: indications of compositional changes in DOC trends. J. Hydrol. 366, 1–8.
Xing, Y., 2010. Characterization of Dissolved Organic Carbon in Prairie Surface Waters Using Fourier Transform Infrared Spectroscopy (M.Sc. thesis). University of Saskatchewan, Saskatoon, Saskatchewan, Canada.
Zhao, Y.-Y., Boyd, J., Woodbeck, M., Andrews, R., Qin, F., Hrudey, S., Li, X.-F., 2008. Formation of *N*-nitrosamines from eleven disinfection treatments of seven different surface waters. Environ. Sci. Technol. 42, 4857–4862.
Zhao, Z.-Y., Gu, D.-J., Li, H.-B., Li, X.-Y., Leung, K., 2009. Disinfection characteristics of the dissolved organic fractions at several stages of a conventional drinking water treatment plant in South China. J. Hazard. Mat. 172, 1093–1099.
Zularisam, A., Ismail, A., Salim, R., 2006. Behaviours of natural organic matter in membrane filtration for surface water treatment – a review. Desalination 194, 211–231.

CHAPTER 2

Characterization of NOM

Mika Sillanpää*, **Anu Matilainen****, **Tanja Lahtinen**†
*Lappeenranta University of Technology, LUT Faculty of Technology, LUT Chemtech, Laboratory of Green Chemistry, Sammonkatu 12, 50130 Mikkeli, Finland; **Finnish Safety and Chemicals Agency, Kalevantie 2, 33100 Tampere, Finland; †University of Jyväskylä, Department of Chemistry, Organic Chemistry, Survontie 9 B, 40500 Jyväskylä, Finland

Abstract

Worldwide reports over the last few decades have shown that the amount of natural organic matter (NOM) in surface water is continuously increasing, which has an adverse effect on drinking water purification. For many practical and hygienic reasons, the presence of NOM in drinking water is undesirable. Various technologies have been proposed for NOM removal with varying degrees of success. The properties and amount of NOM, however, can significantly affect the process efficiency. To improve and optimize these processes, it is essential to characterize and quantify NOM at various points during purification and treatment. It is also important to be able to understand and predict the reactivity of NOM or its fractions at different stages of the process. Methods used in the characterization of NOM include resin adsorption, size exclusion chromatography, nuclear magnetic resonance (NMR) spectroscopy, and fluorescence spectroscopy. The NOM in water has been quantified with parameters including ultraviolet and visible, total organic carbon, and specific UV-absorbance. More precise methods for determining NOM structures have been developed recently: pyrolysis gas chromatography-mass spectrometry, multidimensional NMR techniques, and Fourier transform ion cyclotron resonance mass spectrometry. This chapter focuses on the methods used for the characterization and quantification of NOM in relation to drinking water treatment.

Keywords: Characterization; Chromatography; Concentration; Fractionation; Natural organic matter (NOM); Spectroscopy.

ABBREVIATIONS

AFM Atomic force microscopy
AOC Assimilable organic carbon
AOPs Advanced oxidation processes
ATR Attenuated total reflectance
BDOC Biodegradable dissolved organic carbon
COD Chemical oxygen demand
DBP Disinfection by-product
DBPFP Disinfection by-product formation potential
DOC Dissolved organic carbon
EEM Excitation–emission matrix

ESI Electrospray ionization
FA Fulvic acids
FIFFF Flow field-flow fractionation
FTICR-MS Fourier transform ion cyclotron resonance mass spectrometry
FTIR Fourier transform infrared
GAC Granular activated carbon
GC Gas chromatography
HA Humic acids
HMW High molecular weight
HMBC Heteronuclear multiple bond correlation
HPLC High performance liquid chromatography
HPSEC High performance size exclusion chromatography
HR-MAS High resolution magic-angle spinning
ICP AES Inductively coupled plasma atomic emission spectroscopy
IR Infrared
LC-MS Liquid chromatography-mass spectrometry
LC-OCD Liquid chromatography-organic carbon detection
MM Molar mass
MWD Molecular weight distribution
NMR Nuclear magnetic resonance
NOM Natural organic matter
OCD Organic carbon detection
PARAFAC Parallel factor
PSS Polystyrene sulfonate
Py-GC-MS Pyrolysis gas chromatography-mass spectrometry
RO Reverse osmosis
RPHPLC Reversed-phase high-performance liquid chromatography
SEC Size exclusion chromatography
SEM Scanning electron microscopy
SHA Slightly hydrophobic acid
SUVA Specific UV-absorbance
TEM Transmission electron microscopy
THM Trihalomethane
TOC Total organic carbon
UV–Vis Ultraviolet and visible
VHA Very hydrophobic acid

2.1 INTRODUCTION

Natural organic matter (NOM) is a complex mixture of organic compounds. Some of this organic matter is negatively charged, and it can possess a wide variety of chemical compositions and molecular sizes (Thurman, 1985; Swietlik et al., 2004).

All disinfection methods (chlorine, ozone, chlorine dioxide, chloramines, and UV radiation) reportedly produce their own suite of disinfection by-products (DBPs) and bioreactive compounds in drinking water (Richardson et al., 2007). Present knowledge and experience show that the hydrophobic and high molecular weight (HMW) compounds of NOM are the most significant precursors to DBP formation (Hua and Reckhow, 2007; Chen et al., 2008). Hydrophilic matter may also play a key role in the formation of new compounds during disinfection, especially in waters with low humic components.

More efficient removal of NOM requires more knowledge of the organic matter present in raw water, and novel methods have been developed for the characterization of these organic compounds. At the same time, existing methods and techniques have been improved. These characterization methods are used to study the composition of NOM prior to treatment and during various stages of the treatment process (Chen et al., 2007; Sarathy and Mohseni, 2007; Her et al., 2008a; Liu et al., 2008; Tercero Espinoza et al., 2009; Zhao et al., 2009; Liu et al., 2010). The diversity of molecules that constitute NOM and the relatively low concentrations of NOM in water often make characterization difficult. There is, therefore, a significant need for methods that can either accurately characterize NOM in these dilute solutions or isolated or concentrated NOM.

The NOM in raw water must be characterized to understand its role in water treatment (Matilainen et al., 2011). Ideally, once the various NOM components and fractions of a raw water source have been determined, the treatment processes that will eliminate the most dominant NOM fractions could be selected. Yet, NOM contains literally thousands of chemical constituents. Thus, it is not realistic to characterize NOM on the basis of individual components, as discussed in the introduction, but it is more practical to identify groups of chemicals with similar properties (Croue et al., 2000).

An alternative approach to the characterization of NOM is to study how it reacts to DBP formation and the occurrence of different DBPs in drinking water (Culea et al., 2006; Kanokkantapong et al., 2006b; Chen et al., 2008; Cooper et al., 2008; Richardson et al., 2008; Blodau et al., 2009). The binding potential of NOM with inorganic and organic micropollutants may also be relevant to drinking water treatment (Gjessing et al., 2007; Laborda et al., 2009; Park, 2009).

Different unit processes remove different NOM fractions during water treatment (Haarhoff et al., 2010). Therefore, more sophisticated characterization techniques are required (Matilainen et al., 2011). Several complementary methods can provide definitive structural or functional information about NOM, and have been found to correlate well, allowing for comprehensive characterization (Jaouadi et al., 2012; Penru et al., 2013). These analytical methods also provide better information for process design and optimization than the conventional dissolved organic carbon (DOC) measurements.

DOC, chemical oxygen demand (COD), UV_{254}, pH, turbidity, and color are common water quality parameters assessed by water treatment facilities in their quality control. Assessment of these parameters does not require sophisticated equipment, is simple and fast to perform, and can be automated. Yet, such assessments offer no information about the characteristics of NOM such as molar mass (MM) or hydrophobicity.

Traditionally, humic substances have been classified into three categories based on their solubility: humic acid (HA), fulvic acid (FA), and humin. This is an operational division, however; the terms do not refer to single compounds but to a wide range of compounds of similar origin (Uyguner-Demirel and Bekbolet, 2011). NOM may be characterized by high performance size exclusion chromatography (HPSEC) analysis, which determines the molecular weight distribution (MWD) of NOM (Chow et al., 2008; Korshin et al., 2009), or by fractionation techniques that use resins to divide the mixture of organic compounds of NOM into hydrophilic and hydrophobic fractions (Sharp et al., 2006a,b). Although these parameters provide useful information regarding the change in organic characteristics during treatment and impact on DBP formation, it has been suggested that their use alone they may not be efficient for predicting DBP formation (Sanly et al., 2007).

The hydrophilic fractions of NOM are composed mostly of aliphatic carbon and nitrogenous compounds such as carboxylic acids, carbohydrates, and proteins. Hydrophobic NOM primarily consists of humic and FAs (humic substances) and is rich in aromatic carbon, phenolic structures, and conjugated double bonds (Swietlik et al., 2004). Humic substances generally account for over half of the total organic carbon (TOC) content in water (Thurman, 1985). Humic substances can be regarded as natural anionic polyelectrolytes, with an anionic charge at pH values higher than 4. Another way to define hydrophobicity is to determine the specific UV-absorbance (SUVA) value (i.e., UV_{254} absorbance divided by the TOC concentration). A high SUVA value indicates that the organic matter is largely composed of hydrophobic, high-MM organic material. A low SUVA value indicates that the water contains organic compounds that are mainly hydrophilic, with a low MM and charge density (Sharp et al., 2006a,b; Edzwald and Tobiason, 1999). Due to their diverse properties, different NOM fractions respond differently to treatment by coagulation, coagulant demand, chlorine and ozone reactivity, and disinfection by-product formation potential (DBPFP) (Fabris et al., 2008; Sharp et al., 2006a,b; Archer and Singer, 2006).

2.2 GENERAL PARAMETERS

In practice, NOM is usually characterized by the measurement of its TOC, DOC, adsorption of UV light (UV_{254}), or COD. NOM is also the major contributor to the brownish-yellow color in water, thus measurement of color can provide some indication of the amount of NOM that the water contains (Uyguner et al., 2007). All these tests are fast and do not require sophisticated sample pretreatment or analytical equipment. Yet, these analyses mostly provide information about the quantity while offering limited information on the qualities of NOM.

2.2.1 Total Organic Carbon/Dissolved Organic Carbon

TOC is the sum of the particulate and DOC when existing inorganic carbon is removed by acidification. A widely accepted operational definition of DOC is the organic carbon in a water sample filtered through a 0.45 μm filter (Danielsson, 1982). TOC and DOC are the most convenient parameters for analyzing the NOM removal of treatment processes. Essentially, all methods for quantifying organic carbon in water involve oxidation. Before TOC analyzers, an oxidizing agent was added to the water and the amount of agent used expressed the concentration of COD present. A variety of oxidation techniques and instruments are currently in use, including burning, radiation, and oxidizing agents. The resulting CO_2 is measured mostly by infrared spectroscopy (IR). Yet, TOC and DOC measurements alone do not provide much insight into the behavior of NOM during water treatment (Haarhoff et al., 2010).

2.2.2 Specific UV-Absorbance

SUVA (UV-adsorption $A_{254\,nm(1\,cm)}\,mgC^{-1}L^{-1}$) is defined as the UV absorbance of a given sample at 254 nm divided by the DOC concentration of the sample. This ratio describes the hydrophobicity and hydrophilicity of NOM in water; a SUVA >4 indicates mainly hydrophobic and especially aromatic material, whereas a SUVA <3 illustrates mainly hydrophilic material (Edzwald and Tobiason, 1999). Several researchers have emphasized that NOM removable by coagulation is likely to have a high SUVA value (Archer and Singer, 2006; Bose and Reckhow, 2007). Accordingly, SUVA and A_{254}/A_{204} correlate relatively well because both are associated with the amount of aromatic carbon present in NOM (Hur et al., 2006). Jung and Son (2008) also reported a clear correlation between the SUVA value and trihalomethane (THM) formation potential. However, the latter relationship has not been confirmed by others (Ates et al., 2007; Wei et al., 2008). Methods used to characterize different features of NOM are presented in Table 2.1.

Table 2.1 Methods used to characterize different features of natural organic matter (NOM), with their advantages and disadvantages

Methods	Detected features	Advantages	Disadvantage	References
Bulk parameters				
TOC	Total organic carbon content in water	Easy to use; analytical equipment relatively inexpensive; can be used as on-line method	Only gives information on quantity of NOM, not quality	Koprivnjak et al. (2006), Liu et al. (2007), Seredynska-Sobecka et al. (2007), and Spencer et al. (2007)
DOC	Dissolved organic carbon in water, after filtration through 0.45 lm filter			Koprivnjak et al. (2006), Liu et al. (2007), Seredynska-Sobecka et al. (2007), and Spencer et al. (2007)
SUVA	High SUVA value >4 refers to hydrophobic, aromatic compounds, while low SUVA <3 indicates mainly hydrophilic material	Easy to determine; simple instrumentation	High nitrate content in low-DOC waters may interfere with measurement	Edzwald and Tobiason (1999), Archer and Singer (2006), Bose and Reckhow (2007), Ates et al. (2007), Jung and Son (2008), and Wei et al. (2008)

2.3 ISOLATION AND CONCENTRATION

A variety of NOM isolation and concentration methods are available, of which the most commonly used are reverse osmosis (RO), evaporation under reduced pressure (vacuum distillation), and freeze drying. Some sorption methods also exist (Aiken, 1985; Aiken et al., 1992; Croue, 2004). The disadvantage of most of these techniques is that they also concentrate the salts that are always present in natural waters (Croue, 2004). RO has been used in many NOM isolation studies (Croue, 2004; Koprivnjak et al., 2006; Song et al., 2009), since it has a relatively small alteration effect on the physicochemical properties of NOM (Song et al., 2009; Alasonati et al., 2010). In the RO method, the water is passed through a cation exchanger (mostly Na- or H-saturated) prior to RO filtration. Recovery of NOM from waters with a high SUVA value (mass recoveries 80–99%) is more efficient than for waters with low SUVA. Recovery is also affected by the ambient pH (Song et al., 2009).

2.4 FRACTIONATION

NOM molecules are generally unique, but share many common properties. One of the major analytical difficulties in the characterization of aqueous NOM is the lack of a suitable fractionation procedure to yield pure components that can subsequently be characterized by standard techniques (Leenher, 1985). To characterize NOM effectively, it must first be isolated and fractionated into separate groups of molecules possessing similar chemical or physical properties. Commonly used chemical fractionation methods include precipitation, solvent extraction, and adsorption chromatography. Available physical fractionation methods include electrophoresis, ultrafiltration, size exclusion chromatography (SEC), reversed-phase high-performance liquid chromatography (RPHPLC), field-flow fractionation (FFF), and ultracentrifugation (Leenher, 1985; Leenher and Croué, 2003; Chow et al., 2005; Matilainen et al., 2011).

2.4.1 Resin and Membrane Fractionation

2.4.1.1 Resin Fractionation

The most common way of distinguishing between hydrophobic and hydrophilic NOM is to determine whether or not it is adsorbed onto Amberlite XAD resin. This is a commercially available synthetic resin introduced by Malcolm et al. at the U.S. Geological Survey in Denver for the isolation of humic substances from water. Many researchers have developed and modified resin adsorption methods (Leenher, 1981; Thurman and Malcolm, 1981; Malcolm and MacCarthy, 1992; Marhaba et al., 2003), and these are still used in NOM fractionation studies (Swietlik et al., 2004; Kanokkantapong et al., 2006a,b; Sharp et al., 2006a; Her et al., 2008a; Jung and Son, 2008; Wei et al., 2008; Zhang et al., 2008). This technique was adopted by the International Humic Substances Society (IHSS) as a standard method for isolating FA and HA. Yet, FA and HA do not correspond to Svein Odén's classical definition (Odén, 1919). The XAD method involves the Amberlite resins XAD-8 and XAD-4. The XAD-8 resin favors the isolation of hydrophobic, HMW, and aromatic NOM. The hydrophilic fraction does not adsorb onto either of the resins. XAD-4 resin adsorbs weakly hydrophobic acid fractions, commonly defined as transphilic NOM (Sharp et al., 2006a; Croue, 2004; Wei et al., 2008). The IHSS Standard Method for FA is based on the matter adsorbed onto XAD-8 after reducing the pH to 2 and prefiltering with a 0.45 μm filter. The hydrophilic fraction can be separated further using, for example, AG-MP-50 cationic resin and WA-10 a weak anionic resin (Marhaba et al., 2003).

The rapid resin fractionation technique, based on XADs (Chow et al., 2004), separates DOC into four fractions based on character and molecular weight (MW). Fractions are defined as very hydrophobic acids (VHA), slightly hydrophobic acids (SHA), charged hydrophilics, and neutral hydrophilics (NEU). According to Chow et al. (2004), VHA is composed of HMW HAs, whereas FAs constitute SHA, and the NEU fraction contains substances that do not adsorb onto any of the applied resins. A variety of studies use this method (Fabris et al., 2008; Liu et al., 2008, 2010; Soh et al., 2008). A new fractionation technique based on NOM polarity in combination with an XAD-8 adsorption chromatography technique involving gradient elution has also been shown to separate different hydrophobic and hydrophilic moieties (Li et al., 2009).

Although fractionation methods based on various resins have been extensively used and accepted, they still have disadvantages. These discrepancies include possible chemical or physical alterations of NOM due to extreme pH levels and changes in pH during fractionation, irreversible adsorption of NOM compounds onto the resin, contamination from resin bleeding, and size-exclusion effects. Different operational conditions during fractionation may also influence the results (Song et al., 2009). The proposed compositions of NOM fractions separated by fractionation are presented in Table 2.2.

Table 2.2 Proposed compositions of natural organic matter (NOM) fractions separated using fractionation techniques

Fraction	Organic compound class	References
Humic acid	Portion of humic substances precipitated at pH 1	Peuravuori et al. (1997)
Hydrophobic acid	Soil fulvic acids, C5–C9 aliphatic carboxylic acids, 1- and 2-ring aromatic carboxylic acids, 1- and 2-ring phenols	Leenher (1981), Aiken et al. (1992), Marhaba et al. (2000), and Barber et al. (2001)
Hydrophobic base	1- and 2-ring aromatics (except pyridine), proteinaceous substances	Leenher (1981), Marhaba et al. (2000), and Barber et al. (2001)
Hydrophobic neutral	Mixture of hydrocarbons, >C5 aliphatic alcohols, amides, aldehydes, ketones, esters, >C9 aliphatic carboxylic acids and amines, >3 ring aromatic carboxylic acids and amines	Leenher (1981), Marhaba et al. (2000), and Barber et al. (2001)
Hydrophilic acid	Mixtures of hydroxy acids, <C5 aliphatic carboxylic acids, polyfunctional carboxylic acids	Leenher (1981), Aiken et al. (1992), Marhaba et al. (2000), and Barber et al. (2001)
Hydrophilic base	Pyridine, amphoteric proteinaceous material (i.e., aliphatic amino acids, amino sugars, <C9 aliphatic amines, peptides, and proteins)	Leenher (1981), Marhaba et al. (2000), and Barber et al. (2001)
Hydrophilic neutral	<C5 aliphatic alcohols, polyfunctional alcohols, short-chain aliphatic amines, amides, aldehydes, ketones, esters, cyclic amides, polysaccharides, and carbohydrates	Leenher (1981), Marhaba et al. (2000), and Barber et al. (2001)

Adapted from Swietlik et al. (2004).

2.4.1.2 Membrane Filtration

Filtration is prevalent in water research. The current international definition of matter dissolved in water is "the matter that passes a filter with a cut-off of 0.45 μm" and RO is the most used research method for isolating NOM from water. As early as 1970, it was demonstrated that NOM (aquatic humus) could be fractionated by membranes with different cut-off ranges (Gjessing, 1970). Further development of this technique confirms that

NOM is a mixture of organic matter with different sized molecules (Aiken, 1984; Liu et al., 2007; Lankes et al., 2008; Wei et al., 2008; Zhao et al., 2009).

Separation by membranes may not only be dependent on the size of NOM, but also on its molecular structures (Assemi et al., 2004). Methodological conditions also influence the separation of molecules according to size and shape by membrane filtration. One problem associated with this method is the adherence of macromolecules to the sides of membrane pores, affecting permeability. Results may also be affected by flow restriction, also known as concentration polarization, in which the deposition of macromolecules results in a gel layer that becomes the dominant resistance to flow (Amy et al., 1987). Generally, this fractionation technique correlates reasonably well with SEC in terms of size and MW (Gjessing, 1973). It should be emphasized, however, that the reproducibility of the characterization method is more important than accuracy regarding size and shape.

2.5 CHROMATOGRAPHIC METHODS

2.5.1 Liquid Chromatography

2.5.1.1 Fractionation of NOM by SEC

SEC is a fractionation method based on molecular size: the larger the molecules, the shorter the retention time. Yet, molecular shape and some interaction characteristics may influence the results (Lankes et al., 2009). SEC fractionation of NOM has been utilized since the early 1960s (Posner, 1963; Gjessing, 1965); however, the soft gels used had some major disadvantages (Gjessing, 1973), such as poor separation. To overcome this, SEC column fractionation was applied to high-performance liquid chromatography (HPLC), developing a technique known as HPSEC (Fukano et al., 1978). HPSEC has become particularly useful for NOM characterization during various stages of drinking water treatment (Matilainen et al., 2006; Allpike et al., 2007; Fabris et al., 2008; Chow et al., 2009a,b; Tercero Espinoza et al., 2009; Zhao et al., 2009).

2.5.1.2 Eluents and Columns Used for HPSEC Measurements

In NOM studies using HPSEC, the choice of a proper eluent is important, since its ionic strength and pH significantly affect results. The surface charge characteristics of the gel, the charge and structure of the NOM, and NOM–gel interactions are all affected by the eluent (Specht and Frimmel, 2000). Studies of NOM characterization by HPSEC have used various eluents as a mobile phase, including a phosphate buffer solution with ionic strength adjustment using sodium chloride and sodium acetate solution. The columns have different silica-based and polymer-based supports (Swietlik et al., 2004; Matilainen et al., 2006; Sarathy and Mohseni, 2007; Wu et al., 2007a,b; Fabris et al., 2008; Her et al., 2008a; Liu et al., 2008, 2010; Soh et al., 2008; Zhang et al., 2008; Chow et al., 2009b; Korshin et al., 2009; Zhao et al., 2009). According to Specht and Frimmel (2000), however, both polymer-based and silica-based columns have specific interactions with NOM.

2.5.1.3 Detectors Used for HPSEC Measurements

Detectors used in different HPSEC systems include Fourier transform infrared (FTIR) (Allpike et al., 2007), on-line DOC analyzers (Her et al., 2008a), and excitation emission fluorescence detection (Wu et al., 2007a). The detectors most widely used in HPSEC fractionation are the single and variable wavelength ultraviolet and visible light (UV–Vis) detectors or diode array detectors (Matilainen et al., 2006; Ates et al., 2007; Sarathy and Mohseni, 2007; Wu et al., 2007b; Fabris et al., 2008; Her et al., 2008a,b; Liu et al., 2008, 2010; Zhang et al., 2008; Chow et al., 2009b; Korshin et al., 2009). UV–Vis detectors are simple and quick to use and are available in most laboratories. However, they only respond to or detect analytes that absorb on the wavelength at which they are operating. Since NOM structures contain a range of chromophores with varying molar absorptivities, the MW calculated may not involve all the NOM compounds present (O'Loughlin and Chin, 2001; Wu et al., 2007b; Soh et al., 2008; Tercero Espinoza et al., 2009). Reducing the wavelength results in an increase in the absorbance (O'Loughlin and Chin, 2001; Lankes et al., 2009). On the other hand, low molecular weight (LMW) compounds reportedly are not easily detected by UV–Vis due to their lack of conjugated double bonds.

Despite the success of on-line specific Ex/Em fluorescence detection in the characterization of NOM fractions, this only provides limited information and somewhat poorer separation resolution compared to UV–Vis absorbance detection (Wu et al., 2007a). Yet, in contrast to UV–Vis, on-line 3D excitation–emission matrix (EEM) fluorescence detection gives more insight into the chemical qualitative and structural features of NOM besides molecular size (Wu et al., 2003, 2007a).

On-line organic carbon detectors (OCD) may offer better quantification than UV–Vis detectors (Her et al., 2002; Soh et al., 2008; Lankes et al., 2009; Tercero Espinoza et al., 2009; Simon et al., 2013). Liquid chromatography-organic carbon detection (LC-OCD) is a combination of SEC separation and continuous analysis to quantify organic carbon, nitrogen, and UV absorbance (Penru et al., 2013). During the last 20 years, an LC-OCD combination has been developed that both quantifies NOM and characterizes its compounds (Huber and Frimmel, 1992a,b). Using LC-OCD, it is also possible to evaluate the content of HMW polysaccharides and biopolymers in the sample, which are not generally visible or traceable by UV detectors. Recently, LC-OCD has been successfully used to characterize NOM fractions and assess the impact of pretreatments on them (Simon et al., 2013). Huber and Frimmel (1994) used an HPSEC coupled with an LC-OCD to fractionate NOM into five parts: (1) biopolymers (such as polypeptides, polysaccharides, proteins and amino sugars), (2) humic substances (fulvic and humic acids), (3) building blocks (hydrolysates of humic substances), (4) LMW humic substances and acids, and (5) LMW neutrals (such as alcohols, aldehydes, ketones and amino acids). High-resolution mass spectrometry (MS) has also been coupled with HPSEC to provide more information about the molecules in complex mixtures of NOM fractions.

To achieve better quantitative and qualitative NOM evaluation, an increasing number of reports couple multiple detectors on-line, such as UV–Vis and spectrofluorometer (Wu et al., 2007a), or UV–Vis and OCD (Her et al., 2002; Allpike et al., 2007; Soh et al., 2008; Baghoth et al., 2009; Lankes et al., 2009; Tercero Espinoza et al., 2009). Accordingly, a SEC set-up that uses a UV detector followed by a TOC module and finally an MS as the detector has been described (Warton et al., 2008). The results of these combined detector systems look promising.

2.5.1.4 Determination of MW by HPSEC

The size and the MW of the NOM may be specified by comparison with known molecular mass standards. The selection of proper standards to characterize the MWs of organic substances in NOM using HPSEC is based mostly on the hypothesized structure of NOM. Polystyrene sulfonate (PSS) standards have been widely used for this (Zhou et al., 2000; Swietlik et al., 2004; Ates et al., 2007; Wu et al., 2007a; Fabris et al., 2008; Chow et al., 2009b; Korshin et al., 2009; Zhao et al., 2009; Liu et al., 2010). Yet, NOM in water can be more branched and cross linked than PSS. Therefore, the resulting MWs are approximations. Polyethylene glycols are also used to determine the MWD of NOM (Her et al., 2008a; Zhang et al., 2008). The MW magnitudes determined by HPSEC reportedly correlate well with other MW measuring techniques (Egeberg et al., 2002; Assemi et al., 2004).

2.5.1.5 HPSEC Determinations in Water Treatment Monitoring

HPSEC can be used as a tool to characterize the MW profile of NOM as well as to model and predict NOM removal (Xing et al., 2012). SEC profiles provide valuable information in the study of the removal mechanism of HMW NOM during coagulation. The advantages of SEC include ease of operation, minimal sample pretreatment, and rapid analysis (Kristiana et al., 2010). Several studies compare changes in both the quality of NOM during different stages of the treatment process (Matilainen et al., 2006; Fabris et al., 2008; Baghoth et al., 2009; Chow et al., 2009b; Zhao et al., 2009) and the actual degree of its removal (Wu et al., 2007b). These SEC profiles do not, however, contain sufficient information to evaluate the DBPFP of the remaining NOM molecules. Nevertheless, Korshin et al. (2009) reported a good correlation between DBPFP and HPSEC multi-wavelength UV results based on spectroscopic index calculations (Korshin et al., 2009).

SEC determinations by Haarhoff et al. (2010) revealed that rapid sand filtration targeted the low and high molecular fractions of NOM, granular activated carbon targeted the low to intermediate fractions, and ultrafiltration (UF) targeted the high molecular weight (HMW) fraction.

2.5.1.6 Reversed-Phase High-Performance Liquid Chromatography

RPHPLC is rather limited in application to NOM fractionation, especially compared to HPSEC (Stenson, 2008; Reemtsma, 2009). RPHPLC is based on a nonpolar stationary phase and an aqueous, moderately polar mobile phase, thus fractionating NOM according to its polarity. Xing et al. (2012) investigated RPHPLC as a rapid assessment of the hydrophobicity or hydrophilicity of NOM. It was found that the reduction of total RPHPLC peak area correlated well with DOC and UV absorbance removal efficiency. The ratio between the hydrophobic and hydrophilic peak areas could be used to quantify the treatability of NOM (Xing et al., 2012).

2.5.1.7 Liquid Chromatography-Mass Spectrometry

Liquid chromatography-mass spectrometry (LC-MS, or alternatively HPLC-MS) is a technique that combines the physical separation capabilities of liquid chromatography with the mass analysis capabilities of MS. LC-MS is one of the main recent innovations in NOM structure research. Coupling LC with MS by an advanced technique called electrospray ionization (ESI) made the derivatization of NOM molecules unnecessary (Reemtsma, 2009). The LC-MS system is used in combination with SEC in describing

LMW FAs (Reemtsma and These, 2005). According to Reemtsma and These (2005), ESI-MS is a more sensitive method than HMW humics for the characterization of LMW FAs. Tandem MS, such as ion trap, triple quadrupoles, or quadrupole time-of-flight (e.g., LC-MS/MS) are also used to investigate NOM structures (Reemtsma and These, 2005; Mawhinney et al., 2009). High-resolution mass spectrometry has been utilized to study marine NOM evolution through a desalination process on a molecular scale (Cortés-Francisco and Caixach, 2013).

2.5.2 Gas Chromatography

2.5.2.1 Pyrolysis Gas Chromatography-Mass Spectrometry

In pyrolysis, the large complex molecules in NOM are broken down into more analytically available fragments by the application of heat and under anoxic conditions. These fragments are then swept into the analytical column for gas chromatography (GC) and detected by MS. Pyrolysis combined with gas chromatography-mass spectrometry (Py-GC-MS) is a useful technique both for quantitative measurements and for producing structural information about the molecular building blocks of NOM (Bruchet et al., 1990; Schulten, 1993; Fabbri et al., 1998; Christy et al., 1999; Leenher and Croué, 2003; Templier et al., 2005; Chow et al., 2009b). The analysis of complex oxygen-containing macromolecules can be challenging. Polar pyrolysis products are poorly GC amenable and undergo undesirable thermal reactions (Templier et al., 2005). Development in the pyrolysis technique has made it possible to overcome some of these analytical limitations (Parsi et al., 2007).

Microscale sealed vessel pyrolysis is a promising new approach to NOM characterization. This technique promotes the reduction of polar NOM moieties and significantly increases the yield of GC amenable products (Greenwood et al., 2006; Berwick et al., 2007, 2010). This analytical method represents an excellent complement to conventional flash pyrolysis techniques, which may have a limited capacity for chromatographic resolution of polar biochemicals. However, the technique needs to be developed further to enable more advanced NOM characterization.

Major pyrolysis fragments, most commonly yielded from aquatic NOM pyrochromatograms, have been listed by Leenher and Croué (2003) and are summarized in Table 2.3. Besides pyrolysis-MS, the molecular-level structural information can also be obtained by chemical degradation methods such as using a reducing agent (e.g., *n*-butylsilane) in combination with MS (Nimmagadda and McRae, 2007).

Table 2.3 The predominant pyrolysis by-products from aquatic natural organic matter (NOM)

Type	Common fragments
Polysaccharides	Methylfuran, furfural, acetylfuran, methylfurfural, levoglucosenone, hydroxypropanone, cyclopentanone, methylcyclopentenone, acetic acid
Amino sugars	Acetamide, *N*-methylacetamide, propionamide, acetic acid
Proteins	Acetonitrile, benzonitrile, phenylacetonitrile, pyridine, methylpyridine, pyrrole, methylpyrrole, indole, methylindole (from tryptophan), toluene, styrene (from phenylalanine), phenol, *p*-cresol (from tyrosine)
Polyphenolic compounds	Phenol; *o*-, *m*-, *p*-cresol; methylphenols; dimethylphenols
Lignins	Methoxyphenols
Tannins	Catechol
DNA	Furfuryl alcohol
Polyhydroxybutyrates	Butenoic acid

Adapted from Leenher and Croué (2003).

2.5.2.2 Flow Field-Flow Fractionation

FFF is similar to chromatography as a method for separating macromolecules, colloids, and particles (Messaud et al., 2009). Flow field-flow fractionation (FIFFF) is a subtechnique of FFF. FIFFF is also a molecular size fractionation method that is used to determine the MWD of NOM (Beckett et al., 1987; Moon et al., 2006; Rosario-Ortiz et al., 2007; Alasonati et al., 2010; Baalousha et al., 2011; Pifer and Fairey, 2012). Detectors used in FIFFF include UV–Vis, fluorescence, inductively coupled plasma atomic emission spectroscopy (ICP AES), and ICP MS (Moon et al., 2006; Alasonati et al., 2010). The results obtained by FIFFF are comparable to those obtained by other techniques such as HPSEC (Assemi et al., 2004). Unfortunately, the technique has not yet matured and published reports are limited (Messaud et al., 2009). Size distributions determined by FIFFF under varying pH and ionic strength should be treated with caution, because the sample recovery of humic substances decreases rapidly with increasing cross flow, lower pH, and greater FIFFF carrier ionic strength, and varies according to the material of the accumulation wall membrane (Baalousha et al., 2011).

2.6 SPECTROSCOPIC METHODS

2.6.1 Ultraviolet and Visible

UV–Vis absorption spectroscopy is a semiquantitative method to determine humic substances in natural waters. It is based on the measurement of the attenuation of a beam of light after it passes through a sample or after its reflection from a sample surface. The concentration of an analyte in a solution can be determined by measuring the absorbance at a certain wavelength, applying the Beer–Lambert law. Absorption measurements can be taken at a single wavelength or over an extended spectral range. The UV absorbance spectra of NOM decrease monotonically with increasing wavelength (Hur et al., 2006). Any wavelength from 220 to 280 nm is appropriate for NOM measurements, although the molar absorptivities will vary due to the range of chromophores in the structure of NOM. The different wavelengths are believed to identify different chromophores: absorbance at 220 nm is associated with both the carboxylic and aromatic chromophores, whereas absorbance at 254 nm is typical for aromatic groups with varying degrees of activation (Korshin et al., 2009). UV_{254} has been identified as a potential surrogate measure for DOC despite its tendency to represent only the aromatic character. Yet, due to the presence of a high number of overlapping groups, humic substances display featureless and monotonic absorbance spectra (Uyguner-Demirel and Bekbolet, 2011 and references therein).

Ratios between two different wavelengths, such as A_{254}/A_{204}, A_{254}/A_{436} or A_{250}/A_{365}, have also been reported to aid NOM characterization (Hur et al., 2006; Spencer et al., 2007; Li et al., 2009). For example, the ratio of A_{253}/A_{203} correlates with the formation of DBPs (Kim and Yu, 2007). Yet, there is a significant possibility of errors due to interference at 200 nm through 230 nm by inorganic ions such as nitrates or sulfates (Her et al., 2008a,b).

2.6.2 Differential Absorption

An alternative approach is to measure differential absorbance, defined as the change in the absorbance of a sample in response to any forcing function, such as chlorination (Korshin et al., 1999; Ates et al., 2007). Differential absorbance spectroscopy can detect subtle changes in NOM structure (Uyguner-Demirel and Bekbolet, 2011 and references therein). The differential absorbance at 272 nm has been proposed for the evaluation of DBP formation. This differential absorbance correlates strongly with total THM and HAA concentrations (Roccaro and Vagliasindi, 2009). Differential absorbance at 254 nm (ΔA_{254}) has no correlation with total THM or HAA formations, however (Ates et al., 2007).

2.6.3 Fluorescence

Fluorescence spectroscopy excites the analyte molecules by irradiation at a certain wavelength, and the emitted radiation is measured at another wavelength. This method has caught the attention of the water industry due to its potential as a monitoring technique: recent research has shown its applicability for long-term monitoring (Sanchez et al., 2013). Fluorescence can provide insight into the chemical characteristics of NOM, because it is a function of molecular structure and functional groups (Hudson et al., 2007). The main advantages of fluorescence over traditional methods such as UV–Vis are increased sensitivity and selectivity (Bieroza et al., 2009; Peiris et al., 2010). The method is also simple, relatively inexpensive, and requires little sample pretreatment (Baghoth et al., 2011). Yet, fluorescence measurements are dependent on environmental factors such as pH, ionic strengths and temperature (Markechova et al., 2013). A particular molecular conformation called fluorophore is characterized by specific excitation and emission wavelengths. These fluorophores are helpful in describing the structural composition of humic materials (Datta et al., 1971; Uyguner et al., 2007; Wu et al., 2007b; Baker et al., 2008; Zhang et al., 2008; Bieroza et al., 2009). Fluorescence has also been recommended as a tool for estimating the presence of biodegradable NOM, since strong correlations between fluorescence and biological oxygen demand values have been demonstrated (Hudson et al., 2008).

Three-dimensional fluorescence EEM spectrophotometry is an increasingly popular technique. An entire review was recently devoted to EEM and parallel factor (PARAFAC) drinking water treatment (Markechova et al., 2013). EEM is a 3D spectrum in which fluorescence intensity can be presented as a function of excitation and emission wavelength (Valencia et al., 2014). The EEM spectrum visualizes a range of different fluorophores with an excitation and emission wavelength ranging from ~200 to ~500 nm, which is more revealing than the traditional single-scan technique (Spencer et al., 2007). EEM fluorescence provides valuable information on the removal of different NOM fractions in water treatment (Jeong et al., 2013; Lee et al., 2013; Sanchez et al., 2014; Valencia et al., 2014). Chen et al. (2003) classified NOM into five different fractions to interpret EEM fluorescence spectra: aromatic proteins (regions I and II); fulvic acidlike and humic acidlike compounds (regions III and V); and microbial by-products (region IV). Typical raw water may contain two major fluorescence peaks, described as humiclike and proteinlike fluorescence maxima (Wu et al., 2007b; Baghoth et al., 2009) or three peaks referred to as tryptophanlike, fulviclike, and humiclike fluorophores (Liu et al., 2007; Seredynska-Sobecka et al., 2007; Spencer et al., 2007; Baker et al., 2008; Hudson et al., 2008). The fluorescence intensity peaks for different NOM fractions of natural waters are listed in Table 2.4 (Coble, 1996; Leenher and Croué, 2003; Spencer et al., 2007; Baghoth et al., 2009). The spectral fluorescent signature technique has been developed by Marhaba (2000) to model these characteristics of NOM. Multivariate data analyses such as principal component analysis of full-fluorescence EEMs have been used to characterize water samples obtained from different sources or stages of the purification process (Marhaba et al., 2009; Peiris et al., 2010).

PARAFAC models have been found to fit on excitation–emission matrices (Sanchez et al., 2013). According to Markechova et al. (2013), PARAFAC models can be applied by initializing the algorithm, establishing the number of components, identifying the fluorophores, and calibrating the model.

Table 2.4 Major fluorescence peaks for water samples

Range of excitation (nm)	Range of emission (nm)	Component type	References
270–280	310–320	Tyrosinelike, proteinlike	Coble (1996) and Baghoth et al. (2009)
270–285 (220–235)	340–360	Tryptophanlike, proteinlike	Coble (1996), Spencer et al. (2007), Baker et al. (2008), Hudson et al. (2008), and Baghoth et al. (2009)
320–350	400–450	Fulviclike	Spencer et al. (2007) and Baker et al. (2008)
310–320	3380–420	Humiclike (marine humiclike)	Coble (1996) and Baghoth et al. (2009)
320–390	410–500	Humiclike	Coble (1996), Stedmon et al. (2003), Stedmon and Markager (2005), Spencer et al. (2007), Murphy et al. (2008), Yamashita and Jaffé (2008), and Baghoth et al. (2009)

They may be utilized to determine the component most easily removed from a treatment process, establish the order of preferred removal, and identify persistent components to be removed (Pifer and Fairey, 2012; Markechova et al., 2013; Kothawala et al., 2013; Shutova et al., 2014). Thus, PARAFAC constitutes a valuable tool for modelling NOM behavior during water treatment. Furthermore, strong linear correlations have been demonstrated between EEMs and DOC, SUVA, and their organic precursors (Valencia et al., 2013).

2.6.4 Fourier Transform Infrared

FTIR has been widely used for the characterization of NOM (Kanokkantapong et al., 2006a; Kim and Yu, 2007; Hay and Myneni, 2007; Her et al., 2008a). Samples exposed to IR light absorb energy corresponding to the vibrational energy of atomic bonds. The resulting absorption spectrum is a unique fingerprint of compounds, allowing the identification of both inorganic and organic functional groups. Severe overlapping of characteristic spectral features due to the complexity and polyfunctionality of NOM can complicate the interpretation of the analysis, however. The main characteristics of NOM obtained from several analyses of FTIR spectra are listed in Table 2.5 (Howe et al., 2002; Kanokkantapong et al., 2006a; Kim and Yu, 2007; Rodriques and Núñez, 2009; Berwick et al., 2010). Advantages and disadvantages of the methods used to characterize different features of NOM are presented in Table 2.6.

Table 2.5 Main characteristics of the Fourier transform infrared (FTIR) spectra of humic substances

Band (cm^{-1})	Assignment
3400	Associated O–H stretch (alcohols, phenols, and carboxylic groups)
2850–2960	C–H stretch (CH_3 and CH_2)
2620	O–H stretch (hydrogen-bonded carboxylic groups)
1720	C=O stretch (carboxylic groups)
1630	C=C stretch (alkenes and aromatic rings)
1540	N–H bend (N–H structures)
1455	C–H bend (CH_3 and CH_2)
1410	O–H bend (carboxylic groups)
1375	C–H bend (CH_3)
1260 and 1220	C–O stretch (carboxylic groups, phenols, aromatic/unsaturated ethers)
1095 and 1030	C–O stretch (alcohols, aliphatic ethers)
805	C–H bend (tri- and tetra-substituted aromatic rings)

Adapted from Rodriques and Núñez (2009).

Table 2.6 Advantages and disadvantages of the methods used to characterize different features of natural organic matter (NOM)

Methods	Detected features	Advantages	Disadvantages	References
Spectroscopic methods				
UV–Vis	Quantitative measurement of all compounds in the sample that adsorb UV light. Conjugated C–C multiple bonds, aromatic carbon, –COOH and –OH increase adsorption	Simple; rapid	Not all compounds of NOM can be detected; wavelength-specific adsorption; sensitive to chemical environment, e.g., pH and ionic strength	Korshin et al. (1999), Hur et al. (2006), Ates et al. (2007), Spencer et al. (2007), Li et al. (2009), and Roccaro and Vagliasindi (2009)
Fluorescence	Molecules of the sample are excited by irradiation at a certain wavelength and the emitted radiation is measured at a different wavelength. Conjugated double bonds and aromatic rings, –OH and $-NH_2$ enhance fluorescence, while –COOH diminishes it. Three major groups: tryptophan-, humic- and fulvic-like fluorophores	Sensitivity; specificity; speed	Sensitivity to chemical environment, e.g., pH	Liu et al. (2007), Seredynska-Sobecka et al. (2007), Spencer et al. (2007), Wu et al. (2007a,b), Baker et al. (2008), Hudson et al. (2008), Zhang et al. (2008), Bieroza et al. (2009), and Peiris et al. (2010)
FTIR	Aromatic and aliphatic hydrocarbon, different bonds and functional groups, e.g., OH in carboxylic and alcoholic groups	Sample can be measured both in liquid and solid phases; compact instruments; reliability; ease of use	Interpretation of the analysis can be difficult due to overlapping of characteristic spectral features due to the complexity and poly-functionality of NOM	Kanokkantapong et al. (2006b), Hay and Myneni (2007), and Her et al. (2008a,b)

^{1}H NMR	Determine the structure with respect to hydrogen-1 nuclei within the molecules of a substance	High sensitivity; powerful tool especially together with ^{13}C NMR in structural analysis	Interpretation of spectra can be limited	Ma et al. (2001), Kim et al. (2006), Kim and Yu (2007), and Mao et al. (2007a)
^{13}C NMR	Carboxylic structures of NOM, as well as aliphatic carbon, *o*-alkyl carbon, aromatic carbon and carbonyl carbon. Can be done on solid or liquid sample	Powerful tool especially together with ^{1}H NMR in structural analysis	Less sensitive to carbon than ^{1}H NMR to hydrogen; NOM sample needs to be isolated in solid-state NMR measurements; parameter sensitive	Chen et al. (2002), Wong et al. (2002), Mao et al. (2007a,b), and Lankes et al. (2008)
^{15}N NMR	Identifies carbons bonded to nitrogen	Broader insight into NOM structure	Rather nuclear insensitive; ^{15}N appears at low percentage in natural nitrogen	Mao et al. (2007b), Lankes et al. (2008), and Thorn and Cox (2009)
2-D NMR	Spreads the spectrum in two dimensions and allows for the observation of cross peaks. Structural environments of, e.g., carboxyls in detail	Multidimensional NMR techniques can be used to resolve contributions that overlap strongly in a 1D NMR experiment	Might not detect very large molecules	Deshmukh et al. (2007) and Mao et al. (2007a,b)

Traditionally, IR spectrometers are used to analyze solids or liquids by transmitting the infrared radiation directly through the sample. In many cases, however, pretreatment is required to achieve a good quality spectrum sample. Dilute aqueous NOM samples are analyzed by drying these to thin films on attenuated total reflectance (ATR) crystals, thereby avoiding the different products that may develop with other concentration methods (Howe et al., 2002; Hay and Myneni, 2007). The ATR-IR technique has revolutionized solid and liquid sample analysis, because it combats the most challenging aspects of IR techniques, namely sample preparation and consequently spectral reproducibility.

2.6.5 Nuclear Magnetic Resonance

Nuclear magnetic resonance (NMR) spectroscopy has been used for decades to study the functional groups in NOM (Barron and Wilson, 1981). The NMR technique is especially useful in combination with elemental composition data, apparent MW data, or IR spectroscopy data on fractionated NOM (Gjessing et al., 1998; Assemi et al., 2004; Templier et al., 2005; Deshmukh et al., 2007; Mao et al., 2007a,b; Lankes et al., 2008; Ritchie and Perdue, 2008; Li et al., 2009; Jeong et al., 2013). The heterogeneity and complexity of NOM limit the practical value of the NMR data. Various fractionation methods (SEC, resin fractionation) prior to analysis have been shown to improve the practical availability of NMR information, although in some cases the prefractionation of NOM has simplified spectra (Wong et al., 2002).

NMR spectroscopy of NOM can be done on both solid and aqueous phase samples. Solution-state NMR is applicable to soluble samples like humic and FAs. Yet, the higher NOM concentration in solid samples enables various ^{13}C NMR experiments, including spectral editing of strongly overlapping bands (Mao et al., 2007a; Lankes et al., 2008). In addition to ^{1}H and ^{13}C NMR techniques, ^{15}N NMR has also been used, for example, in the identification of peptides and other nitrogen containing structures (Thorn and Cox, 2009).

New 2D or multidimensional NMR techniques may have more potential than the traditional 1D method. Two-dimensional and multidimensional NMR spectra are especially useful in determining the structures of complicated molecules such as NOM, and are therefore more appropriate for NOM characterization than either 1D ^{1}H NMR or ^{13}C NMR (Deshmukh et al., 2007; Mao et al., 2007a,b). Other techniques include correlation spectroscopy, nuclear overhauser effect spectroscopy, total correlation spectroscopy, and heteronuclear multiple bond correlation (HMBC) (Cook,

2004; Deshmukh et al., 2007). For example, ^{1}H–^{13}C HMBC reveals couplings between protons and carbon atoms separated by 2–3 bonds, with the potential for examining the structural environments of carboxyls in more detail, although the method still requires improvement to be useful in NOM research (Deshmukh et al., 2007).

Various advanced spectral editing techniques have also been developed to obtain more specific structural information, aiding the understanding of molecular connectivity or functional group identification of NOM compounds. Techniques used include $^{13}C\{^{14}N\}$ SPIDER NMR, which selects signals from carbons bonded to nitrogen (Mao et al., 2007b).

2.6.5.1 *^{1}H NMR*

The ^{1}H NMR spectra provide information on nonexchangeable structural proton distributions in NOM. Structural features of NOM obtained this way are listed in Table 2.7 (Ma et al., 2001; Kim et al., 2006; Kim and Yu, 2007). The ^{1}H NMR results can support elemental data interpretation in the quantification of aromaticity and the degree of humification (Kim et al., 2006). ^{1}H high-resolution magic-angle spinning (HR-MAS) spectroscopy is an innovative NMR technique for examining semisolid samples using liquid-state NMR methods, deployed in the study of NOM in the soil–water interface (Simpson et al., 2006).

Table 2.7 General assignments of the ^{1}H NMR spectra of humic substances

Chemical shift (ppm)	General assignment	References
0.0–1.6	Protons on methyl and methylene carbons directly bonded to other carbons	Ma et al. (2001), Kim et al. (2006), and Kim and Yu (2007)
1.6–3.2	Protons of the methyl and methylene carbons alpha to aromatic rings, carboxyl and carbonyl groups	Ma et al. (2001), Kim et al. (2006), and Kim and Yu (2007)
3.2–4.3	Protons on carbon of hydroxyl, ester and ether and protons on methyl, methylene and methyne carbons directly bonded to oxygen and nitrogen	Ma et al. (2001), Kim et al. (2006), and Kim and Yu (2007)
6.0–8.5	Protons attached to unsaturated carbons. Aromatic protons including quinones, phenols, oxygen containing heteroaromatics	Ma et al. (2001), Kim et al. (2006), and Kim and Yu (2007)

Table 2.8 Chemical shift regions for the integration of peak areas in solid-state ^{13}C NMR spectra of natural organic matter and their respective assignments (Chen et al., 2002; Mao et al., 2007b)

Chemical shift (ppm)	Carbon assignment
0–50	Aliphatic or paraffinic carbon chains
50–75	Methoxyl groups
75–112	Carbohydrate RC–OH or RC–OR functional groups
112–145	Aromatic carbons
145–163	Phenolic groups
163–180	Carboxylic, carbonyl, amine and ester carbons

2.6.5.2 ^{13}C NMR

Solid-state ^{13}C NMR is a conventional tool for studying NOM structure. Several specific pulse techniques are available for improving the quality and quantity of ^{13}C NMR spectra of NOM. This technique is useful for determining the details of carbon structures (Peuravuori et al., 2003). Cross polarization magic-angle spinning solid-state nuclear magnetic resonance (CP/MAS ^{13}C NMR) spectroscopy is used to determine the functional groups of carbon in NOM (Chen et al., 2002; Wong et al., 2002; Mao et al., 2007b). According to Peuravuori et al. (2003), quantitative analysis with solid-state ^{13}C NMR is difficult, but qualitative analysis is possible, so the relative compositions of NOM can be determined. Although this technique is effective concerning the gross chemical compositions, specific compounds are rarely identified (Peuravuori et al., 2003). As NOM is chemically and structurally heterogeneous, the chemical shift dispersion of the NMR spectrum can be relatively large. The obtained ^{13}C chemical shift range is divided into aliphatic, alkoxyl, unsaturated, and carbonyl carbon regions (Wong et al., 2002; Lankes et al., 2008). The general assignments for structural features of NOM obtained by solid-state ^{13}C NMR are shown in Table 2.8 (Chen et al., 2002; Mao et al., 2007b).

2.7 BIOLOGICAL TESTING

Several biological tests have been developed to assess the level of biodegradable organic matter in water. These bioassays are generally based on two concepts: (1) the assimilable organic carbon (AOC), which measures the growth of bacterial inoculum in response to the amount of nutrients in the water and (2) the biodegradable dissolved organic carbon (BDOC), which measures the fraction of DOC assimilated and mineralized by heterotrophic microbes (Escobar and Randall, 2001).

AOC is believed to be composed mostly of LMW compounds, which are the most difficult to remove. These are also readily degradable, thus indicating the potential for bacterial growth in the distribution system (Escobar and Randall, 2001; Kang et al., 2006; Matilainen et al., 2006; Chen et al., 2007; Chow et al., 2009a). BDOC methods have been developed and used to determine the biological stability of raw and treated water (Escobar and Randall, 2001; Chen et al., 2007; Soh et al., 2008; Zappia et al., 2008; Yapsakli and Cecen, 2009; Haarhoff et al., 2010). BDOC fractionation provides further insight into existing oxidation and biofiltration processes (Haarhoff et al., 2010).

2.8 OTHER CHARACTERIZATION METHODS

Many other NOM characterization methods are rather complex and time consuming, requiring access to sophisticated analytical instrumentation. Even though these techniques are not used routinely, they may provide further insight into the quality and structural features of NOM at various stages in the water treatment process, promoting a better understanding of its behavior.

The elemental composition of NOM is often used to calculate atomic ratios such as O/C, H/C, or N/C. In addition to the classic Elemental Analyzer (Li et al., 2009; Zhao et al., 2009), elemental composition can be determined by ICP AES (Blodau et al., 2009). Another structural analysis method is to measure the size and conformation of NOM under different environmental conditions with photon correlation spectroscopy, transmission electron microscopy, or atomic force microscopy (AFM) (Lead et al., 2000; Baalousha et al., 2006; Abu-Lail et al., 2007). Recent advances in AFM are providing better chemical and structural information on NOM to determine to bacteria–NOM adhesion forces (Abu-Lail et al., 2007). Scanning electron microscope images are also used to study adsorbent layers and foulants on membrane surfaces during drinking water treatment (Kim et al., 2010). Determination of the electrical character and surface charge of its fractions may also provide valuable information about the nature of NOM (Sharp et al., 2006a).

Fourier transform ion cyclotron resonance mass spectrometry (FTICR-MS) is the latest approach to NOM research. This method separates thousands of ions from each other and determines molecular formulae with reasonable accuracy. Reemtsma (2009) attributes two key inventions to the primacy of MS in NOM research. The first development was ESI, which allowed the infusion of aqueous solutions into the MS, and furthered the coupling of MS with LC. This circumvented the need to deconstruct and derivatize NOM molecules, allowing their direct analysis by MS. The second discovery was ultrahigh resolution in the form of Fourier transform ion cyclotron resonance.

As a result, electrospray ionization Fourier transform ion cyclotron resonance mass spectrometry (ESI-FTICR-MS) has recently been developed for NOM characterization (Reemtsma, 2009; Zhang et al., 2012a, b). More than 4000 NOM components could be resolved in one sample using this technique. Interestingly, in practical application, NOM molecules with a low degree of oxidation were found to be more reactive toward chlorine than those with a high O/C ratio (Zhang et al., 2012a), and NOM molecules with a high degree of oxidation were preferentially removed during coagulation by PACl (Zhang et al., 2012b). FTICR-MS is expected to further the understanding of NOM transformations on a molecular level during water treatment processes and provide useful evidence for improving their performance (Zhang et al., 2012b). More specific information is obtained by coupling ESI-FTICR-MS with SEC (Reemtsma et al., 2006, 2008; Reemtsma, 2009) and RPHPLC (Stenson, 2008; Reemtsma, 2009), thereby producing different fractions of NOM, which are easier to determine with ESI-FTICR-MS. An application of the universal algorithm for automatic treatment of mass spectra, based on total mass difference statistics, has been also developed and implemented to support the interpretation of data on complex NOM structures (Kunenkov et al., 2009).

Methods used to characterize different features of NOM with their pros and cons are presented in Table 2.9.

2.9 CONCLUSIONS

Surface water is the main drinking water source in many regions of the world. Due to contact with the soil surface, fresh water will always contain some inorganic and organic matter. The quality of all fresh water thus depends on the nature of the catchment. In the northern hemisphere, surface water often has a high NOM content. Knowledge of this matter has considerably increased with the last three or four decades of international research, leading to the general conclusion that for both practical and hygienic reasons, NOM should be removed from drinking water. The nature, character, and amount of NOM in raw water all depend on geology, topography, and climate; and climate change clearly has an effect on these

Table 2.9 Advantages and disadvantages of natural organic matter (NOM) characterization methods

Methods	Detected features	Advantages	Disadvantages	References
Chromatographic methods				
GC–MS	Structural and molecular properties of compounds in NOM	Sensitive; specific; rapid	Unwanted thermal reactions; data interpretation is complicated, due to the complexity of NOM	Templier et al. (2005) and Parsi et al. (2007)
FIFFF	A separation technique similar to chromatography. Fractionate NOM by size and MWD	No stationary phase	Choice of proper standards	Moon et al. (2006), Messaud et al. (2009), and Alasonati et al. (2010)
FTICR-MS	Molecular species, exact masses, and molecular formulas of compounds on NOM	The full scale of the molecular complexity of NOM can be revealed; can be combined with SEC or RPHPLC	Data interpretation is difficult, due to the complexity of NOM	Reemtsma et al. (2006, 2008), Stenson (2008), Kunenkov et al. (2009), and Reemtsma (2009)
HP-SEC	Fractionates NOM on basis of molecular sizes of organic compounds present	Rapid; sensitive; no preextraction needed; equipment relatively simple; inexpensive	Charge effects during measurement (column/NOM, eluent/NOM), choice of detector and proper standards	Wu et al. (2007a), Chow et al. (2005), Fabris et al. (2008), Her et al. (2008a,b), Zhang et al. (2008), and Korshin et al. (2009)

parameters, as is demonstrated by a number of reports from across the globe testifying to a significant increase in the both the amount, and probably also the quality, of NOM in water sources.

Water purification should be optimized by controlling water quality change during the treatment process. Many advanced characterization methods to describe the nature of NOM have been developed over the last few years, which are an asset in optimizing NOM removal from raw water. The main aim is to produce quality water that in turn will create fewer problems at later stages, such as disinfection and alkalization during transport to the tap.

Isolation, concentration, and fractionation of NOM are often implemented before analytical procedures are performed. Resin isolation and fractionation are widely accepted pretreatment methods to HPSEC or FTIR analysis. HPSEC has proved highly popular for size fractionation of NOM in conjunction with a number of different detection procedures. NMR spectroscopy has also been used extensively during the last few decades to investigate the structures and other properties of NOM.

Coupling MS with HPLC, allowing direct analysis without prior destruction or derivatization of the NOM molecules, has increased the significance of MS as a technique. More recently, the development of FTICR-MS has facilitated the molecular-level analysis of NOM, enabling the separation of molecular species from very complex NOM structures. Multidimensional NMR techniques together with FTICR-MS are new methods that appear promising for characterizing the major fractions of NOM. These techniques shed new light on the structural and functional features of NOM and fractionated NOM. Finally, reproducible characterization methods are more important than chemical structural analysis in promoting better knowledge of the practical, hygienic, and ecological impact of NOM.

REFERENCES

Abu-Lail, L., Liu, Y., Atabek, A., Camesano, T., 2007. Quantifying the adhesion and interaction forces between *Pseudomonas aeruginosa* and natural organic matter. Environ. Sci. Technol. 41, 8031–8037.

Aiken, G., 1984. Evaluation of ultrafiltration for determining molecular weight of fulvic acid. Environ. Sci. Technol. 18, 978–981.

Aiken, G., 1985. Isolation and concentration techniques for aquatic humic substances. In: Aiken, G., McKnight, D., Wershaw, R., MacCarthy, P. (Eds.), Humic Substances in Soil, Sediment and Water. John Wiley & Sons, pp. 363–385.

Aiken, G., McKnight, D., Thorn, K., Thurman, E., 1992. Isolation of hydrophilic organic acids from water using non-ionic macroporous resins. Org. Geochem. 18 (4), 567–573.

Alasonati, E., Slaveykova, V., Gallard, H., Croué, J.-P., Benedetti, M., 2010. Characterization of the colloidal organic matter from Amazonian basin by asymmetrical flow field-flow fractionation and size exclusion chromatography. Water Res. 44, 223–231.
Allpike, B., Heitz, A., Joll, C., Kagi, R., 2007. A new organic carbon detector for size exclusion chromatography. J. Chromatogr. A 1157, 472–476.
Amy, G., Collins, M., Kuo, C., King, P., 1987. Comparing gel permeation chromatography and ultrafiltration for the molecular weight characterization of aquatic organic matter. J. Am. Water Works Assoc. 79 (1), 43–49.
Archer, A., Singer, P., 2006. An evaluation of the relationship between SUVA and NOM coagulation using ICR database. J. Am. Water Works Assoc. 98 (7), 110–123.
Assemi, S., Newcombe, G., Hepplewhite, C., Beckett, R., 2004. Characterization of natural organic matter fractions separated by ultrafiltration using flow field-flow fractionation. Water Res. 38, 1467–1476.
Ates, N., Kitis, M., Yetis, U., 2007. Formation of chlorination by-products in waters with low SUVA – correlations with SUVA and differential UV spectroscopy. Water Res. 41, 4139–4148.
Baalousha, M., Motelica-Heino, M., Le Coustumer, P., 2006. Conformation and size of humic substances: effects of major cation concentration and type, pH, salinity, and residence time. Colloids Surf. A 272, 48–55.
Baalousha, M., Stolpe, B., Lead, J.R., 2011. Flow field-flow fractionation for the analysis and characterization of natural colloids and manufactured nanoparticles in environmental systems: a critical review. J. Chromatogr. A 1218, 4078–4103.
Baghoth, S., Dignum, M., Grefte, A., Kroesbergen, J., Amy, G., 2009. Characterization of NOM in a drinking water treatment process train with no disinfectant residual. Water Sci. Technol. 9 (1), 379–386.
Baghoth, S., Sharma, S., Amy, G., 2011. Tracking natural organic matter (NOM) in a drinking water treatment plant using fluorescence excitation–emission matrices and PARAFAC. Water Res. 45, 797–809.
Baker, A., Tipping, E., Thacker, S., Gondar, D., 2008. Relating dissolved organic matter fluorescence and functional properties. Chemosphere 73, 1765–1772.
Barber, L., Leenher, J., Noyes, T., Stiles, E., 2001. Nature and transformation of dissolved organic matter in treatment wetlands. Environ. Sci. Technol. 35, 4805–4816.
Barron, P., Wilson, M., 1981. Humic soil and coal structure study with magic-angle spinning ^{13}C CP NMR. Nature 289, 275–276.
Beckett, R., Jue, Z., Giddings, J.C., 1987. Determination of molecular weight distributions of fulvic and humic acids using flow field-flow fractionation. Environ. Sci. Technol. 21, 289–295.
Berwick, L., Greenwood, P., Kagi, R., Croué, J.-P., 2007. Thermal release of nitrogen organics from natural organic matter using micro scale sealed vessel pyrolysis. Org. Geochem. 38, 1073–1090.
Berwick, L., Greenwood, P., Smernik, R., 2010. The use of MSSV pyrolysis to assist the molecular characterization of aquatic natural organic matter. Water Res. 44, 3039–3054.
Bieroza, M., Baker, A., Bridgeman, J., 2009. Relating freshwater organic matter fluorescence to organic carbon removal efficiency in drinking water treatment. Sci. Total Environ. 407, 1765–1774.
Blodau, C., Bauer, M., Regenspurg, S., Macalady, D., 2009. Electron accepting capacity of dissolved organic matter as determined by reaction with metallic zinc. Chem. Geol. 260, 186–195.
Bose, P., Reckhow, D., 2007. The effect of ozonation on natural organic matter removal by alum coagulation. Water Res. 41, 1516–1524.
Bruchet, A., Rousseau, C., Mallevialle, J., 1990. Pyrolysis-GC-MS for investigating high-molecular-weight THM precursors and other refractory organics. J. Am. Water Works Assoc. 82, 66–74.

Chen, J., Gu, B., LeBoef, E.J., Pan, H., Dai, S., 2002. Spectroscopic characterization of the structural and functional properties of natural organic matter fractions. Chemosphere 48, 59–68.
Chen, W., Westerhoff, P., Leenheer, J., Booksh, K., 2003. Fluoresence exitation-emission matrix regional integration to quantify spectra for dissolved organic matter. Environ. Sci. Technol. 37, 5701–5710.
Chen, C., Zhang, X., He, W., Lu, W., Han, H., 2007. Comparison of seven kinds of drinking water treatment processes to enhance organic material removal: a pilot test. Sci. Total Environ. 382, 93–102.
Chen, C., Zhang, X., Zhu, L., Liu, J., He, W., Han, H., 2008. Disinfection by-products and their precursors in a water treatment plant in North China: seasonal changes and fraction analysis. Sci. Total Environ. 397, 140–147.
Chow, C., Fabris, R., Drikas, M., 2004. A rapid fractionation technique to characterize natural organic matter for the optimization of water treatment processes. J. Water Supply Res. Technol. Aqua 53, 85–92.
Chow, A.T., Dahlgren, R.A., Gao, S., 2005. Physical and chemical fractionation of dissolved organic matter and trihalomethane precursors: a review. J. Water Supply Res. Technol. 54, 475–507.
Chow, C., Fabris, R., van Leeuwen, J., Wang, D., Drikas, M., 2008. Assessing natural organic matter treatability using high performance size exclusion chromatography. Environ. Sci. Technol. 42, 6683–6689.
Chow, C., Kuntke, P., Fabris, R., Drikas, M., 2009a. Organic characterization tools for distribution system management. Water Sci. Technol. 9, 1–8.
Chow, C., von Leeuwen, J., Fabris, R., Drikas, M., 2009b. Optimized coagulation using aluminium sulfate for the removal of dissolved organic carbon. Desalination 245, 120–134.
Christy, A., Bruchet, A., Rybacki, D., 1999. Characterization of natural organic matter by pyrolysis/GC-MS. Environ. Int. 25, 181–189.
Coble, P.G., 1996. Characterization of marine and terrestrial DOM in seawater using excitation–emission matrix spectroscopy. Mar. Chem. 51, 325–346.
Cook, R.L., 2004. Coupling NMR to NOM. Anal. Bioanal. Chem. 376, 1484–1503.
Cooper, W., Song, W., Gonsior, M., Kalnina, D., Peake, B., Mezyk, S., 2008. Recent advances in structure and reactivity of dissolved organic matter in natural waters. Water Sci. Technol. 8, 615–623.
Cortés-Francisco, N., Caixach, J., 2013. Molecular characterization of dissolved organic matter through a desalination process by high resolution mass spectrometry. Environ. Sci. Technol. 47, 9619–9627.
Croue, J.-P., Korshin, G.V., Benjamin, M.M., 2000. Characterization of Natural Organic Matter in Drinking Water. American Water Works Association. pp. 324.
Croue, J.-P., 2004. Isolation of humic and non-humic NOM fractions: structural characterization. Environ. Monit. Assess. 92, 193–207.
Culea, M., Cozar, O., Ristoiu, D., 2006. Methods validation for the determination of trihalomethanes in drinking water. J. Mass Spectrom. 41, 1594–1597.
Danielsson, L., 1982. On the use of filters for distinguishing between dissolved and particulate fractions in natural waters. Water Res. 16, 179–182.
Datta, C., Ghosh, K., Mukherjee, S., 1971. Fluorescence excitation spectra of different fractions of humus. J. Ind. Chem. Soc. 48, 279–287.
Deshmukh, A., Pacheco, C., Hay, M., Myneni, S., 2007. Structural environments of carboxyl groups in natural organic molecules from terrestrial systems. Part 2: 2D NMR spectroscopy. Geochim. Cosmochim. Acta 71, 3533–3544.
Edzwald, J., Tobiason, J., 1999. Enhanced coagulation: USA requirements and a broader view. Water Sci. Technol. 40, 63–70.

Egeberg, P., Christy, A., Eikenes, M., 2002. The molecular size of natural organic matter (NOM) determined by diffusivimetry and seven other methods. Water Res. 36, 925–932.

Escobar, I., Randall, A., 2001. Assimilable organic carbon (AOC) and biodegradable dissolved organic carbon (BDOC): complementary measurements. Water Res. 35, 4444–4454.

Fabbri, D., Mongardi, M., Mintanari, L., Galletti, G., Chiavari, G., Scotti, R., 1998. Comparison between CP/MS ^{13}C-NMR and pyrolysis-GC/MS in the structural characterization of humins and humic acids of soil and sediments. J. Anal. Chem. 362, 299–306.

Fabris, R., Chow, C., Drikas, M., Eikebrokk, B., 2008. Comparison of NOM character in selected Australian and Norwegian drinking waters. Water Res. 42, 4188–4196.

Fukano, K., Komiya, K., Sasaki, H., Hashimoto, T., 1978. Evaluation of new supports for high-pressure aqueous gel permeation chromatography: TSK-GEL SW type columns. J. Chromatogr. A 166, 47–54.

Gjessing, E., 1965. Use of Sephadex gels for estimation of humic substances in natural water. Nature 208, 1091–1092.

Gjessing, E., 1970. Ultrafiltration of aquatic humus. Environ. Sci. Technol. 4, 437–438.

Gjessing, E.T., 1973. Gel- and ultramembrane filtration of aquatic humus: a comparison of the two methods. Schweiz. Z. Hydrol. 35, 286–294.

Gjessing, E., Alberts, J., Bruchet, A., Egeberg, P., Lydersen, E., McGown, L., Mobed, J., Munster, U., Pempkowiak, J., Perdue, M., Ratnawerra, H., Rybacki, D., Takacs, M., Abbt-Braun, G., 1998. Multi-method characterization of natural organic matter isolated from water: characterization of reverse osmosis-isolates from water of two semi-identical dystrophic lakes basins in Norway. Water Res. 32, 3108–3124.

Gjessing, E., Steiro, C., Becher, G., Christy, A., 2007. Reduced analytical availability of polychlorinated biphenyls (PCB's) in colored surface water. Chemosphere 66, 644–649.

Greenwood, P., Leenher, J., McIntyre, C., Berwick, L., Franzmann, P., 2006. Bacterial biomarkers thermally released from dissolved organic matter. Org. Geochem. 37, 597–609.

Haarhoff, J., Kubare, M., Mamba, B., Krause, R., Nkambule, T., Matsebula, B., Menge, J., 2010. NOM characterization and removal at six Southern African water treatment plants. Drink. Water Eng. Sci. 3, 53–61.

Hay, M., Myneni, S., 2007. Structural environments of carboxyl groups in natural organic molecules from terrestrial systems. Part 1: infrared spectroscopy. Geochim. Cosmochim. Acta 71, 3518–3532.

Her, N., Amy, G., Foss, D., Cho, J., Yoon, Y., Kosenka, P., 2002. Optimization of method for detecting and characterizing NOM by HPLC-size exclusion chromatography with UV and on-line DOC detection. Environ. Sci. Technol. 36, 1069–1076.

Her, N., Amy, G., Chung, J., Yoon, J., Yoon, Y., 2008a. Characterizing dissolved organic matter and evaluating associated nanofiltration membrane fouling. Chemosphere 70, 495–502.

Her, N., Amy, G., Sohn, J., Gunten von, U., 2008b. UV absorbance ratio index with size exclusion chromatography (URI-SEC) as an NOM property indicator. J. Water Supply Res. Technol. Aqua 57, 35–44.

Howe, K., Ishida, K., Clark, M., 2002. Use of ATR/FTIR spectrometry to study fouling of microfiltration membranes by natural waters. Desalination 147, 251–255.

Hua, G., Reckhow, D., 2007. Characterization of disinfection byproduct precursors based on hydrophobicity and molecular size. Environ. Sci. Technol. 41, 3309–3315.

Huber, S., Frimmel, F., 1992a. A liquid chromatographic system with multi-detection for direct analysis of hydrophilic organic compounds in natural waters. J. Anal. Chem. 342, 198–200.

Huber, S., Frimmel, F., 1992b. A new method for the characterization of organic carbon in aquatic systems. Int. J. Environ. Anal. Chem. 49, 49–57.

Huber, S., Frimmel, F., 1994. Direct gel chromatic characterization and quantification of marine dissolved organic carbon using high-sensitivity DOC detection. Environ. Sci. Technol. 28, 1194–1197.

Hudson, N., Baker, A., Reynolds, D., 2007. Fluorescence analysis of dissolved organic matter in natural, waste and polluted waters—a review. River Res. Appl. 23, 631–649.

Hudson, N., Baker, A., Ward, D., Reynolds, D., Brunsdon, C., Carliell-Marquet, C., Browning, S., 2008. Can fluorescence spectrometry be used as a surrogate for the biochemical oxygen demand (BOD) test in water quality assessment? An example from South West England. Sci. Total Environ. 391, 149–158.

Hur, J., Williams, M., Schlautman, M., 2006. Evaluating spectroscopic and chromatographic techniques to resolve dissolved organic matter via end member mixing analysis. Chemosphere 63, 387–402.

Jaouadi, M., Amdouni, N., Duclaux, L., 2012. Characteristics of natural organic matter extracted from the waters of Medjerda dam (Tunisia). Desalination 305, 64–71.

Jeong, S., Kim, S.-J., Kim, C., Vigneswaran, S., Nguyen, T., Shon, H.-K., Kandasamy, J., Kim, I., 2013. A detailed organic matter characterization of pretreated seawater using low pressure microfiltration hybrid systems. J. Membr. Sci. 428, 290–300.

Jung, C., Son, H., 2008. The relationship between disinfection by-products formation and characteristics of natural organic matter in raw water. Korean J. Chem. Eng. 25, 714–720.

Kang, M.-G., Ku, Y.-H., Cho, Y.-K., Yu, M.-J., 2006. Variation of dissolved organic matter and microbial regrowth potential through drinking water treatment processes. Water Sci. Technol. 6, 57–66.

Kanokkantapong, V., Marhaba, T., Panyapinyophol, B., Pavasant, P., 2006a. FTIR evaluation of functional groups involved in the formation of haloacetic acids during the chlorination of raw water. J. Hazard. Mater. B136, 188–196.

Kanokkantapong, V., Marhaba, T., Pavasant, P., Panyapinyophol, B., 2006b. Characterization of haloacetic acid precursors in source water. J. Environ. Manage. 80, 214–221.

Kim, H.-C., Yu, M.-J., 2007. Characterization of aquatic humic substances to DBPs formation in advanced treatment processes for conventionally treated water. J. Hazard. Mater. 143, 486–493.

Kim, H.-C., Yu, M.-J., Han, I., 2006. Multi-method study of the characteristic chemical nature of aquatic humic substances isolated from the Han River, Korea. Appl. Geochem. 21, 1226–1239.

Kim, J., Cai, Z., Benjamin, M., 2010. NOM fouling mechanisms in a hybrid adsorption/membrane system. J. Membr. Sci. 349, 35–43.

Koprivnjak, J., Perdue, E., Pfromm, P., 2006. Coupling reverse osmosis with electrodialysis to isolate natural organic matter from fresh waters. Water Res. 40, 3385–3392.

Korshin, G., Benjamin, M., Li, C.-W., 1999. Use of differential spectroscopy to evaluate the structure and reactivity of humics. Water Sci. Technol. 40, 9–16.

Korshin, G., Chow, C., Fabris, R., Drikas, M., 2009. Absorbance spectroscopy-based examination of effects of coagulation on the reactivity of fractions of natural organic matter with varying apparent molecular weights. Water Res. 43, 1541–1548.

Kothawala, D.N., Stedmon, C.A., Müller, R.A., Weyhenmeyer, G.A., Köhler, S.J., Tranvik, L.J., 2013. Controls of dissolved organic matter quality: evidence from a large-scale boreal lake survey. Glob. Chang. Biol. 20, 1101–1114.

Kristiana, I., Allpike, B., Joll, C., Heitz, A., Trolio, R., 2010. Understanding the behaviour of molecular weight fractions of natural organic matter to improve water treatment processes. Water Sci. Technol. Water Supply 10, 59–68.

Kunenkov, E., Kononikhin, A., Perminova, I., Hertkorn, N., Gaspar, A., Scmitt-Kopplin, P., Popov, I., Garmash, A., Nikolaev, E., 2009. Total mass difference statistics algorithm: a new approach to identification of high-mass building blocks in electrospray ionization Fourier transform ion cyclotron mass spectrometry data of natural organic matter. Anal. Chem. 81, 10106–10115.

Laborda, F., Ruiz-Beguería, S., Bolea, E., Castillo, J., 2009. Functional speciation of metal-dissolved organic matter complexes by size exclusion chromatography coupled to inductively coupled plasma mass spectrometry and deconvolution analysis. Spectrochim. Acta B 392, 392–398.
Lankes, U., Lüdemann, H.-D., Frimmel, F., 2008. Search for basic relationships between "molecular size" and "chemical structure" of aquatic natural organic matter – answers from ^{13}C and ^{15}N CPMAS NMR spectroscopy. Water Res. 42, 1051–1060.
Lankes, U., Müller, M., Weber, M., Frimmel, F., 2009. Reconsidering the quantitative analysis of organic carbon concentrations in size exclusion chromatography. Water Res. 43, 915–924.
Lead, J.R., Wilkinson, K.J., Balnois, E., Cutak, B.J., Larive, C.K., Assemi, S., Beckett, R., 2000. Diffusion coefficients and polydispersities of the Suwannee river fulvic acid: comparison of fluorescence correlation spectroscopy, pulsed-field gradient nuclear magnetic resonance, and flow field-flow fractionation. Environ. Sci. Technol. 34, 3508–3513.
Lee, D.-J., Cheng, Y.-L., Wong, R.-J., 2013. Characteristics of natural organic matters in raw and treated drinking water: comparison study. J. Taiwan Inst. Chem. Eng. 44, 103–110.
Leenher, J., 1981. Comprehensive approach to preparative isolation and fractionation of dissolved organic carbon from natural waters and wastewaters. Environ. Sci. Technol. 15, 578–587.
Leenher, J.A., 1985. Fractionation techniques for aquatic humic substances. In: Aiken, G., McKnight, D., Wershaw, R., MacCarthy, P. (Eds.), Humic Substances in Soil, Sediment and Water. John Wiley & Sons, pp. 409–429.
Leenher, J., Croué, J.-P., 2003. Characterizing dissolved aquatic organic matter. Environ. Sci. Technol. 37, 18A–26A.
Li, A., Hu, J., Li, W., Zhang, W., Wang, X., 2009. Polarity based fractionation of fulvic acids. Chemosphere 77, 1419–1426.
Liu, R., Lead, J., Baker, A., 2007. Fluorescence characterization of cross flow ultrafiltration derived freshwater colloidal and dissolved organic matter. Chemosphere 68, 1304–1311.
Liu, S., Lim, M., Fabris, R., Chow, C., Chiang, K., Drikas, M., Amal, R., 2008. Removal of humic acid using TiO_2 photocatalytic process – fractionation and molecular weight characterization studies. Chemosphere 72, 263–271.
Liu, S., Lim, M., Fabris, R., Chow, C., Drikas, M., Amal, R., 2010. Comparison of photocatalytic degradation of natural organic matter in two Australian surface waters using multiple analytical techniques. Org. Geochem. 41, 124–129.
Ma, H., Allen, H., Yin, Y., 2001. Characterization of isolated fractions of dissolved organic matter from natural waters and a wastewater effluent. Water Res. 35, 985–996.
Malcolm, R., MacCarthy, P., 1992. Quantitative evaluation of XAD-8 and XAD-4 resins used in tandem for removing organic solutes from water. Environ. Int. 18, 597–607.
Mao, J., Cory, R., McKnight, D., Schmidt-Rohr, K., 2007a. Characterization of a nitrogen-rich fulvic acid and its precursor algae from solid state NMR. Org. Geochem. 38, 1277–1292.
Mao, J., Tremblay, L., Gagné, J.-P., Kohl, S., Rice, J., Schmidt-Rohr, K., 2007b. Humic acids from particulate organic matter in the Saguenay Fjord and St. Lawrence Estuary investigated by advanced solid-state NMR. Geochim. Cosmochim. Acta 71, 5483–5499.
Marhaba, T., 2000. Fluorescence technique for rapid identification of DOM fractions. J. Environ. Eng. Sci. 126, 145–152.
Marhaba, T.F., Van, D., Lippincott, R.L., 2000. Changes in NOM fractions through treatment: a comparison of ozonation and chlorination. Ozone Sci. Technol. 22, 249–266.
Marhaba, T., Pu, Y., Bengraine, K., 2003. Modified dissolved organic matter fractionation technique for natural water. J. Hazard. Mater. 101, 43–53.
Marhaba, T., Borgaonkar, A., Punburananon, K., 2009. Principal component regression model applied to dimensionally reduced spectral fluorescent signature for the determination of organic character and THM formation potential of source water. J. Hazard. Mater. 169, 998–1004.

Markechova, D., Tomokova, M., Sadecka, J., 2013. Fluorescence excitation–emission matrix spectroscopy and parallel factor analysis in drinking water treatment: a review. Pol. J. Environ. Stud. 22, 1289–1295.
Matilainen, A., Iivari, P., Sallanko, J., Heiska, E., Tuhkanen, T., 2006. The role of ozonation and activated carbon filtration in the natural organic matter removal from drinking water. Environ. Technol. 27, 1171–1180.
Matilainen, A., Gjessing, E.T., Lahtinen, T., Hed, L., Bhatnagar, A., Sillanpää, M., 2011. An overview of the methods used in the characterization of natural organic matter (NOM) in relation to drinking water treatment. Chemosphere 83, 1431–1442.
Mawhinney, D., Rosario-Ortiz, F., Baik, S., Vanderford, B., Snyder, S., 2009. Characterization of fulvic acids by liquid chromatography-quadrupole time-of-flight mass spectrometry. J. Chromatogr. A 1216, 1319–1324.
Messaud, F., Sanderson, R., Runyon, J., Otte, T., Pasch, H., Williams, S., 2009. An overview on field-flow fractionation techniques and their applications in the separation and characterization of polymers. Prog. Polym. Sci. 34, 351–368.
Moon, J., Kim, S., Cho, J., 2006. Characterizations of natural organic matter as nano particle using flow field-flow fractionation. Colloids Surf. A Physicochem. Eng. Aspects 287, 232–236.
Murphy, K., Stedmon, C., Waite, T., Ruiz, G., 2008. Distinguishing between terrestrial and autochthonous organic matter sources in marine environments using fluorescence spectroscopy. Mar. Chem. 108, 40–58.
Nimmagadda, R., McRae, C., 2007. Characterization of the backbone structures of several fulvic acids using a novel selective chemical reduction method. Org. Geochem. 38, 1061–1072.
O'Loughlin, E., Chin, Y.-P., 2001. Effect of detector wavelength on the determination of the molecular weight of humic substances by high-pressure size exclusion chromatography. Water Res. 35, 333–338.
Odén, S., 1919. Die Huminsaure. Kolloidchem. Beih. 11, 75.
Park, J.-H., 2009. Spectroscopic characterization of dissolved organic matter and its interactions with metals in surface waters using size exclusion chromatography. Chemosphere 77, 485–494.
Parsi, Z., Hartog, N., Górecki, T., Poerschmann, J., 2007. Analytical pyrolysis as a tool for the characterization of natural organic matter – a comparison of different approaches. J. Anal. Appl. Pyrolysis 79, 9–15.
Peiris, R.H., Hallé, C., Budman, H., Moresoli, C., Peldszus, S., Huck, P.M., Legge, R.L., 2010. Identifying fouling events in a membrane-based drinking water treatment process using principal component analysis of fluorescence exitation-emission matrices. Water Res. 44, 185–194.
Penru, Y., Simon, F., Guastalli, A., Esplugas, S., Llorens, J., Baig, Sylvie, 2013. Characterization of natural organic matter from Mediterranean coastal seawater. J. Water Supply Res. Technol. Aqua 62, 42–51.
Peuravuori, J., Pihlaja, K., Välimäki, N., 1997. Isolation and characterization of natural organic matter from lake water: two different adsorption chromatographic methods. Environ. Int. 23, 453–464.
Peuravuori, J., Ingman, P., Pihlaja, K., 2003. Critical comments on accuracy of quantitative determination of natural humic matter by solid state ^{13}C NMR spectroscopy. Talanta 59, 177–189.
Pifer, A.D., Fairey, J.L., 2012. Improving on $SUVA_{254}$ using fluorescence-PARAFAC analysis and asymmetric flow-field flow fractionation for assessing disinfection byproduct formation and control. Water Res. 46, 2927–2936.
Posner, A., 1963. Importance of electrolyte in the determination of molecular weights by sephadex gel with specific reference to humic acids. Nature 198, 1161–1163.

Reemtsma, T., 2009. Determination of molecular formulas of natural organic matter molecules by (ultra-) high-resolution mass spectrometry – status and needs. J. Chromatogr. A 1216, 3687–3701.
Reemtsma, T., These, A., 2005. Comparative investigation of low-molecular-weight fulvic acids of different origin by SEC-Q-TOF-MS: new insights into structure and formation. Environ. Sci. Technol. 39, 3507–3512.
Reemtsma, T., These, A., Springer, A., Linscheid, M., 2006. Fulvic acids as transition state of organic matter: indications from high resolution mass spectrometry. Environ. Sci. Technol. 40, 5839–5845.
Reemtsma, T., These, A., Springer, A., Linscheid, M., 2008. Differences in the molecular composition of fulvic acid size fractions detected by size-exclusion chromatography-on line Fourier transform ion cyclotron resonance (FTICR) mass spectrometry. Water Res. 42, 63–72.
Richardson, S., Plewa, M., Wagner, E., Schoeny, R., DeMarini, D., 2007. Occurrence, genotoxicity, and carcinogenicity of regulated and emerging disinfection by-products in drinking water: a review and roadmap for research. Mutat. Res./Rev. Mutat. Res. 636, 178–242.
Richardson, S., Fasano, F., Ellington, J., Crumley, F., Buettner, K., Evans, J., Blount, B., Silva, L., Waite, T., Luther, G., McKague, A., Miltner, R., Wagner, E., Plewa, M., 2008. Occurrence and mammalian cell toxicity of iodinated disinfection by-products in drinking water. Environ. Sci. Technol. 42, 8330–8338.
Ritchie, J., Perdue, E., 2008. Analytical constraints on acidic functional groups in humic acids. Org. Geochem. 39, 783–799.
Roccaro, P., Vagliasindi, F., 2009. Differential vs. absolute UV absorbance approaches in studying NOM reactivity in DBPs formation: comparison and applicability. Water Res. 43, 744–750.
Rodriques, F., Núñez, L., 2009. Characterization of aquatic humic substances. Water Environ. J. 23, 1–8.
Rosario-Ortiz, F., Snyder, S., Suffet, I., 2007. Characterization of dissolved organic matter in drinking water sources impacted by multiple tributaries. Water Res. 41, 4115–4128.
Sanchez, N., Skeriotis, A., Miller, C., 2013. Assessment of dissolved organic matter fluorescence PARAFAC components before and after coagulation–filtration in a full scale water treatment plant. Water Res. 47, 1679–1690.
Sanchez, N.P., Skeriotis, A.T., Miller, C., 2014. A PARAFAC-based long-term assessment of DOM in a multi-coagulant drinking water treatment scheme. Environ. Sci. Technol. 48, 1582–1591.
Sanly, Lim M., Chiang, K., Amal, R., Fabris, R., Chow, C., Drikas, M., 2007. A study on the removal of humic acid using advanced oxidation process. Sep. Sci. Technol. 42, 1391–1404.
Sarathy, S., Mohseni, M., 2007. The impact of UV/H_2O_2 advanced oxidation on molecular size distribution of chromophoric natural organic matter. Environ. Sci. Technol. 41, 8315–8320.
Schulten, H., 1993. Analytical pyrolysis of humic substances and soils: geochemical, agricultural and ecological consequences. J. Anal. Appl. Pyrolysis 25, 97–122.
Seredynska-Sobecka, B., Baker, A., Lead, A., 2007. Characterization of colloidal and particulate organic carbon in freshwaters by thermal fluorescence quenching. Water Res. 41, 3069–3076.
Sharp, E., Jarvis, P., Parsons, S., Jefferson, B., 2006a. Impact of fractional character on the coagulation of NOM. Colloids Surf. A Physicochem. Eng. Aspects 286, 104–111.
Sharp, E., Parsons, S., Jefferson, B., 2006b. The impact of seasonal variations in DOC arising from a moorland peat catchment on coagulation with iron and aluminium salts. Environ. Pollut. 140, 436–443.

Shutova, Y., Baker, A., Bridgeman, J., Henderson, R., 2014. Spectroscopic characterization of dissolved organic matter changes in drinking water treatment: from PARAFAC analysis to online monitoring wavelengths. Water Res. 54, 159–169.
Simon, F., Penru, Y., Guastalli, A., Esplugas, S., Llorens, J., Baig, S., 2013. NOM characterization by LC-OCD in a SWRO desalination line. Desalin. Water Treat. 51, 1776–1780.
Simpson, A., Simpson, M., Kingery, W., Lefebvre, B., Moser, A., Williams, A., Kvasha, M., Kvelleher, B., 2006. The application of ^{1}H high-resolution magic-angle spinning NMR for the study of clay-organic associations in natural and synthetic complexes. Langmuir 22, 4498–4503.
Soh, Y., Roddick, F., van Leeuwen, J., 2008. The impact of alum coagulation on the character, biodegradability and disinfection by-product formation potential of reservoir natural organic matter (NOM) fractions. Water Sci. Technol. 58, 1173–1179.
Song, H., Orr, O., Hong, Y., Karanfil, T., 2009. Isolation and fractionation of natural organic matter: evaluation of reverse osmosis performance and impact of fractionation parameters. Environ. Monit. Assess. 153, 307–321.
Specht, C., Frimmel, F., 2000. Specific interactions of organic substances in size-exclusion chromatography. Environ. Sci. Technol. 34, 2361–2366.
Spencer, R., Bolton, L., Baker, A., 2007. Freeze/thaw and pH effects on freshwater dissolved organic matter fluorescence and absorbance properties from a number of UK locations. Water Res. 41, 2941–2950.
Stedmon, C., Markager, S., 2005. Resolving the variability in dissolved organic matter fluorescence in a temperate estuary and its catchment using PARAFAC analysis. Limnol. Oceanogr. 50, 686–697.
Stedmon, C., Markagera, S., Bro, R., 2003. Tracing dissolved organic matter in aquatic environments using a new approach to fluorescence spectroscopy. Mar. Chem. 82, 239–254.
Stenson, A., 2008. Reversed-phase chromatography fractionation tailored to mass spectral characterization of humic substances. Environ. Sci. Technol. 42, 2060–2065.
Swietlik, J., Dabrowska, A., Raczyk-Stanislawiak, U., Nawrocki, J., 2004. Reactivity of natural organic matter fractions with chlorine dioxide and ozone. Water Res. 38, 547–558.
Templier, J., Derenne, S., Croué, J.-P., Largeau, C., 2005. Comparative study of two fractions of riverine dissolved organic matter using various analytical pyrolytic methods and a ^{13}C CP/MAS NMR approach. Org. Geochem. 36, 1418–1442.
Tercero Espinoza, L., ter Haseborg, E., Weber, M., Frimmel, F., 2009. Investigation of the photocatalytic degradation of brown water natural organic matter by size exclusion chromatography. Appl. Catal. B Environ. 87, 56–62.
Thorn, K., Cox, L., 2009. N-15 NMR spectra of naturally abundant nitrogen in soil and aquatic natural organic matter samples of the International Humic Substances Society. Org. Geochem. 40, 484–499.
Thurman, E., 1985. Organic Geochemistry of Natural Waters. Martinus Nijhoff/Dr W. Junk Publishers, Dordrecht, The Netherlands.
Thurman, E., Malcolm, R., 1981. Preparative isolation of aquatic humic substances. Environ. Sci. Technol. 15, 463–466.
Uyguner, C., Suphandag, S., Kerc, A., Bekbolet, M., 2007. Evaluation of adsorption and coagulation characteristics of humic acids preceded by alternative advanced oxidation techniques. Desalination 210, 183–193.
Uyguner-Demirel, C.S., Bekbolet, M., 2011. Significance of analytical parameters for the understanding of natural organic matter in relation to photocatalytic oxidation. Chemosphere 84, 1009–1031.
Valencia, S., Marín, J., Restrepoa, G., Frimmel, F., 2013. Application of excitation–emission fluorescence matrices and UV/Vis absorption to monitoring the photocatalytic degradation of commercial humic acid. Sci. Total Environ. 442, 207–214.

Valencia, S., Marín, J., Restrepo, G., Frimmel, F.H., 2014. Evaluation of natural organic matter changes from Lake Hohloh by three-dimensional excitation–emission matrix fluorescence spectroscopy during TiO_2/UV process. Water Res. 51, 124–133.

Warton, B., Heitz, A., Allpike, B., Kagi, R., 2008. Size-exclusion chromatography with organic carbon detection using a mass spectrometer. J. Chromatogr. A 1207, 186–189.

Wei, Q., Feng, C., Wang, D., Shi, B., Zhang, L., Wei, Q., Tang, H., 2008. Seasonal variations of chemical and physical characteristics of dissolved organic matter and trihalomethane precursors in a reservoir: a case study. J. Hazard. Mater. 150, 257–264.

Wong, S., Hanna, J., King, S., Carroll, T., Eldridge, R., Dixon, D., Bolto, B., Hesse, S., Abbt-Braun, G., Frimmel, F.H., 2002. Fractionation of natural organic matter in drinking water and characterization by ^{13}C cross-polarization magic-angle spinning NMR spectroscopy and size exclusion chromatography. Environ. Sci. Technol. 36, 3497–3503.

Wu, F., Evans, R., Dillon, P., 2003. Separation and characterization of NOM by high-performance liquid chromatography and on-line three-dimensional excitation emission matrix fluorescence detection. Environ. Sci. Technol. 37, 3687–3693.

Wu, F., Evans, R., Dillon, P., Cai, Y., 2007a. Rapid quantification of humic and fulvic acids by HPLC in natural waters. Appl. Geochem. 22, 1598–1605.

Wu, F., Kothawala, D., Evans, R., Dillon, P., Cai, Y., 2007b. Relationships between DOC concentration, molecular size and fluorescence properties of DOM in a stream. Appl. Geochem. 22, 1659–1667.

Xing, L., Fabris, R., Chow, C., van Leeuwen, J., Drikas, M., Wang, D., 2012. Prediction of DOM removal of low specific UV absorbance surface waters using HPSEC combined with peak fitting. J. Environ. Sci. 24, 1174–1180.

Yamashita, Y., Jaffé, R., 2008. Characterizing the interactions between trace metals and dissolved organic matter using excitation–emission matrix and parallel factor analysis. Environ. Sci. Technol. 42, 7374–7379.

Yapsakli, K., Cecen, F., 2009. Use of an enriched inoculum for determination of biodegradable dissolved organic carbon (BDOC) in drinking water. Water Sci. Technol. 9, 149–157.

Zappia, L., Alessadrino, M., Garbin, S., Pringle, P., Heitz, A., Joll, C., Masters, D., Hiller, B., Capewell, S., Franzmann, P., Plumb, J., Von Eckstaedt, S., Cadee, K., 2008. Comparison of methods for the determination of biodegradable dissolved organic carbon in potable water supply: use of a novel biofilm. Water Sci. Technol. 8, 633–641.

Zhang, T., Lu, J., Ma, J., Qiang, Z., 2008. Fluorescence spectroscopic characterization of DOM fractions isolated from a filtered river water after ozonation and catalytic ozonation. Chemosphere 71, 911–921.

Zhang, H., Zhang, Y., Shi, Q., Hu, J., Chu, M., Yu, J., Yang, M., 2012a. Study on transformation of natural organic matter in source water during chlorination and its chlorinated products using ultrahigh resolution mass spectrometry. Environ. Sci. Technol. 46, 4396–4402.

Zhang, H., Zhang, Y., Shi, Q., Ren, S., Yu, J., Ji, F., Luo, W., Yang, M., 2012b. Characterization of low molecular weight dissolved natural organic matter along the treatment trait of a waterworks using Fourier transform ion cyclotron resonance mass spectrometry. Water Res. 46, 5197–5204.

Zhao, Z.-Y., Gu, D.-J., Li, H.-B., Li, X.-Y., Leung, K., 2009. Disinfection characteristics of the dissolved organic fractions at several stages of a conventional drinking water treatment plant in South China. J. Hazard. Mater. 172, 1093–1099.

Zhou, Q., Cabaniss, S., Maurice, P., 2000. Considerations in the use of high-pressure size exclusion chromatography (HPSEC) for determining molecular weights of aquatic chumic substances. Water Res. 34, 3505–3514.

CHAPTER 3

NOM Removal by Coagulation

Mika Sillanpää*, Anu Matilainen**

*Lappeenranta University of Technology, LUT Faculty of Technology, LUT Chemtech, Laboratory of Green Chemistry, Sammonkatu 12, 50130 Mikkeli, Finland; **Finnish Safety and Chemicals Agency, Kalevantie 2, 33100 Tampere, Finland

Abstract

The presence of natural organic matter (NOM) causes various problems in drinking water and drinking water treatment processes, including (1) negative effect on water quality by causing color, taste, and odor problems, (2) increased coagulant and disinfectant doses (which in turn results in increased sludge volumes and production of harmful disinfection by-products), (3) promoted biological growth in distribution system, and (4) increased levels of complexed heavy metals and adsorbed organic pollutants. NOM can be removed from drinking water through the use of several treatment options, of which the most common and economically feasible processes are considered to be coagulation and flocculation followed by sedimentation/flotation and sand filtration. Most of the NOM can be removed by coagulation, although the hydrophobic fraction and high molar mass compounds of NOM are removed more efficiently than hydrophilic fraction and the low molar mass compounds. Thus, enhanced and/or optimized coagulation, as well as new process alternatives for better removal of NOM by coagulation, have been suggested. In this chapter, we present an overview of the recent research dealing with coagulation and flocculation in the removal of NOM from drinking water.

Keywords: Coagulation; Drinking water; Natural organic matter; NOM; Water treatment.

ABBREVIATIONS

APAMs Anionic polyacrylamides
BDOC Biodegradable dissolved organic carbon
CNT Carbon nanotube
CPAMs Cationic polyacrylamides
DBP Disinfection by-product
DOC Dissolved organic carbon
FP Formation potential
HAAs Haloacetic acids
HFCVD Hot filament chemical vapor deposition
HMM High molar mass
HPAC Composite polyaluminum chloride
IPFs Inorganic polymer flocculants
LMM Low molar mass

NOM Natural organic matter
PAA Polyacrylamide
PACl Polyaluminum chloride
PAFC Polymer-polyaluminum ferric chloride
PAF-SiC Polyaluminum ferric silicate chloride
PAM Polyacrylamides
PAS Polyaluminum sulfate
PASiC Polyaluminum silicate chloride
PDADMAC Polydiallyldimethyl ammonium chloride
PFC Polyferric chloride
PFS Polyferric sulfate
PFSiS Polyferric silicate sulfate
PSF Polysilicate ferric
PTSS Polytitanium silicate sulfate
RSM Response surface method
SUVA Specific UV absorbance
THM Trihalomethane
TOC Total organic carbon

3.1 INTRODUCTION

Generally, the most common and economically feasible standard processes for the removal of natural organic matter (NOM) (and particulate matter) are considered to be coagulation and flocculation, followed by sedimentation/flotation and sand filtration (Jacangelo et al., 1995). Coagulation tends to balance cost and NOM removal efficiency well. Thus, enhanced and/or optimized coagulation appears to be the best option for better control of NOM and disinfection by-product (DBP) formation.

This chapter presents the options for NOM removal from drinking water using various coagulants and coagulation techniques. In the coagulation stage, solid particles are destabilized by a coagulant. This may include the formation of small aggregates by Brownian motion. The subsequent process by which large flocs are formed is known as *flocculation*. Chemical coagulation is achieved by the addition of inorganic coagulants (Duan and Gregory, 2003). Coagulants are most often iron or aluminum salts. When added to water, aqueous Al(III) and Fe(III) salts are dissociated to their respective trivalent ions, i.e., Al^{3+} and Fe^{3+}. They are then hydrolyzed and form several soluble complexes possessing high positive charges, thus adsorbing onto the surface of the negative colloids (Edzwald and Tobiason, 1999; Pham et al., 2006). According to the literature, at pH higher than the minimum solubility of the coagulants (e.g., 5.8 and 6.3 for ferric chloride and aluminum chloride, respectively), the hydrolysis products are high molar mass (HMM) polymers or colloidal/precipitated species, while at a pH slightly lower than the minimum solubility, the hydrolysis products are mainly medium polymers or monomers (Yan et al., 2008a). The degree of polymerization can be determined by ferron-timed spectroscopy. The different species can be classified into three categories: the mononuclear (Al_a and Fe_a), medium polynuclear (Al_b and Fe_b), and precipitated/colloidal (Al_c and Fe_c) species, in which the Al_b and Fe_b are considered to be the most efficient species for dissolved organic carbon (DOC) removal (Duan and Gregory, 2003; Zouboulis and Moussas, 2008; Yan et al., 2009a).

Coagulation is a physical–chemical process where the repulsive potential of electrical double layer of colloids is reduced in such a way that microparticles can be produced. These microparticles collide with each other to form larger structures, called flocs in the flocculation process. The use of coagulants for NOM removal was reviewed a few years ago (Matilainen et al., 2010) and this chapter is an update. Traditionally, coagulation has been employed in water treatment to decrease turbidity, resolve color problems, and remove pathogens. It must be emphasized, however, that optimal turbidity or color removal is not always achieved under the same conditions as NOM removal (Budd et al., 2004; Yu et al., 2007; Yan et al., 2008a; Vasyukova et al., 2012). For example, rapid changes in raw water composition can occur. Seasonal variations also play a key role (Vasyukova et al., 2012). In baseline coagulation, the coagulation conditions are optimized for turbidity removal, whereas optimized coagulation refers to dose and pH conditions optimized especially for organic matter reduction. Enhanced coagulation uses a higher amount of coagulant than would be needed for baseline coagulation, with accompanying changes in pH, order of chemical addition, or use of alternative coagulant chemicals. The goal is more efficient total organic carbon (TOC) and thus NOM removal (Budd et al., 2004; Bell-Ajy et al., 2000). Little work has been done on changes in water chemistry in NOM removal even by conventional ferric chloride and aluminum sulfate, but recently it has been discovered that if, during a treatment train, pH changes, removal of organic matter is affected (Slavik et al., 2012).

The aggregation mechanisms through which NOM is removed include a combination of charge neutralization, entrapment, adsorption, and complexation with coagulant metal ions into insoluble particulate aggregates (Figure 3.1) (Jarvis et al., 2004). Because of the variable composition of NOM, the removal mechanism will be different for the specific types of NOM molecules in water (Sharp et al., 2006a; Jarvis et al., 2004); however, the charge-driven mechanisms are most evident when pH is optimized

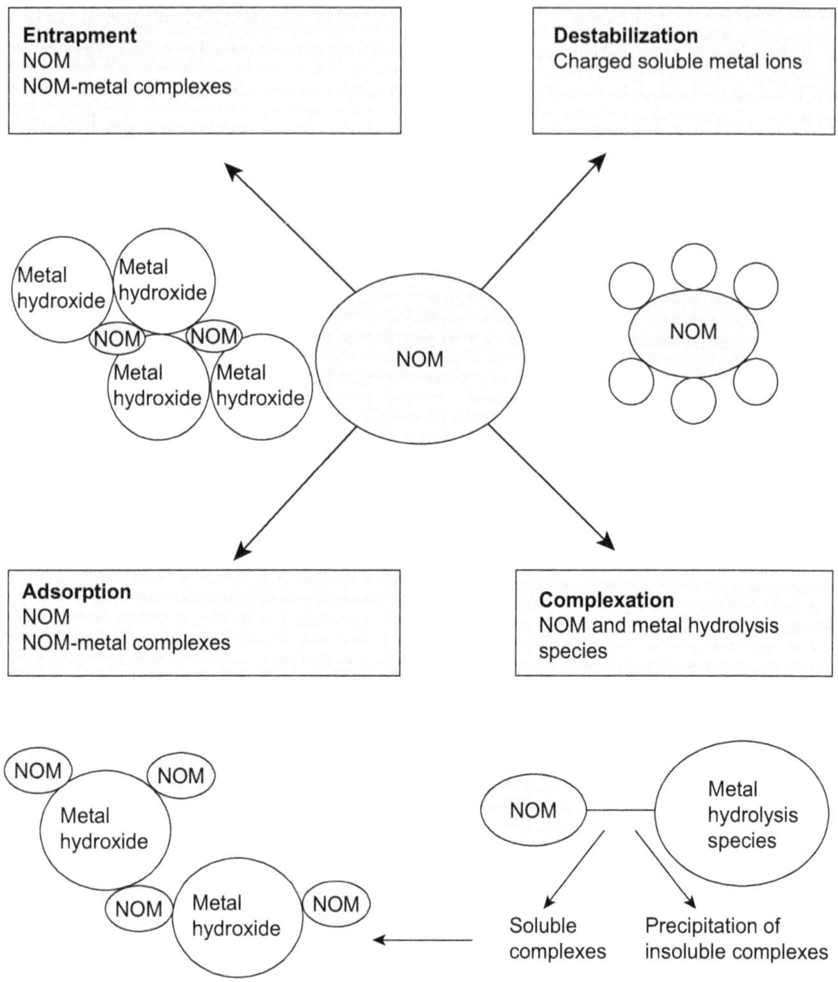

Figure 3.1 The possible removal mechanisms of natural organic matter (NOM) during coagulation. *Adapted from Jarvis et al. (2004).*

(Sharp et al., 2006a; Zhao et al., 2008; Bond et al., 2009). Moreover, the HMM polymer species (e.g., Al_c) efficiently remove particles through bridging or sweep flocculation, while medium polymer or monomer species appear to have a high ability to remove DOC by complexation, adsorption, charge neutralization, or coprecipitation (Yan et al., 2008a). The flocs formed during various coagulation mechanisms have different properties, including size, structure, and strength (Li et al., 2006).

The ability of coagulation to remove NOM and particles depends on several factors, including coagulant type and dosage, mixing conditions, pH, temperature, particle and NOM properties (such as size, functionality, charge, and hydrophobicity) and the presence of divalent cations and concentrations of destabilizing anions including bicarbonate, chloride and sulfate (Sharp et al., 2006a; Duan and Gregory, 2003; Kim et al., 2006; Sharp et al., 2006b; Qin et al., 2006). Ferric chloride, ferric sulfate, and alum are acidic, thus lowering the pH after application. Low-alkalinity waters therefore may require the addition of base, while high alkalinity may sometimes need to be lowered by adding acid to reach an efficient pH range (Jacangelo et al., 1995). The buffer capacity of high-alkalinity waters is higher than that of low-alkalinity waters against pH decrease (Budd et al., 2004).

The nature of NOM has a significant effect on the coagulant dose. The optimal coagulant dose for water with HMM NOM is expected to be low because the removal mechanism is mostly charge neutralization. If NOM consists of low molar mass (LMM) or non-humic substances, however, the removal mechanism is expected to be adsorption onto metal hydroxide surfaces and the optimal coagulant dose is thus much larger. The hydrophobic fraction of NOM is generally removed more efficiently than the hydrophilic fraction by coagulation. Furthermore, HMM compounds are more readily removed than LMM compounds (Korshin et al., 2009; Sharp et al., 2006a; Bond et al., 2009; Soh et al., 2008; Chow et al., 2009; Szlachta and Adamski, 2009) most likely because HMM matter is quite hydrophobic in nature, consisting more of aromatic compounds (Swietlik et al., 2004). The ability of high specific UV absorbance (SUVA) values to reflect the better reactivity of organic compounds in water for removal by coagulation has generated a great deal of discussion (Edzwald and Tobiason, 1999; Archer and Singer, 2006; Bose and Reckhow, 2007). The suggested relationship between SUVA and DOC is illustrated in Table 3.1 (Edzwald and Tobiason, 1999). Compared to DOC, UV_{254} has been noted to reduce more, suggesting that aromatic materials are removed more effectively than other NOM fractions (Uyguner et al., 2007a; Uyak and Toroz, 2007; Uyak et al., 2007). Hydrophobic NOM has been observed to carry high levels of negative charge due to the presence of ionized groups such as carboxylic and phenolic groups. Thus, the hydrophobic fraction dominates the specific colloidal charge character of water, and fractions with higher charge are more amenable to removal (Sharp et al., 2006c; Bond et al., 2009; Bose and Reckhow, 2007). NOM removal can be monitored by the removal of negative charge from the suspension. Maximum NOM removal has been achieved in a zeta potential window between −10 and +5 mV (Sharp et al., 2006c; Bond et al., 2009; Henderson et al., 2006).

Table 3.1 Relationship between specific UV absorbance (SUVA) and dissolved organic carbon (DOC) removal during coagulation

SUVA	Composition	Coagulation	DOC removals
>4	Mostly aquatic humics, high hydrophobicity, high molar mass (MM) compounds	Natural organic matter (NOM) controls, good DOC removals	>50% for alum, little greater for ferric
2–4	Mixture of aquatic humics and other NOM, mixture of hydrophobic and hydrophilic NOM, mixture of MMs	NOM influences, DOC removals should be fair to good	25–50% for alum, little greater for ferric
<2	Mostly nonhumics, low hydrophobicity, low MM compounds	NOM has little influence, poor DOC removals	<25% for alum, little greater for ferric

Adapted from Edzwald and Tobiason (1999).

The most widely used coagulant in drinking water production has been aluminum sulfate (alum), but recently the use of ferric salts has become more popular (Duan and Gregory, 2003). This may be due to a suggestion that aluminum in drinking water may increase the risk of Alzheimer's disease (Flaten, 2001). Many investigations have been made to compare the effectiveness of one coagulant over another. According to these studies, ferric-based coagulants have been generally noted to remove NOM, measured as DOC and UV_{254}, more effectively than aluminum-based ones (Sharp et al., 2006c; Budd et al., 2004; Bell-Ajy et al., 2000; Uyguner et al., 2007a; Uyak and Toroz, 2007) and were found to be less sensitive to low temperatures than alum with regard to turbidity removal (Yu et al., 2007), although contradictory results have also been obtained in some cases (Uyguner et al., 2007b).

It has been proposed that the removal mechanisms of ferric and alum coagulants are different (Sharp et al., 2006c; Jarvis et al., 2005; Lei et al., 2009). Ferric-based coagulants have proved more effective, especially in removing NOM with intermediate MM (1000–4000 $g\,mol^{-1}$). This fraction may be largely composed of fulvic acids and more efficiently removed by ferric salts due to the higher charge density of ferric coagulants (Lindqvist et al., 2002, 2004). Moreover, the flocs formed during ferric coagulation are noted to be both larger and more numerous than those formed during aluminum coagulation, therefore increasing the likelihood of collisions and subsequent removal (Lindqvist et al., 2002, 2004; Ratnaweera et al., 1999).

The choice of proper coagulant ultimately depends on the characteristics of the water to be treated. In addition, Cheng et al. (2010) discovered that intensive and prolonged rapid mixing enhances the NOM removal.

New inorganic polymer flocculants (IPFs) have been developed during the past few years based on inorganic salts. Prehydrolyzed coagulants are formed by the forced hydrolysis of a simple coagulant, such as alum. Concerning their coagulation–flocculation properties, IPFs are considered to be somewhere between traditional inorganic salts and organic flocculants (Moussas and Zouboulis, 2009). Besides aluminum and ferric-based polymeric flocculants, a combination of the two has also been tested: polyaluminum ferric silicate chloride (PAF-SiC) coagulant (Cheng et al., 2008). It was observed that flocs formed during coagulation were relatively large but nonsettleable, causing high turbidity when the dosage was insufficient, although turbidity was effective at high coagulant doses.

3.2 ALUMINUM-BASED COAGULANTS

The most commonly used aluminum-based coagulant is alum ($Al_2(SO_4)_3$), with other coagulants including aluminum chloride ($AlCl_3$). Yet relatively high aluminum residuals may remain in the finished water, which can cause possible health hazards or other problems in the distribution system, e.g., spontaneous flocculation, especially during periods of cold temperatures or at low pH levels (Chow et al., 2009; Flaten, 2001; Matsukawa et al., 2006; Bérubê and Dorea, 2008). This can be avoided by pH control, but aluminum-based coagulants also affect turbidity and NOM removal (Matsukawa et al., 2006; Bérubê and Dorea, 2008). Table 3.2 presents pros and cons of aluminum-based coagulants.

Increasing the alum dose increases NOM removal up to a certain point. At very high doses (>100 $mg L^{-1}$), NOM removal is not significantly improved, suggesting the presence of components of NOM recalcitrant to removal by coagulation. Larger MM components are removed more efficiently, whereas LMM compounds are more resistant to removal (Chow et al., 2008, 2009; Soh et al., 2008). The hydrophilic neutral NOM fraction becomes difficult to remove with the addition of alum at any concentration. Furthermore, this fraction has been observed to have the highest biodegradable dissolved organic carbon (BDOC) contribution to the treated water, thus potentially supporting biofilm formation in the distribution system, and contributing to trihalomethane (THM) formation (Soh et al., 2008). The recalcitrant fraction of NOM is indicated to be polysaccharides and

Table 3.2 Advantages and disadvantages of aluminum-based coagulants

Coagulant type	Features	Advantages	Disadvantages	References
Alum, aluminum chloride	Trivalent aluminum ions are released into a solution from the respective salt. They are hydrolyzed and form soluble complexes possessing high positive charges. Coagulation efficiency depends on, e.g., coagulant dose, mixing, pH, temperature, particle and natural organic matter (NOM) properties. During coagulation, the most effective range of pH is 5–6.5.	Stable, easily handled, readily soluble. Better turbidity removal than with ferric salts in many cases. Can be more effective than ferric in low doses. Higher color removal efficiency.	Relatively high coagulant residuals in the finished water in some cases. Possibly increases risk of Alzheimer's disease. Ferric salts were found to be better than aluminum salts for NOM removal. High alkalinity consumption. Sulfate and/or chloride in finished water increases corrosivity. Net increase in the dissolved constituents.	Smith and Kamal (2009), Chow et al. (2008), Yu et al. (2007), Uyguner et al. (2007a), Qin et al. (2006), Soh et al. (2008), Chow et al. (2009), Uyak and Toroz (2007), Flaten (2001), Lindqvist et al. (2004), Matsukawa et al. (2006), Bérubê and Dorea (2008), Shin et al. (2008), Yan et al. (2008a), Rizzo et al. (2008), Yan et al. (2006)

their derivatives (Chow et al., 2009). Coagulation pH has a significant effect on coagulation efficiency, and the observed optimal pH for alum coagulation is in the range of 5.0–6.5 (with alum doses between 5 and 100 mgAl L^{-1}), resulting in an average of 25–67%, 44–77%, 25–66% and 97% of DOC, UV_{254}, THMFP and turbidity removal, respectively (Smith and Kamal, 2009; Yu et al., 2007; Uyguner et al., 2007a; Qin et al., 2006; Soh et al., 2008; Chow et al., 2009; Uyak and Toroz, 2007; Shin et al., 2008; Rigobello et al., 2011).

3.3 FERRIC-BASED COAGULANTS

Ferric salts commonly used in coagulation processes include ferric chloride ($FeCl_3$) and ferric sulfate ($Fe_2(SO_4)_3$) (Budd et al., 2004). The observed optimal pH level for ferric-based coagulation is pH 4.5–6, with NOM removal at about 29–70% as DOC (Budd et al., 2004; Yu et al., 2007; Bell-Ajy et al., 2000; Bond et al., 2009; Uyak and Toroz, 2007; Abbaszadegan et al., 2007; Meyn et al., 2008; Park and Yoon, 2009; Rigobello et al., 2011). DOC removal and THMFP mitigation increase with increasing coagulant dose up to 100 mgL^{-1} (Uyak et al., 2007). At low coagulant doses, the anionic humic network reorganizes upon association with cationic coagulant species to yield more compact structures; when the coagulant dose is increased, the aggregation of humic acid with hydrolyzed-Fe species has been ascribed to competition between the humic network reconformation rate and collision rate of destabilized colloids (Siéliéchi et al., 2008). Table 3.3 presents pros and cons of iron-based coagulants.

3.4 INORGANIC POLYMER FLOCCULANTS

Prehydrolyzed aluminum coagulants such as polyaluminum chloride (PACl) and polyaluminum sulfate (PAS) have been developed in the past decade (Yan et al., 2008b, 2009a; Hu et al., 2006; Iriarte-Velasco et al., 2007; Ji et al., 2008; Li et al., 2008; Wang et al., 2009; Zhan et al., 2011; Trinh and Kang, 2011; Amin et al., 2012). PACl has been made by partially neutralizing $AlCl_3$ to different basicity ratios, and its use is becoming more widespread. Prehydrolyzing the $AlCl_3$ can enhance the amount of Al_{13} ($AlO_4Al_{12}(OH)_{24}{}^{7+}$) in the coagulation process, making it the most efficient known Al-species for contaminant removal (Hu et al., 2006; Dongsheng et al., 2006; Yan et al., 2008c). These coagulant species (Al_{13} or Al_b) are considered to be the most efficient Al-species due to their larger size and higher positive charges. The

Table 3.3 Advantages and disadvantages of iron-based coagulants

Coagulant type	Features	Advantages	Disadvantages	References
Ferric chloride, ferric sulfate	Ferric salts hydrolyze similarly to aluminum salts when added to water and form different hydrolysis products. Coagulation effectiveness is dependent on the same factors as for alum salts. The most effective pH range is 4.5–6.	Many studies found ferric salts better than aluminum salts at removing natural organic matter (NOM), especially the middle size NOM fractions. Not as sensitive to temperature changes as alum. Larger flocs than those produced by Al coagulants.	Ferric-based coagulant acid strength and associated optimized flocculation pH ranges can produce purified water with less buffering capacity and require greater chemical addition for stabilization and corrosion control. High alkalinity consumption. Sulfate and/or chloride in finished water increases corrosion. Net increase in the dissolved constituents.	Sharp et al. (2006a), Budd et al. (2004), Yu et al. (2007), Bell-Ajy et al. (2000), Uyguner et al. (2007a), Uyak and Toroz (2007), Lindqvist et al. (2002), Lindqvist et al. (2004), Rizzo et al. (2008), Rigobello et al. (2011), Zhao et al. (2011)

properties of flocs formed under various aluminum coagulant species have been observed to differ (Hu et al., 2006; Li et al., 2008; Wang et al., 2009; Dongsheng et al., 2006; Shi et al., 2007), and the destabilization mechanisms of different aluminum species have also been investigated (Zhao et al., 2008; Yan et al., 2008c; Kazpard et al., 2006; Sarpola et al., 2007). Prehydrolyzed polymer coagulants have reported advantages over conventional coagulants, including reduced temperature or pH dependence, as well as lower alkalinity consumption (Iriarte-Velasco et al., 2007; Wang et al., 2009; Dongsheng et al., 2006; Shi et al., 2007; Yan et al., 2009b), but the characteristics of the water to be treated (e.g., alkalinity, pH, NOM content) play a major role in the choice of a proper coagulant (Hu et al., 2006; Yan et al., 2008b). Consequently, prehydrolyzed coagulants do not appear to enhance the removal efficiency of NOM in all cases (Hu et al., 2006; Shi et al., 2007). Lou et al. (2012) suggested that the drawback of using some polymeric salts is that the preformed species are stable and cannot be further hydrolyzed during coagulation. Thus, these coagulants may be relatively inefficient in removing a highly hydrophobic NOM (Lou et al., 2012).

Activated silica is an efficient coagulant aid and enhances the bridging ability of PACl (Yan et al., 2008a). Polymeric aluminum coagulant with supplemented activated SiO_2 has a larger molecular weight and smaller surface charge than PACl (Yan et al., 2008a; Cheng et al., 2008; Zouboulis and Tzoupanos, 2009). Tubić et al. (2013) studied the role of various NOM fractions in changes in THM and haloacetic acid formation during coagulation with $FeCl_3$ and a combination of PACl and $FeCl_3$. Based on their hydrophobicity, they fractionated the DOC in the raw water into four parts after coagulation and observed significant changes in the reactivities of the hydrophilic acidic and nonacidic fractions.

Polyaluminum silicate chloride (PASiC) can enhance the efficiency of NOM and turbidity removal, proving more efficient than PACl in the treatment of low turbidity waters (Cheng et al., 2008). Improved coagulation properties depend on both the basicity and silica content of a polymer and the preparation technique. PASiC is more resistant to pH variation (the effective pH range is 6.5–9) than PACl, and the flocs formed under optimal conditions are larger, contributing to greater removal efficiency (Zouboulis and Tzoupanos, 2009).

Instead of processing one factor at a time—which is time consuming and incapable of revealing the optimal combination of factors as it ignores the interrelation—Trinh and Kang (2011) applied the response surface method (RSM) to PACl coagulation. RSM includes the impacts of various individual factors and their interactions. The authors successfully optimized the coagulation conditions for turbidity and DOC removal.

More attention has recently been paid to inorganic polymeric iron coagulants, including polyferric sulfate (PFS), polymeric iron chloride (PICl), or polyferric chloride (PFC) (Jarvis et al., 2005; Zouboulis et al., 2008; Liu and Chin, 2009; Cao et al., 2011; Wang et al., 2008; Lei et al., 2009; Zhan et al., 2010, 2011). These coagulants may have some advantages over monomeric forms of ferric salts, such as a wider working pH range, lower sensitivity to water temperature, reduced amounts of coagulant, and lower residual iron concentrations (Park and Yoon, 2009; Liu and Chin, 2009), although they are still in the development stage and the applications are limited. Prehydrolyzed coagulants are prepared by partial neutralization of metal salts, e.g., PFS from ferric sulfate. The prime advantage of these coagulants is that they produce only the most favorable hydrolysis and polymeric species for coagulation. The performance of the prehydrolyzed coagulants depends on the nature of the polymeric species formed in the process of preparation (Lei et al., 2009; Moussas and Zouboulis, 2009). Possible dominant removal mechanisms during PFC coagulation are adsorption, entrapment, and complexation rather than charge neutralization (Wei et al., 2009a). Zhan et al. (2011) reported the lack of research on coagulation performance and floc properties for NOM removal under various low-shear flows at the flocculation stage, especially at low NOM levels. They compared PFC and PACl as coagulants and observed that the G values in the slow-mixing phase had greater influence on the PFC flocs.

Polymeric coagulant efficiency can be improved by adding other components to the polymer such as polysilic acid, to produce new composite coagulants such as polysilicate ferric (PSF) and polyferric silicate sulfate (PFSiS) (Zouboulis and Moussas, 2008; Li et al., 2008; Fu and Yu, 2007; Fu et al., 2009; Moussas and Zouboulis, 2008; Xu et al., 2009). These coagulants work across a broad pH range and exhibit a markedly better coagulation performance than, e.g., ferric sulfate. Table 3.4 presents pros and cons of inorganic polymer flocculants.

3.5 ORGANIC POLYELECTROLYTES

A polymer is termed a *polyelectrolyte* when its monomeric unit contains ionizable groups (e.g., carboxyl, amino or sulphonic). Thus, polyelectrolytes are characterized by their charge, charge density, and molecular weight (MW). Consequently, polyelectrolytes can be cationic, anionic, or ampholytic,

Table 3.4 Advantages and disadvantages of inorganic polymer flocculants

Coagulant type	Features	Advantages	Disadvantages	References
PACl	Prepared by partially neutralized (prehydrolyzed) aluminum chloride. Enhanced amounts of high-charged, moderate MM hydrolysis species, e.g., Al_{13}.	Less temperature and pH dependent than alum salts. Lower alkalinity consumption. Better natural organic matter (NOM) removal capacity than alum in many cases. Lower dose requirement and less sludge produced. Lower residual aluminum in treated water.	Coagulant effectiveness is significantly affected by coagulant hydrolysis species speciation. Preformed Al species are stable and cannot be further hydrolyzed during coagulation. Might not be so efficient in removing HMM and highly hydrophobic NOM.	Yan et al. (2009a), Cheng et al. (2008), Yan et al. (2008b), Dongsheng et al. (2006), Yan et al. (2006, 2008a), Zhan et al. (2011), Trinh and Kang (2011), Amin et al. (2012)
PICl, PFC	Prepared by partially neutralized (prehydrolyzed) ferric chloride. Enhanced amounts of high-charged, moderate MM hydrolysis species.	Wider pH range, lower sensitivity to temperature, reduced amounts of coagulant, lower residual iron concentration. Less corrosive than ferric chloride. Lower sludge production.	Still in the development stage. Not so widely used.	Liu and Chin (2009), Wei et al. (2009a), Zhan et al. (2010, 2011), Cao et al. (2011)

Continued

Table 3.4 Advantages and disadvantages of inorganic polymer flocculants—cont'd

Coagulant type	Features	Advantages	Disadvantages	References
PFS	Made by partially neutralized (prehydrolyzed) ferric sulfate. Enhanced amounts of high-charged, moderate MM hydrolysis species.	Broader pH range, lower sensitivity to temperature, reduced amounts of coagulants, lower residual iron concentration. Less corrosive than ferric chloride.	Hydrolysis conditions have impact on speciation of hydrolysis/polymeric species. Still in the development stage. Not so widely used.	Lei et al. (2009), Zouboulis et al. (2008)
PASiC, PAF-SiC, PFSiS, PSF	Combination of aluminum and/or ferric-based polymeric flocculants with polysilicate.	Enhances bridging ability of coagulant. Relatively large flocs formed. More resistant to pH variations. Less corrosive. Wider pH range. Better performance. Produces little sludge. Effective in hydrophobic NOM removal. Only most favorable hydrolysis and polymeric species are produced in some cases.	Flocs' difficulty in settling creates high turbidity values at low dosages. Still in the development stage (especially ferric-based coagulants). High costs.	Yan et al. (2008b), Zouboulis and Moussas (2008), Cheng et al. (2008), Li et al. (2008), Zouboulis and Tzoupanos (2009), Fu and Yu (2007), Fu et al. (2009), Moussas and Zouboulis (2008), Xu et al. (2009), Lou et al. (2012)

HMM, high molar mass; PACl, polyaluminum chloride; PICl, polymeric iron chloride; PFC, polyferric chloride; PFS, polyferric sulfate; PASiC, polyaluminum silicate chloride; PAF-SiC, polyaluminum ferric silicate chloride; PFSiS, polyferric silicate sulfate; PSF, polysilicate ferric.

while polymers without ionizable groups are termed nonionic. Cationic polyelectrolytes include polydiallyldimethyl ammonium chloride (PDADMAC), cationic polyacrylamides (CPAMs) and natural cationic polymers (e.g. chitosan, starch), whereas anionic polyelectrolytes include anionic PAMs or natural biopolymers. Nonionic polyelectrolytes include polyacrylamide (PAA) (Bolto and Gregory, 2007).

Synthetic or natural polymers are used as both primary coagulants and coagulant or flocculant aids. Cationic polyelectrolytes are most commonly used for NOM removal because NOM in water is likely to be negatively charged, favoring NOM removal by charge neutralization over bridging. The cationic polymer PDADMAC has been used as primary coagulant for NOM removal, since it is highly charged and has HMM, although its high cost has been a limiting factor (Bolto and Gregory, 2007; Gao et al., 2008). Polymer is used in combination with a conventional coagulant to lower the process costs by reducing coagulant demand, resulting in larger aggregates from already formed microflocs (Bolto and Gregory, 2007; Jarvis et al., 2006). Polyelectrolytes are more resistant to pH changes than conventional coagulants, and have achieved better floc characteristics in terms of size, strength, settleability and filterability, thus producing a lower volume of sludge (Edzwald and Tobiason, 1999; Jarvis et al., 2004; Li et al., 2006; Bolto and Gregory, 2007; Barbot et al., 2008). Moreover, cationic polymers have proved efficient coagulation aids in improving PACl coagulation performance by enhancing the neutralization ability of PACl (Yan et al., 2008a) and improving the ability of PFS to remove DOC and turbidity (Wei et al., 2009a). In some cases, however, no improvement in floc characteristics or NOM removal efficiency was reported (Jarvis et al., 2006; Rizzo et al., 2008). The NOM removal capacity of different polyelectrolytes has varied from case to case (Yu et al., 2007; Yan et al., 2008a), so polyelectrolytes may not always be suitable coagulants, e.g., for turbidity removal (Yu et al., 2007). Cationic organic polyelectrolytes in particular are considered to be quite toxic to aquatic organisms and several countries have thus restricted their use in water purification (Bolto and Gregory, 2007; Rizzo et al., 2008). Table 3.5 presents the advantages and disadvantages of various organic polyelectrolytes, which were discussed above.

3.6 COMPOSITE COAGULANTS

The combination of organic polymers or inorganics into IPFs to create a composite coagulant is an innovation in coagulation. These composites are synthesized by adding inorganic or organic modifiers to the respective coagulants (Zhu et al., 2012). The aim is to combine the advantages of both

Table 3.5 The advantages and disadvantages of organic polyelectrolytes

Coagulant type	Features	Advantages	Disadvantages	References
Organic polyelectrolytes				
Cationic (e.g., PDADMAC, CPAMs, chitosan)	Cationic polyelectrolytes are the most commonly used polyelectrolytes due to their charge density. Charge neutralization is probably the main NOM removal mechanism.	High charge density polyelectrolytes remove hydrophobic NOM effectively. Produces smaller amounts of sludge, which is easier to dewater.	Higher the polymer dose required, higher the cost. Toxic effects. Formation of smaller flocs due to charge neutralization.	Yan et al. (2008b), Lindqvist et al. (2004), Wei et al. (2009a), Bolto and Gregory (2007), Gao et al. (2008), Rizzo et al. (2008)
Anionic (e.g., APAMs)	High MW carboxylic acid polymers. NOM removal mainly by bridging mechanism.	Improves low molecular weight (LMW) matter removal in some cases.	Not as effective as cationic polymers in every case.	Lindqvist et al. (2004); Bolto and Gregory (2007)
Inorganic salt + polyelectrolyte	Coagulation of NOM with inorganic salts using polyelectrolytes as a coagulant and/or flocculant aid, to maximize the benefits of both.	Larger and stronger flocs are formed than with either coagulant alone. Lower coagulant dose requirements. Smaller volume of sludge. Cost savings.	Higher the polymer dose required, higher the cost. Toxic effects.	Lindqvist et al. (2004); Bolto and Gregory (2007)

NOM, natural organic matter; PDADMAC, polydiallyldimethyl ammonium chloride; CPAM, cationic polyacrylamides; APAM, anionic polyacrylamides; MW, molecular weight.

components and further enhance coagulation efficiency (Tzoupanos and Zouboulis, 2011). Several researchers have studied the composition of PFC-PDADMAC (Wei et al., 2009a; Bolto and Gregory, 2007; Wang et al., 2008; Wei et al., 2009b,c), as well as PFS-PAA (Lindqvist et al., 2004). PFC-PDADMAC has efficiently removed NOM and created high levels of Fe_b species, the most active components in ferric salt coagulants (Gao et al., 2008; Wang et al., 2008; Wei et al., 2009b). Another novel composite coagulant is polytitanium silicate sulfate (PTSS) (Huang et al., 2014). Composites may be more suitable for surface water treatment than conventional metal coagulants or organic polymers, as they are efficient for DOC, SUVA, and turbidity removal through stronger charge neutralization (Wei et al., 2009a). PFS-PAA is particularly effective, generating flocs that are larger and have a broader size distribution range than those produced by a PSF + PAA mixture or either component alone (Moussas and Zouboulis, 2009).

A highly efficient composite polyaluminum chloride (HPAC) has been developed from PACl and other organic and inorganic additives such as active silicates and PDADMAC. HPAC has exhibited about 30% more efficiency on NOM removal than conventional coagulants or PACl (Yan et al., 2006, 2007, 2008a, 2009b), and has proved more efficient than conventional coagulants in basic conditions, thus being effective in waters with high pH or alkalinity (Yan et al., 2008a, 2009b). Polymer polyaluminum ferric chloride (PAFC) and composite $FeCl_3$/PAFC coagulate more efficiently than $FeCl_3$ (Lou et al., 2012). This composite could integrate the effects of adsorption, charge neutralization, entrapment, and complexation.

Bioflocculants are nontoxic and biodegradable (Yang et al., 2009) but have a relatively low flocculating capacity (Li et al., 2009). The compound bioflocculant-aluminum sulfate dual coagulant was found to exhibit a similar floc growth rate and strength to aluminum sulfate alone in the coagulation of kaolin-humic acid solution (Bo et al., 2012). Table 3.6 presents advantages and disadvantages of various composite coagulants, which have been discussed above.

3.7 NOVEL COAGULANTS

Al and Fe coagulants meet the need for improved NOM removal. In addition to Fe and Al salts, the inorganic coagulant Zr may prove effective due to its increased positive charge (Jarvis et al., 2008, Jarvis et al., 2012). Zr provides a significantly lower DOC residual, lower THMs, and larger and stronger flocs than Fe. This was verified in pilot test runs in which flocs

Table 3.6 Advantages and disadvantages of composite coagulants

Coagulant type	Features	Advantages	Disadvantages	References
$FeCl_3$/PAFC inorganic-bioflocculant composites	Composite bioflocculant-aluminum sulfate	Nontoxic and biodegradable.	Relatively low flocculating capacity.	Bo et al. (2012)
HPAC	Composition of organic polymer (PDADMAC) and inorganic polymeric coagulant (PACl) with additives such as active silicates.	Effective in waters with high alkalinity or pH. Wider working pH range.	Flocs formed can be small or incompact and thus unfavorable to sedimentation, although flotation is much more efficient.	Yan et al. (2008b), Yan et al. (2009a), Yan et al. (2006), Yan et al. (2007)
PFC-PDADMAC	Composition of organic polymer (PDADMAC) and inorganic polymeric coagulant (PFC). High content of Fe_b hydrolysis species.	Stronger charge neutralization, hence efficient DOC, SUVA, and turbidity removal capacity.	Still in the development stage. Not so widely used.	Wei et al. (2009a), Gao et al. (2008), Wang et al. (2008), Wei et al. (2009b), Wei et al. (2009c)

PFC, polyferric chloride; PDADMAC, polydiallyldimethyl ammonium chloride; DOC, dissolved organic carbon; SUVA, specific UV absorbance; PAFC, polymer-polyaluminum ferric chloride; HPAC, composite polyaluminum chloride.

formed by the Zr coagulant were better removed by filtration and flotation processes. Yet, redestabilization occurred, so careful control of coagulation conditions is necessary (Jarvis et al., 2012). Zr was proposed as a good substitute when additional NOM removal is required, particularly due to its nontoxicity.

Aluminum formate was found to be as efficient in coagulation as conventional polyaluminum chloride and aluminum sulfate (Väänänen et al., 2012; Mikola et al., 2013). This coagulant is advantageous because it does not contribute harmful sulfate and chloride ions to waters. It was observed that aluminum formate functions selectively toward some NOM fractions. The composite coagulant polymeric aluminum ferric sulfate (PAFS) also proved effective (Zhu et al., 2012).

Carbon nanotubes (CNT) have demonstrated their efficiency as coagulants, although they lack dispersion and solubility properties (Simate et al., 2012a). Attempts have been made to overcome these challenges (Zhang et al., 2002, 2009; Pei et al., 2008). Due to their cost and potential toxicity, CNTs need to be separated by membrane filtration or ultracentrifugation and regenerated after use (Simate et al., 2012b). Another novel nanostructured coagulant that may prove useful for NOM removal is nanocellulose (Suopajärvi et al., 2013).

3.8 CONCLUSIONS

This chapter has reviewed recent research in the field of NOM removal from drinking water by coagulation. The character and concentration of NOM in raw water sources change both regionally and seasonally (Fabris et al., 2008; Sharp et al., 2006d). In addition, an overall increase in NOM content has been observed in various regions worldwide (Eikebrokk et al., 2004). Coagulation, particularly when enhanced or optimized, is a major treatment option to address raised NOM levels. Enhanced coagulation is an especially good option for high-alkalinity waters (Yan et al., 2008a).

Enhanced coagulation removed more NOM than baseline coagulation in the majority of the reported studies, yet some of the LMM compounds and nonhumic material in NOM still remained in the treated water. Applying increased coagulant dosages and controlling the raw water pH by the addition of acid or base are two common options for enhancing NOM removal by coagulation. Yet, these solutions increase the proportion of rapidly precipitated material as sludge or flocs, forming looser aggregates and thus exhibiting inferior dewatering properties (Verrelli et al., 2009, 2010). Sludge can be dewatered by thickening, centrifugation, or filtration. Fenton's reagent may also improve sludge dewaterability, and is considered a good option for conditioning an alum-based sludge (Tony et al., 2008). Other ways of handling the sludge formed include atomization and use in the cement industry (Rodriguez et al., 2010). Enhanced coagulation also increases the iron corrosion of the finished water due to the consumption of alkalinity and introduction of sulfate, so water discoloration may result (Shi and Taylor, 2007). Instead of increasing coagulant doses, NOM removal may be better enhanced by recycling backwash streams that contain destabilized particles, especially in low-turbidity waters (Gottfried et al., 2008). Since enhanced coagulation aims to further NOM removal, the aesthetic quality of the water must be preserved, which includes efficient microbial and virus deterioration (Abbaszadegan et al., 2007).

Conventional coagulation has been performed with inorganic salts, including alum, ferric chloride, and ferric sulfate. A recent improvement is the development of prehydrolyzed ferric and alum coagulants called IPFs (e.g., PFC and PACl). Their advantages over conventional coagulants include a broader working pH range, lower sensitivity to water temperature, and lower residual coagulant concentrations (Dongsheng et al., 2006; Liu and Chin, 2009). Synthetic and natural organic polyelectrolytes have long been used in combination with metal coagulants to achieve better coagulation efficiency and floc characteristics (Bolto and Gregory, 2007). A new application combines IPFs and organic polymers, such as PFC-PDADMAC and PFS-PAA, with efficient NOM removal rates (Moussas and Zouboulis, 2009; Wei et al., 2009c). Other new coagulants include Mg/Al hydrotalcite (Santosa et al., 2008) and Zr^{4+}, which creates significantly larger flocs than Fe^{3+} coagulation, achieving a higher DOC and recalcitrant NOM removal efficiency; this technique is still in development (Jarvis et al., 2008). A highly promising new approach to removing organics from water is nanotechnology, which has introduced new types of material such as dendrimers, hyperbranched polymers, carbon nanotubes, cyclodextrins (Upadhyayula et al., 2009), and nanocellulose (Suopajärvi et al., 2013). These nanomaterials have exceptional adsorption properties and are able to entrap a variety of organic molecules. Nanomaterials are particularly effective in removing DBPs and other organic pollutants, and have also been observed to remove NOM (Lu and Su, 2007).

REFERENCES

Abbaszadegan, M., Mayer, B.K., Ryu, H., Nwachuku, N., 2007. Efficacy of removal of CCL viruses under enhanced coagulation conditions. Environ. Sci. Technol. 41, 971–977.
Amin, M., Faraji, M., Momenbeik, F., Hasanzadeh, A., Sadani, M., 2012. Performance evaluation of an anaerobic migrating blanket reactor in the biodegradation of perchloroethylene from industrial wastewaters. Int. J. Environ. Health Eng. 1, 1–8.
Archer, A., Singer, P., 2006. An evaluation of the relationship between SUVA and NOM coagulation using the ICR database. J. Am. Water Works Assoc. 98, 110–123.
Barbot, E., Moustier, S., Bottero, J., Moulin, P., 2008. Coagulation and ultrafiltration: understanding of the key parameters of the hybrid process. J. Membr. Sci. 325, 520–527.
Bell-Ajy, K., Abbaszadegan, M., Ibrahim, E., Verges, D., LeChevallier, M., 2000. Conventional and optimized coagulation for NOM removal. J. Am. Water Works Assoc. 92, 44–58.
Bérubê, D., Dorea, C., 2008. Optimizing alum coagulation for turbidity, organics, and residual Al reductions. Water Sci. Technol. Water Supply 8, 505–511.
Bo, X., Gao, B., Peng, N., Wang, Y., Yue, Q., Zhao, Y., 2012. Effect of dosing sequence and solution pH on floc properties of the compound bioflocculant–aluminum sulfate dual-coagulant in kaolin–humic acid solution treatment. Bioresour. Technol. 113, 89–96.
Bolto, B., Gregory, J., 2007. Organic polyelectrolytes in water treatment. Water Res. 41, 2301–2324.
Bond, T., Goslan, E.H., Parsons, S.A., Jefferson, B., 2009. Disinfection by-product formation of natural organic matter surrogates and treatment by coagulation, MIEX and nanofiltration. Water Res. 44, 1645–1653.
Bose, P., Reckhow, D., 2007. The effect of ozonation on natural organic matter removal by alum coagulation. Water Res. 41, 1516–1524.
Budd, G., Hess, A., Shorney-Darby, H., Neemann, J., Spencer, C., Bellamy, J., Hargette, P., 2004. Coagulation applications for new treatment goals. J. Am. Water Works Assoc. 92, 102–113.
Cao, B., Gao, B., Liu, X., Wang, M., Yang, Z., Yue, Q., 2011. The impact of pH on floc structure characteristic of polyferric chloride in a low DOC and high alkalinity surface water treatment. Water Res. 45, 6181–6188.
Cheng, W., Chi, F., Li, C., Yu, R., 2008. A study on the removal of organic substances from low-turbidity and low-alkalinity water with metal-polysilicate coagulants. Colloids Surf. A Physicochem. Eng. Aspects 312, 238–244.
Cheng, Y.-L., Wong, R.-J., Lin, J., Huang, C., Lee, D.-J., Mujumdar, A., 2010. Water coagulation using electrostatic patch coagulation (EPC) mechanism. Drying Technol. 28, 850–857.
Chow, C., Fabris, R., van Leeuwen, J., Wang, D., Drikas, M., 2008. Assessing natural organic matter treatability using high performance size exclusion chromatography. Environ. Sci. Technol. 42, 6683–6689.
Chow, C., van Leeuwen, J., Fabris, R., Drikas, M., 2009. Optimized coagulation using aluminium sulfate for the removal of dissolved organic carbon. Desalination 245, 120–134.
Dongsheng, W., Hong, L., Chunhua, L., Hongxiao, T., 2006. Removal of humic acid by coagulation with nano-Al_{13}. Water Sci. Technol. Water Supply 6, 59–67.
Duan, J., Gregory, J., 2003. Coagulation by hydrolysing metal salts. Adv. Colloid Interface Sci. 100–102, 475–502.
Edzwald, J., Tobiason, J., 1999. Enhanced coagulation: US requirements and a broader view. Water Sci. Technol. 40, 63–70.
Eikebrokk, B., Vogt, R., Liltved, H., 2004. NOM increase in Northern European source waters: discussion of possible causes and impacts on coagulation/contact filtration processes. Water Sci. Technol. Water Supply 4, 47–54.
Fabris, R., Chow, C., Drikas, M., Eikebrokk, B., 2008. Comparison of NOM character in selected Australian and Norwegian drinking waters. Water Res. 42, 4188–4196.

Flaten, T., 2001. Aluminium as a risk factor in Alzheimer's disease, with emphasis on drinking water. Brain Res. Bull. 55, 187–196.

Fu, Y., Yu, S., Han, C., 2009. Morphology and coagulation performance during preparation of poly-silicic-ferric (PSF) coagulant. Chem. Eng. J. 149, 1–10.

Fu, Y., Yu, S., 2007. Characterization and coagulation performance of solid poly-silicic–ferric (PSF) coagulant. J. Non-Cryst. Solids 353, 2206–2213.

Gao, B.-Y., Wang, Y., Yue, Q.-Y., Wei, J.-C., Li, Q., 2008. The size and coagulation behaviour of a novel composite inorganic–organic coagulant. Sep. Purif. Technol. 62, 544–550.

Gottfried, A., Shepard, A., Hardiman, K., Walsh, M., 2008. Impact of recycling filter backwash water on organic removal in coagulation–sedimentation processes. Water Res. 42, 4683–4691.

Henderson, R., Sharp, E., Jarvis, P., Parsons, S., Jefferson, B., 2006. Identifying the linkage between particle characteristics and understanding coagulation performance. Water Sci. Technol. Water Supply 6, 31–38.

Hu, J., Liu, H., Qu, J., Wang, D., Ru, J., 2006. Coagulation behaviour of aluminum salts in eutrophic Water: significance of Al_{13} species and pH control. Environ. Sci. Technol. 40, 325–331.

Huang, X., Gao, B., Wang, Y., Yue, Q., Li, Q., Zhang, Y., 2014. Coagulation performance and flocs properties of a new composite coagulant: polytitanium-silicate-sulfate. Chem. Eng. J. 245, 173–179.

Iriarte-Velasco, U., Àlvarez-Uriarte, J., Gonzáles-Velasco, J., 2007. Enhanced coagulation under changing alkalinity-hardness conditions and its implications on trihalomethane precursors removal and relationship with UV absorbance. Sep. Purif. Technol. 55, 368–380.

Jacangelo, J., DeMarco, J., Owen, D., Randtke, S., 1995. Selected processes for removing NOM: an overview. J. Am. Water Works Assoc. 87, 64–77.

Jarvis, P., Banks, J., Molinder, R., Stephenson, T., Parsons, S., Jefferson, B., 2008. Processes for enhanced NOM removal: beyond Fe and Al coagulation. Water Sci. Technol. Water Supply 8, 709–716.

Jarvis, P., Jefferson, B., Parsons, S., 2004. Characterizing natural organic matter flocs. Water Sci. Technol. Water Supply 4, 79–87.

Jarvis, P., Jefferson, B., Parsons, S., 2005. Breakage, regrowth, and fractal nature of natural organic matter flocs. Environ. Sci. Technol. 39, 2307–2314.

Jarvis, P., Jefferson, B., Parsons, S., 2006. Floc structural characteristics using conventional coagulation for a high doc, low alkalinity surface water source. Water Res. 40, 2727–2737.

Jarvis, P., Sharp, E., Pidou, M., Molinder, R., Parsons, S., Jefferson, B., 2012. Comparison of coagulation performance and floc properties using a novel zirconium coagulant against traditional ferric and alum coagulants. Water Res. 46, 4179–4187.

Ji, Q., Liu, H., Hu, C., Qu, J., Wang, D., Li, J., 2008. Studies on the treatment of brilliant green solution by combination microwave induced oxidation with $CoFe_2SO_4$. Sep. Purif. Technol. 62, 464–469.

Kazpard, V., Lartiges, B., Frochot, C., d'Espinose de la Caillerie, J., Viriot, M., Portal, J., Görner, T., Bersillon, J., 2006. Fate of coagulant species and conformational effects during the aggregation of a model of a humic substance with Al_{13} polycations. Water Res. 40, 1965–1974.

Kim, H., Lee, S., Byun, S., Yu, M., 2006. Application of improved rapid mixing for enhanced removal of dissolved organic matter and DBPFP (disinfection by-product formation potential) control. Water Sci. Technol. Water Supply 6, 49–57.

Korshin, G., Chow, C., Fabris, R., Drikas, M., 2009. Absorbance spectroscopy-based examination of effects of coagulation on the reactivity of fractions of natural organic matter with varying apparent molecular weights. Water Res. 43, 1541–1548.

Lei, G., Ma, X., Guan, X., Song, A., Cui, Y., 2009. Effect of basicity on coagulation performance of polyferric chloride applied in eutrophicated raw water. Desalination 247, 518–529.

Li, F., Akira, Y., Yuka, A., 2008. Characterization of micro-flocs of NOM coagulated by PACl, alum and polysilicate-iron in terms of molecular weight and floc size. Water Sci. Technol. 57, 83–90.

Li, T., Zhu, Z., Wang, D., Yao, C., Tang, H., 2006. Characterization of floc size, strength and structure under various coagulation mechanisms. Powder Technol. 168, 104–110.

Li, Z., Zhong, S., Lei, H.-Y., Chen, R.-W., Yu, Q., Li, H.-L., 2009. Production of a novel bioflocculant by *Bacillus licheniformis* X14 and its application to low temperature drinking water treatment. Bioresour. Technol. 100, 3650–3656.

Lindqvist, N., Jokela, J., Tuhkanen, T., 2004. In: Proceedings of the 11th International Gothenburg Symposium on Chemical Treatment of Water and Wastewater, November 8–11, Orlando, Florida, USA.

Lindqvist, N., Korhonen, S., Jokela, J., Tuhkanen, T., 2002. In: Hahn, H., Hofmann, E., Odegaard, H. (Eds.), Chemical Water and Wastewater Treatment VIII. IWA Publishing, London, p. 133.

Liu, T.-K., Chin, C., 2009. Improved coagulation performance using preformed polymeric iron chloride (PICl). Colloids Surf. A Physicochem. Eng. Aspects 339, 192–198.

Lou, I., Gong, S., Huang, X., Liu, Y., 2012. Coagulation optimization for low temperature and low turbidity source water using combined coagulants: a case study. Desalin. Water Treat. 46, 107–114.

Lu, C., Su, F., 2007. Adsorption of natural organic matter by carbon nanotubes. Sep. Purif. Technol. 58, 113–121.

Mikola, M., Rämö, J., Sarpola, A., Tanskanen, J., 2013. Removal of different NOM fractions from surface water with aluminium formate. Sep. Purif. Technol. 118, 842–846.

Matilainen, A., Vepsäläinen, M., Sillanpää, M., 2010. Natural organic matter removal by coagulation during drinking water treatment: a review. Adv. Colloid Interface Sci. 159, 189–197.

Matsukawa, S., Itoho, S., Habutnu, S., Aizawa, T., 2006. Water Sci. Technol. Water Supply 6, 67.

Meyn, T., Bahn, A., Leiknes, T., 2008. Significance of flocculation for NOM removal by coagulation–ceramic membrane microfiltration. Water Sci. Technol. Water Supply 8, 691–700.

Moussas, P., Zouboulis, A., 2008. A study on the properties and coagulation behaviour of modified inorganic polymeric coagulant—Polyferric silicate sulphate (PFSiS). Sep. Purif. Technol. 63, 475–483.

Moussas, P., Zouboulis, A., 2009. A new inorganic–organic composite coagulant, consisting of Polyferric Sulphate (PFS) and Polyacrylamide (PAA). Water Res. 43, 3511–3524.

Park, S., Yoon, T., 2009. Effects of iron species and inert minerals on coagulation and direct filtration for humic acid removal. Desalination 239, 146–158.

Pei, X., Hu, L., Liu, W., Hao, J., 2008. Synthesis of water-soluble carbon nanotubes via surface initiated redox polymerization and their tribological properties as water-based lubricant additive. Eur. Polym. J. 44, 2458–2464.

Pham, A., Rose, A., Feitz, A., Waite, T., 2006. Kinetics of Fe(III) precipitation in aqueous solutions at pH 6.0–9.5 and 25 °C. Geochim. Cosmochim. Acta 70, 640–650.

Qin, J.-J., Oo, M., Kekre, K., Knops, F., Miller, P., 2006. Impact of coagulation pH on enhanced removal of natural organic matter in treatment of reservoir water. Sep. Purif. Technol. 49, 295–298.

Ratnaweera, H., Gjessing, E., Oug, E., 1999. Influence of physical–chemical characteristics of natural organic matter (NOM) on coagulation properties: an analysis of eight Norwegian water sources. Water Sci. Technol. 40, 89–95.

Rigobello, E., Dantas, A., Di Bernardo, L., Vieira, E., 2011. Influence of the apparent molecular size of aquatic humic substances on colour removal by coagulation and filtration. Environ. Technol. 32, 1767–1777.

Rizzo, L., Di Gennaro, A., Gallo, M., Belgiorno, V., 2008. Coagulation/chlorination of surface water: a comparison between chitosan and metal salts. Sep. Purif. Technol. 62, 79–85.

Rodriguez, N., Ramirez, S., Blanco Varela, M., Guillem, M., Puig, J., 2010. Re-use of drinking water treatment plant (DWTP) sludge: characterization and technological behaviour of cement mortars with atomized sludge additions. Cem. Concr. Res. 40, 778–786.

Santosa, S.J., Kunarti, E.S., Karmanto, 2008. Synthesis and utilization of Mg/Al hydrotalcite for removing dissolved humic acid. Appl. Surf. Sci. 254, 7612–7617.

Sarpola, A., Hellman, H., Hietapelto, V., Jalonen, J., Jokela, J., Rämö, J., Saukkoriipi, J., 2007. Hydrolysis products of water treatment chemical aluminium sulfate octadecahydrate by electrospray ionization mass spectrometry. Polyhedron 26, 2851–2858.

Sharp, E., Jarvis, P., Parsons, S., Jefferson, B., 2006a. Impact of fractional character on the coagulation of NOM. Colloids Surf. A Physicochem. Eng. Aspects 286, 104–111.

Sharp, E., Jarvis, P., Parsons, S., Jefferson, B., 2006b. The impact of zeta potential on the physical properties of ferric–NOM flocs. Environ. Sci. Technol. 40, 3934–3940.

Sharp, E., Parsons, S., Jefferson, B., 2006c. The impact of seasonal variations in DOC arising from a moorland peat catchment on coagulation with iron and aluminium salts. Environ. Pollut. 140, 436–443.

Sharp, E., Parsons, S., Jefferson, B., 2006d. Seasonal variations in natural organic matter and its impact on coagulation in water treatment. Sci. Total Environ. 363, 183–194.

Shi, B., Taylor, J., 2007. Potential impact of enhanced coagulation on corrosion by-product release in a distribution system. Desalination 208, 260–268.

Shi, B., Wei, Q., Wang, D., Zhu, Z., Tang, H., 2007. Coagulation of humic acid: the performance of preformed and non-preformed Al species. Colloids Surf. A Physicochem. Eng. Aspects 296, 141–148.

Shin, J., Spinette, R., O'Melia, C., 2008. Stoichiometry of coagulation revisited. Environ. Sci. Technol. 42, 2582–2589.

Siéliéchi, J.-M., Lartiges, B., Kayem, G., Hupont, S., Frochot, C., Thieme, J., Ghanbaja, J., d'Espinose de la Caillerie, J., Barrés, O., Kamga, R., Levitz, P., Michot, L., 2008. Changes in humic acid conformation during coagulation with ferric chloride: implications for drinking water treatment. Water Res. 42, 2111–2123.

Simate, G., Iyuke, S., Ndlovu, S., Heydenrych, M., Walubita, L., 2012a. Human health effects of residual carbon nanotubes and traditional water treatment chemicals in drinking water. Environ. Int. 39, 38–49.

Simate, G., Iyuke, S., Ndlovu, S., Heydenrych, M., 2012b. The heterogeneous coagulation and flocculation of brewery wastewater using carbon nanotubes. Water Res. 46, 1185–1197.

Slavik, I., Müller, S., Mokosch, R., Azongbilla, J., Uhl, W., 2012. Impact of shear stress and pH changes on floc size and removal of dissolved organic matter (DOM). Water Res. 46, 6543–6553.

Smith, E., Kamal, Y., 2009. Optimizing treatment for reduction of disinfection by-product (DBP) formation. Water Sci. Technol. Water Supply 9, 191–198.

Soh, Y., Roddick, F., van Leeuwen, J., 2008. The impact of alum coagulation on the character, biodegradability and disinfection by-product formation potential of reservoir natural organic matter (NOM) fractions. Water Sci. Technol. 58, 1173–1179.

Suopajärvi, T., Liimatainen, H., Hormi, O., Niinimäki, J., 2013. Coagulation–flocculation treatment of municipal wastewater based on anionized nanocelluloses. Chem. Eng. J. 231, 59–67.

Swietlik, J., Dabrowska, A., Raczyk-Stanislawiak, U., Nawrocki, J., 2004. Reactivity of natural organic matter fractions with chlorine dioxide and ozone. Water Res. 38, 547–558.
Szlachta, M., Adamski, W., 2009. Effects of natural organic matter removal by integrated processes: alum coagulation and PAC-adsorption. Water Sci. Technol. 59, 1951–1957.
Tony, M., Zhao, Y., Fu, J., Tayeb, A., 2008. Conditioning of aluminium-based water treatment sludge with Fenton's reagent: effectiveness and optimizing study to improve dewaterability. Chemosphere 72, 673–677.
Trinh, T., Kang, L., 2011. Response surface methodological approach to optimize the coagulation–flocculation process in drinking water treatment. Chem. Eng. Res. Des. 89, 1126–1135.
Tzoupanos, N., Zouboulis, A., 2011. Preparation, characterization and application of novel composite coagulants for surface water treatment. Water Res. 45, 3614–3626.
Upadhyayula, V., Deng, S., Mitchell, M., Smith, G., 2009. Application of carbon nanotube technology for removal of contaminants in drinking water: a review. Sci. Total Environ. 408, 1–13.
Uyak, V., Toroz, I., 2007. Disinfection by-product precursors reduction by various coagulation techniques in Istanbul water supplies. J. Hazard. Mater. 141, 320–328.
Uyak, V., Yavuz, S., Toroz, I., Ozaydin, S., Genceli, E., 2007. Disinfection by-products precursors removal by enhanced coagulation and PAC adsorption. Desalination 216, 334–344.
Uyguner, C., Bekbolet, M., Selcuk, H., 2007a. A comparative approach to the application of a physico-chemical and advanced oxidation combined system to natural water samples. Sep. Sci. Technol. 42, 1405–1419.
Uyguner, C., Suphandag, S., Kerc, A., Bekbolet, M., 2007b. Evaluation of adsorption and coagulation characteristics of humic acids preceded by alternative advanced oxidation techniques. Desalination 210, 183–193.
Vasyukova, E., Uhl, W., Braga, F., Simõnes, C., Baylão, T., Neder, K., 2012. Drinking water production from surface water sources in the tropics: Brasília DF, Brazil. Environ. Earth Sci. 65, 1587–1599.
Verrelli, D., Dixon, D., Scales, P., 2009. Effect of coagulation conditions on the dewatering properties of sludges produced in drinking water treatment. Colloids Surf. A Physicochem. Eng. Aspects 348, 14–23.
Verrelli, D., Dixon, D., Scales, P., 2010. Assessing dewatering performance of drinking water treatment sludges. Water Res. 44, 1542–1552.
Väänänen, M., Kupiainen, L., Rämö, J., Sarpola, A., Tanskanen, J., 2012. Speciation and coagulation performance of novel coagulant – aluminium formate. Sep. Purif. Technol. 86, 242–247.
Wang, Y., Gao, B.-Y., Xu, X.-M., Xu, W.-Y., Xu, G.-Y., 2009. Characterization of floc size, strength and structure in various aluminum coagulants treatment. J. Colloid Interface Sci. 332, 354–359.
Wang, Y., Gao, B., Yue, Q., Wei, J., Li, Q., 2008. The characterization and flocculation efficiency of composite flocculant iron salts–polydimethyldiallylammonium chloride. Chem. Eng. J. 142, 175–181.
Wei, J., Gao, B., Yue, Q., Wang, Y., Li, W., Zhu, X., 2009b. Comparison of coagulation behavior and floc structure characteristic of different polyferric-cationic polymer dual-coagulants in humic acid solution. Water Res. 43, 724–732.
Wei, J., Gao, B., Yue, Q., Wang, Y., Lu, L., 2009a. Performance and mechanism of polyferric-quaternary ammonium salt composite flocculants in treating high organic matter and high alkalinity surface water. J. Hazard. Mater. 165, 789–795.
Wei, J., Gao, B., Yue, Q., Wang, Y., 2009c. Strength and regrowth properties of polyferric-polymer dual-coagulant flocs in surface water treatment. J. Hazard. Mater. 175, 949–954.
Xu, X., Yu, S.-I., Shi, W., Jiang, Z., Wu, C., 2009. Effect of acid medium on the coagulation efficiency of polysilicate-ferric (PSF)—a new kind of inorganic polymer coagulant. Sep. Purif. Technol. 66, 486–491.

Yan, M., Liu, H., Wang, D., Ni, J., Qu, J., 2009a. Natural organic matter removal by coagulation: effect of kinetics and hydraulic power. Water Sci. Technol. Water Supply 9, 21–30.
Yan, M., Wang, D., Ni, J., Qu, J., Chow, C.W.K., Liu, H., 2008c. Mechanism of natural organic matter removal by polyaluminum chloride: effect of coagulant particle size and hydrolysis kinetics. Water Res. 42, 3361–3370.
Yan, M., Wang, D., Ni, J., Qu, J., Ni, W., Van Leeuwen, J., 2009b. Natural organic matter (NOM) removal in a typical North-China water plant by enhanced coagulation: targets and techniques. Sep. Purif. Technol. 68, 320–327.
Yan, M., Wang, D., Qu, J., Ni, J., Chow, C.W.K., 2008a. Enhanced coagulation for high alkalinity and micro-polluted water: the third way through coagulant optimization. Water Res. 42, 2278–2286.
Yan, M., Wang, D., Shi, B., Wang, M., Yan, Y., 2007. Effect of pre-ozonation on optimized coagulation of a typical North-China source water. Chemosphere 69, 1695–1702.
Yan, M., Wang, D., You, S., Qu, J., Tang, H., 2006. Enhanced coagulation in a typical North-China water treatment plant. Water Res. 40, 3621–3627.
Yan, M., Wang, D., Yu, J., Ni, J., Edwards, M., Qu, J., 2008b. Enhanced coagulation with polyaluminum chlorides: role of pH/alkalinity and speciation. Chemosphere 71, 1665–1673.
Yang, Z.-H., Huang, J., Zeng, G.-M., Ruan, M., Zhou, C.-S., Li, L., Rong, Z.-G., 2009. Optimization of flocculation conditions for kaolin suspension using the composite flocculant of MBFGA1 and PAC by response surface methodology. Biores. Technol. 100, 4233–4239.
Yu, J., Wang, D., Yan, M., Ye, C., Yang, M., Ge, X., 2007. Optimized coagulation of high alkalinity, low temperature and particle water: pH adjustment and polyelectrolytes as coagulant aids. Environ. Monit. Assess. 131, 377–386.
Zhan, X., Gao, B., Yue, Q., Wang, Y., Gao, B., 2010. Coagulation behavior of polyferric chloride for removing NOM from surface water with low concentration of organic matter and its effect on chlorine decay model. Sep. Purif. Technol. 75, 61–68.
Zhan, X., Gao, B., Wang, Y., Yue, Q., 2011. Influence of velocity gradient on aluminum and iron floc property for NOM removal from low organic matter surfacewater by coagulation. Chem. Eng. J. 166, 116–121.
Zhang, N., Xie, J., Varadan, V., 2002. Functionalization of carbon nanotubes by potassium permanganate assisted with phase transfer catalyst. Smart Mater. Struct. 11, 962–965.
Zhang, L., Ni, Q.-Q., Fu, Y., Natsuki, T., 2009. One-step preparation of water-soluble single-walled carbon nanotubes. Appl. Surf. Sci. 255, 7095–7099.
Zhao, Y., Gao, B., Shon, H., Cao, B., Kim, J.-H., 2011. Coagulation characteristics of titanium (Ti) salt coagulant compared with aluminum (Al) and iron (Fe) salts. J. Hazard. Mater. 185, 1536–1542.
Zhao, H., Hu, C., Liu, H., Zhao, X., Qu, J., 2008. Role of aluminum speciation in the removal of disinfection byproduct precursors by a coagulation process. Environ. Sci. Technol. 42, 5752–5758.
Zhu, G., Zheng, H., Chen, W., Fan, W., Zhang, P., Tshukudu, T., 2012. Preparation of a composite coagulant: polymeric aluminum ferric sulfate (PAFS) for wastewater treatment. Desalination 285, 315–323.
Zouboulis, A., Moussas, P., Vasilakou, F., 2008. Polyferric sulphate: preparation, characterization and application in coagulation experiments. J. Hazard. Mater. 155, 459–468.
Zouboulis, A., Moussas, P., 2008. Polyferric silicate sulphate (PFSiS): preparation, characterization and coagulation behaviour. Desalination 224, 307–316.
Zouboulis, A., Tzoupanos, N., 2009. Polyaluminium silicate chloride—a systematic study for the preparation and application of an efficient coagulant for water or wastewater treatment. J. Hazard. Mater. 162, 1379–1389.

CHAPTER 4

NOM Removal by Electrochemical Methods

Mika Sillanpää*, Heikki Särkkä*, Mikko Vepsäläinen†
*Lappeenranta University of Technology, LUT Faculty of Technology, LUT Chemtech, Laboratory of Green Chemistry, Sammonkatu 12, 50130 Mikkeli, Finland, †CSIRO Mineral Resources Flagship, Box 312, Clayton South, VIC 3169, Australia

Abstract

Electrochemical techniques such as electrocoagulation (EC) and electrooxidation (EO) have proved their efficiency in the removal of humic acid (HA), coliform, and algae from surface waters. Many investigations have also been conducted with synthetic wastewaters. EC combined with membrane filtration hybrid systems can increase natural organic matter (NOM) removal rates remarkably. In EO technology, electrolysis efficiency is strongly linked to electrode composition. Efficiency could be increased by changing the reactor design, using commercial electrodes, and exploring the semiconducting properties of oxide mixtures. Electrochemical methods may present an attractive alternative to other NOM removal techniques, such as conventional coagulation and chemical oxidation methods, for natural waters. Surface water treatment with EC can produce high-quality water for either potable or industrial use. This technology appears to remove some toxic pollutants from wastewater and could be used as a pretreatment in combination with some other purification technology. Boron-doped diamond (BDD) anodes have proved effective in HA removal from aqueous solutions and potentially their total mineralization.

Keywords: Electrochemical methods; Electrocoagulation; Electrooxidation; Natural organic matter (NOM); Water treatment.

ABBREVIATIONS

AOPs Advanced oxidation processes
BDD Boron-doped diamond
BIEC Doubled EC
COD Chemical oxygen demand
DBP Disinfection by-product
DOC Dissolved organic carbon
DSA Dimensional stable anode
EC Electrocoagulation
EEM Excitation-emission matrix
EF Electroflotation

EO Electrooxidation
EMT-BIEC Electromagnetic treatment-doubled EC
eMBR Electro-MBR
EPS Extracellular polymeric substance
HA Humic acids
HFCVD Hot filament chemical vapor deposition
NF Nanofiltration
NOM Natural organic matter
ORP Oxidation reduction potential
ROS Reactive oxygen species
SHE Standard hydrogen electrode
SPEF Solar photoelectro-Fenton
TOC Total organic carbon

4.1 INTRODUCTION

Electrochemical technologies for water purification and disinfection have been the subject of growing interest in recent years (Chen, 2004). Techniques such as electrocoagulation (EC), also known as electroflocculation, electroflotation (EF), and electrooxidation (EO) can easily be distributed (Mollah et al., 2004). Conventional water purification techniques such as chemical coagulation, biological treatment, or UV oxidation are not effective against some toxic and refractory organic pollutants, and electrochemical techniques can provide a more efficient means of treating these. Electrochemical methods are innovative, inexpensive, and effective in purifying industrial wastewaters before discharge into water systems or circulation back into processes (Mollah et al., 2001, 2004; Holt et al., 2005). They could also be regarded as "green technologies" because little or no chemicals are needed to facilitate water treatment.

EC has a long history: the first plant was built in London in 1889 for the treatment of sewage (Moreno et al., 2009; Vik et al., 1984). Despite some promising results, this technology has enjoyed limited success. Scientific, economic, and environmental interest in this technology has revived in recent years, however, due the demand for alternative water treatment technologies. EC understandably has several similarities to chemical coagulation but also significant differences, such as side reactions.

A relatively new technology, used especially for the disinfection of drinking water or treatment of wastewaters, is the electrochemical mineralization of organic compounds (Patermarakis and Fountoukidis, 1990; Comninellis et al., 2008; Chen, 2004). An electrochemical oxidation mechanism involves the production of •OH-radicals at the active sites of the anode. The activity of these electrogenerated •OH radicals is strongly linked to their interaction with the anode surface (Comninellis et al., 2008). Electrochemical oxidation over anodes made of graphite, Pt, TiO_2, IrO_2, PbO_2, several Ti-based alloys and more recently, boron-doped diamond (BDD) electrodes in the presence of a suitable electrolyte (typically NaCl) has been employed for the decontamination of various inorganic and organic pollutants (Comninellis et al., 2008; Klavarioti et al., 2009; Särkkä et al., 2009), as well as microbial inactivation (Bergmann et al., 2009; Jeong et al., 2009).

4.2 PRINCIPLES OF EC AND EO TECHNOLOGIES

4.2.1 Theory of EC

In the EC system, multiple electrochemical reactions occur simultaneously at the anodes and cathodes. These can be divided into the main mechanisms that cause the destabilization of pollutants, and side reactions, such as hydrogen formation. Electrodes that produce coagulants in water are made from either iron or aluminum. In addition, there can be inert electrodes, typically cathodes, that are sometimes used in the system as counterelectrodes. Iron and aluminum cations dissolve from the anodes according to Eqns (4.1) and (4.2):

$$Fe\,(s) \rightarrow Fe^{n+}\,(aq) + ne^- \tag{4.1}$$

$$Al\,(s) \rightarrow Al^{3+}\,(aq) + 3e^- \tag{4.2}$$

In typical aqueous environments and conditions of the EC process, iron can dissolve in divalent Fe(II) and trivalent Fe(III) forms, whereas aluminum dissolves only in trivalent form Al(III). Fe(II) can further oxidize to Fe(III) if oxidation reduction potential and pH conditions are suitable. Oxygen has to be present and pH has to be neutral or alkaline to achieve a reasonable reaction rate (Morgan and Lahav, 2007).

Moreno et al. (2009) studied the electrochemical reactions of iron electrodes in an EC system. According to their results and the thermodynamic data they presented, the potential of the iron anode in an EC system is in the region where Fe(III) iron is produced. Some authors suggest, however, that the potential of the cathode is in the region where Fe(III) is reduced to Fe(II) form and that, therefore, both forms exist in the EC system (Parga et al., 2005; Gomes et al., 2007). Iron is produced on the anodes mainly in the form of Fe(II) (Ben-Sasson et al., 2009). At low pH, the chemical dissolution of iron can be significant and total iron concentration can be higher than would theoretically be expected. Ben-Sasson et al. (2009) also studied the oxidation of produced Fe(II). According to their results, no significant oxidation occurred at pH 5, the oxidation rate was moderate at pH 6, and very rapid oxidation occurred at pH 7–9.

The amount of metal cations dissolved during the reactions at the anode can be calculated according to Faraday's law (Eqn (4.3)):

$$m = \frac{ItMw}{zF} \tag{4.3}$$

where I is the current, t is the operation time (s), Mw is molecular weight of the substance (g mol^{-1}), F is the Faraday constant (96,485 C mol^{-1}), z is the number of electrons involved in the reaction (2 for Fe^{2+} and 3 for Fe^{3+} and Al^{3+}), and m is the quantity of metal dissolved (g). Several studies have reported current yields of over 100% for the dissolution of aluminum electrodes (Picard et al., 2000). It seems that aluminum also dissolves from the cathodes. This occurs when the pH on the surface of the cathode decreases due to the formation of HO^- (Eqn (4.4)) or by the consumption of hydronium ions/protons (Eqn (4.5)):

$$2H_2O + 2e^- \rightarrow H_2\,(g) + 2HO^-\,(aq) \tag{4.4}$$

$$2H^+\,(aq) + 2e^- \rightarrow H_2\,(g) \tag{4.5}$$

At high pH, aluminum dissolves as aluminate (Eqn (4.6)):

$$2Al\,(s) + 6H_2O + 2HO^-\,(aq) \rightarrow 2\left[Al(OH)_4\right]^-\,(aq) + 3H_2\,(g) \tag{4.6}$$

Picard et al. (2000) studied cathodic dissolution during the EC process. They compared the hydrogen evolution from stainless-steel cathodes and aluminum cathodes. Hydrogen produced at the stainless-steel cathodes followed Faraday's law. In tests with the aluminum cathodes, hydrogen production was higher due to the chemical dissolution of aluminum according to Eqn (4.6). Hydrogen formation and aluminum dissolution from the cathodes increased exponentially with current intensity.

In contrast to chemical coagulation, EC treatment raises the pH of the solution when it is acidic, neutral, or slightly alkaline, and lowers the pH when the solution is highly alkaline. This change of pH during EC treatment affects the speciation of aluminum and iron hydroxides. At highly acidic pH (pH 2) the alkalinity produced during EC is not sufficient to increase the pH of the solution, whereas at initial pH 3 and above, pH rises during the treatment (Mouedhen et al., 2008). This is easily understandable, as the concentration of hydronium ions increases by a factor of 10 when pH decreases from pH 3 to pH 2. When initial pH is significantly alkaline (pH > 9), pH probably decreases due to the formation of aluminate [$Al(OH)_4$], which is an alkalinity consumer (Chen et al., 2000; Kobya et al., 2006). It seems that the pH change rate and final steady-state pH depend on the concentration of anions in the solution. According to Trompette and Vergnes (2009), pH increases more in sulfate solution than in chloride solution. Sulfates can replace hydroxyl ions in the hydroxide precipitates, and therefore fewer hydroxyl ions are bound to hydroxides.

4.2.2 Treatment Parameters in EC

Various parameters affect EC efficiency in removing the pollutants from water. Parameters that are known to have an effect are as follows:

- Electrode materials can be iron, aluminum and/or inert material (typically cathodes). Iron and aluminum ions and hydroxides have different chemistries and applications.
- Solution pH has an effect on the speciation of metal hydroxides in the solution, the ζ-potential of the colloidal particles, and the dissolution of aluminum cathodes.
- Current density is proportional to the amount of electrochemical reactions taking place on the electrode surface.
- Treatment time or electric charge added per volume is proportional to the amount of coagulants produced in the EC system and other reactions taking place in the system.
- Electrode potential defines which reactions occur on the electrode surface.
- Pollutant concentration affects removal efficiency because coagulation does not follow zero-order reaction kinetics but rather pseudosecond- or first-order kinetics.
- The concentration of anions, such as sulfate or fluoride, affects the composition of hydroxides, because they can replace hydroxide ions in the precipitates.
- Temperature affects flocs formation, reaction rates, and conductivity. Depending on the pollutant, increasing temperature can have a negative or a positive effect on removal efficiency.

Other parameters, such as hydrodynamic conditions and interelectrode distance, may affect treatment efficiency and electricity consumption. These parameters either have not been widely studied or are found to have a negligible effect on overall removal efficiency.

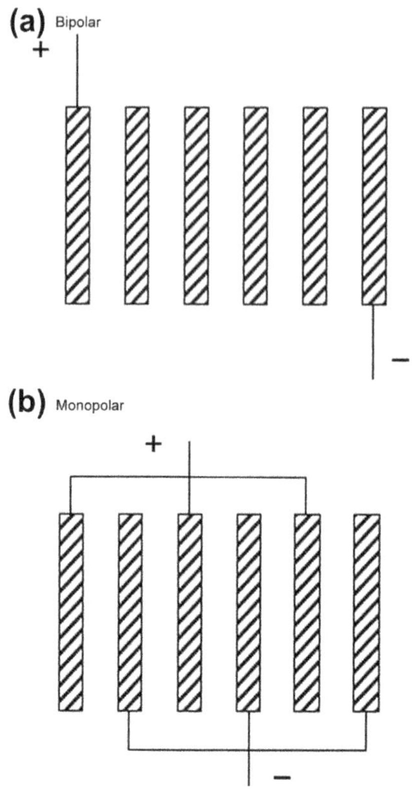

Figure 4.1 (a) Bipolar and (b) monopolar electrode arrangements. *Adapted from Chen (2004).*

4.2.3 Reactor Design in EC

EC systems are typically constructed of plate electrodes, and water flows through the space between the electrodes (Chen, 2004). Electrodes can be arranged in the EC system in either a monopolar or bipolar way. In the monopolar systems (Figure 4.1), all anodes are connected to each other and similarly all cathodes are also connected to each other. In the bipolar systems (Figure 4.1), the outermost electrodes are connected to a power source and current passes through the other electrodes, thus polarizing them. In the bipolar systems, the side of the electrode facing the anode is negatively polarized and vice versa on the other side facing the cathode. Water flow between the electrodes can follow a vertical or horizontal direction.

4.2.4 Theory of EO

Thermodynamically, the electrochemical degradation of any soluble organic compound in water should be achieved at low potentials, before the thermodynamic potential of water oxidation to molecular oxygen (1.23V per standard hydrogen electrode (SHE) under standard conditions), as indicated in formula (4.7) (Kapalka et al., 2010):

$$2H_2O \rightarrow O_2 + 4H^+ + 4e^- \quad (4.7)$$

In acidic media, water can be discharged on the electrode, producing highly oxidative absorbed hydroxyl radicals (reaction (4.8)):

$$H_2O + M \rightarrow M\,(^*OH) + H^+ + e^- \quad (4.8)$$

where M is the electrode surface.

These radicals are physisorbed on the anode surface where the organic pollutant R can be oxidized as follows (reaction (4.9)):

$$R_{(aq)} + M(^*OH)_{n/2} \rightarrow M + \text{Oxidation products} + \frac{n}{2}H^+ + \frac{n}{2}e^-, \quad (4.9)$$

where n is the number of electrons involved in the oxidation reaction of R. The reaction of organics with electrogenerated hydroxyl radicals (Eqn (4.9)) is in competition with the side reaction of the anodic discharge of these radicals to oxygen (reaction (4.10)):

$$M\,(^*OH) \rightarrow M + \frac{1}{2}O_2 + H^+ + e^- \quad (4.10)$$

Anodic activity by electrodes depends on their overpotential for oxygen evolution (Chen, 2004). Platinum electrodes have a much lower potential value for oxygen evolution reactions (1.3V vs SHE) than, for example, SnO_2 electrodes (1.9V vs SHE) or BDD electrodes (Ti/BDD, 2.7V vs SHE). This signifies that anodic oxidation by hydroxyl radicals can take place on a Ti/BDD electrode surface at a significantly higher current density with a minimal oxygen evolution side reaction.

Electrochemical oxidation can also occur indirectly through the formation of oxidants such as chlorine, hypochlorous acid, and hypochlorite (Rajkumar and Kim, 2006; Rajkumar et al., 2007; Kraft et al., 1999; Bergmann and Koparal, 2005; Chatzisymeon et al., 2006; Sakalis et al., 2006; Gotsi et al., 2005) or hydrogen peroxide/ozone (Polcaro et al., 2007;

Dhaouadi et al., 2009; Chu et al., 2012; Guinea et al., 2010) at the electrodes by the following reactions (4.11)–(4.17):

$$2Cl^- \rightarrow Cl_2 + 2e^- \tag{4.11}$$

$$Cl_2 + H_2O \rightarrow HOCl + H^+ + Cl^- \tag{4.12}$$

$$HOCl \rightarrow H^+ + OCl^- \tag{4.13}$$

$$H_2O \rightarrow {}^*OH + H^+ + e^- \tag{4.14}$$

$$2\,{}^*OH \rightarrow H_2O_2 \tag{4.15}$$

$$H_2O_2 \rightarrow O_2 + 2H^+ + 2e^- \tag{4.16}$$

$$O_2 + {}^*O \rightarrow O_3 \tag{4.17}$$

At acidic pH, chlorine is present in the solution in the form of hypochlorous acid, which has a higher oxidation potential (1.49V) than hypochlorite (0.94V). Under alkaline conditions, hypochlorite is the dominant species (Rajkumar et al., 2007). Under higher pH conditions, more chlorate or perchlorate is formed than chlorine/hypochlorite, which decreases oxidation efficiency. Higher initial chloride concentration in the electrolyte solution naturally encourages greater chlorine/hypochlorite production (Rajkumar and Kim, 2006; Rajkumar et al., 2007; Kraft et al., 1999). A similar effect can be observed when the temperature of the electrolyte solution is lowered, but this depends on the electrode material used (Kraft et al., 1999). Higher current density also increases chlorine production (Rajkumar et al., 2007; Kraft et al., 1999).

Electrochemical production of reactive oxygen species, such as ozone and hydrogen peroxide, has proved effective in water disinfection without the mediation of active chlorine products (Polcaro et al., 2007). In an electro-Fenton oxidation process, H_2O_2 is continuously generated in acidic solutions by the reduction of O_2 at the graphite cathodes or carbon felt (Dhaouadi et al., 2009). When Fe^{2+} is introduced as the catalyst, a Fenton reaction takes place in the solution, generating hydroxyl radicals to decompose pollutants or disinfect bacteria. Chu et al. (2012) effectively degraded 4-nitrophenol by employing a dual-cathode system to generate H_2O_2 and Fe^{2+} simultaneously at two cathodes to encourage an electro-Fenton reaction in the solution together with anodic oxidation by hydroxyl radicals at the anode. Guinea et al. (2010) achieved almost total mineralization of enrofloxacin solutions by solar photoelectro-Fenton treatment using an O_2-diffusion cathode.

Efficient organic pollutant removal can also be achieved through generating Fe^{2+} in situ by an EC process in which an iron anode is dissolved into a solution, causing a Fenton reaction when hydrogen peroxide is added to it (Martins et al., 2006; Zhao et al., 2012). Martins et al. (2006) were able to remove 95% of nonylphenol polyethoxylate in 5 min from aqueous solution and in 10 min from wastewater, respectively.

4.2.5 Electrodes and Reactors Used in EO

Several electrodes have been used for water and wastewater treatment by EO. Anodes used include lead and lead dioxide (El-Ashtoukhy et al., 2009; Hamza et al., 2011; Awad and Abo Galwa, 2005; Flox et al., 2009), dimensional stable anode (DSA) electrodes (Tavares et al., 2012; Wang et al., 2007b; Qu et al., 2012; Miwa et al., 2006), graphite (Patel and Suresh, 2008; Wang et al., 2007a; Ma et al., 2007), and BDD electrodes (Flox et al., 2009; Montilla et al., 2002; Kraft et al., 2003; Rodrigo et al., 2010; Panizza and Cerisola, 2005; Chen and Chen, 2006, 2011; Faouzi et al., 2007; Migliorini et al., 2011).

Lead and lead dioxide are effective anodes due to their stability, low cost, and high oxygen evolution potential, which delays O_2 evolution in favor of Cl_2 evolution (El-Ashtoukhy et al., 2009). Hamza et al. (2011) completely mineralized 1,3,5-trimethoxybenzene in acid media at a Ta/PbO_2 anode. They discovered that carboxylic acids finally oxidized all oxidation products to CO_2. Awad and Abo Galwa (2005) found that the electrocatalytic activity of a lead dioxide electrode depends on the conductive electrolyte. They concluded that electrode poisoning occurred in the presence of an H_2SO_4 electrolyte, since an adherent film was formed on the anode surface. The dissolution of toxic Pb^{2+} ions also hinders the use of lead and lead dioxide as anodes (Chen, 2004).

DSA electrodes are catalytic oxide electrodes that can effectively generate active hydroxyl radicals and active chloride species (Wang et al., 2007b). They also have a relatively high overpotential for oxygen evolution. Efficient degradation of paper mill wastewater has been achieved using 3D electrodes ($Ti/Co/SnO_2$–Sb_2O_5) combined with activated carbon treatment (Wang et al., 2007a). This result was mainly due to the fact that their large specific surface area in comparison to conventional 2D electrodes can substantially increase the conversion rate within an electrochemical reactor. "Nonactive" electrodes, such as SnO_2, form hydroxyl radicals on their surface more easily, which can result in the complete oxidation of the organic molecules to CO_2 (Miwa et al., 2006). With "active electrodes," such as RuO_2 and IrO_2, only selective oxidation of the organic species in the solution occurs.

Effective removal of chemical oxygen demand (COD) (>96%) was achieved when the electrochemical degradation process was catalyzed by transition metals (Co and Cu; Wang et al., 2007a,b) or molybdenum and phosphate (Mo–P) modified kaolin with graphite as the anode and cathode (Ma et al., 2007). Pollutants were adsorbed on the surface of the kaolin, where they were oxidized by hydroxyl radicals produced at the graphite cathode by the reaction of hydrogen peroxide and transition metals (Wang et al., 2007a,b). This process is similar to the electro-Fenton process.

The potential of conducting diamond films for water treatment has recently been recognized. These films have an inert surface with low adsorption properties, remarkable corrosion stability even in strong acidic media, and an extremely wide potential window in aqueous and nonaqueous media (Kraft et al., 2003; Panizza and Cerisola, 2005). They also have the highest oxygen evolution overpotential value (Chen, 2004; Kraft et al., 2003), meaning that more hydroxyl radicals are formed on the anode surface during treatment. BDD electrodes can also degrade refractory organic pollutants completely, and the nature of the pollutant does not significantly affect the efficiency of the process (Rodrigo et al., 2010). Furthermore, besides hydroxyl radical formation on the electrode surface, diamond electrodes also increase mediated oxidation by other electrochemically formed compounds such as persulfate, perphosphate, percarbonate, or hypochlorite, depending on the electrolyte used.

The low pressure conversion of carbon to diamond crystals has made it possible to grow a thin layer of diamond film on suitable substrates such as silicon, niobium, tungsten, molybdenum, and titanium (Chen and Chen, 2011). The hot filament chemical vapor deposition (HFCVD) technique has been applied primarily to fabricate active and stable BDD electrodes, mainly using titanium as a substrate material (Chen and Chen, 2006, 2011; Migliorini et al., 2011). Migliorini et al. (2011) used an additional H_2 gas flux passing through a bubbler containing a solution of B_2O_3 dissolved in CH_3OH with a B/C ratio of 30,000 ppm during the HFCVD coating of diamond films. Two different fluxes were used to produce heavily BDD films.

The main advantages and disadvantages of different electrodes in EO treatment are presented in Table 4.1.

Table 4.1 Comparison of electrodes in EO treatment

Electrode	Advantages	Disadvantages	Compared to other electrodes
Ti	Stable	Passive, expensive	
Pt	Inert, low oxygen. Evolution overpotential	Expensive	Poor efficiency in anodic oxidation of organic compounds
PbO_2	Good current efficiency, cheap, effective in oxidizing pollutants, high oxygen evolution overpotential, easy to prepare	Corrosive, toxic Pb^{2+}-ions could be released	
Dimensional stable anode electrodes	Supports indirect oxidation, good current efficiency, high oxygen evolution overpotential, lower cost, higher availability	Short lifespan, lack of electrochemical stability	
Boron-doped diamond	Inert under tough conditions, high oxygen evolution overpotential and electrochemical stability, good current efficiency, high corrosion stability, good conductivity	Very expensive	Higher activity

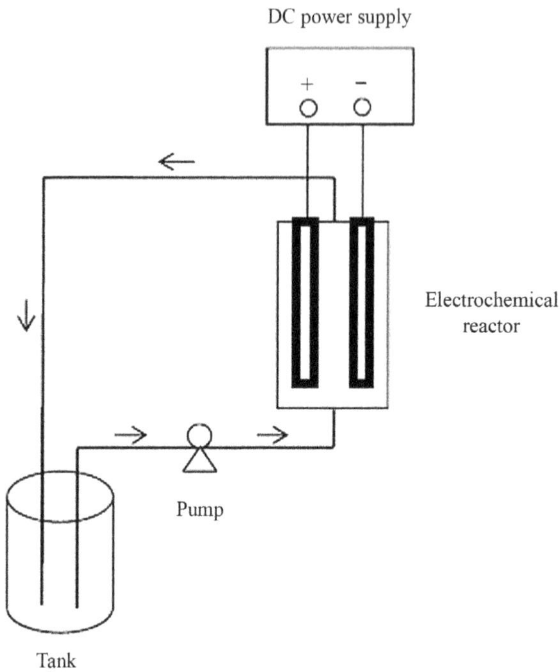

Figure 4.2 Typical electrooxidation set-up. *Adapted from Zhu et al. (2011).*

The simplest EO reactor design is the bipolar cell. Besides plane electrodes, the cylindrical electrodes can also be employed (Chen, 2004). Typical reactor design for EO treatment is presented in Figure 4.2. It includes a tank, a pump, a DC power supply, and an electrochemical reactor with electrodes (Zhu et al., 2011).

4.3 EC AND EO TECHNOLOGIES IN NOM REMOVAL

4.3.1 EC in NOM Removal

EC has been proposed as a promising alternative to chemical coagulation for removing various pollutants from freshwater and wastewater (Kobya et al., 2011; Meas et al., 2010; Kabdaslı et al., 2009). When iron or aluminum is used as a sacrificial anode, it dissolves and produces Al^{+3}, Fe^{+2} and Fe^{+3} ions in the solution. These ions form metal hydroxides, which dissolve less readily, such as $Al(OH)_3$, $Fe(OH)_2$ and $Fe(OH)_3$, by linking hydroxides ions in water. Metal hydroxide species that adsorb readily are effective in the destabilization of suspended pollutants. They can remove pollutants from water either by complexation or by electrostatic attraction, followed by coagulation, sedimentation, or flotation.

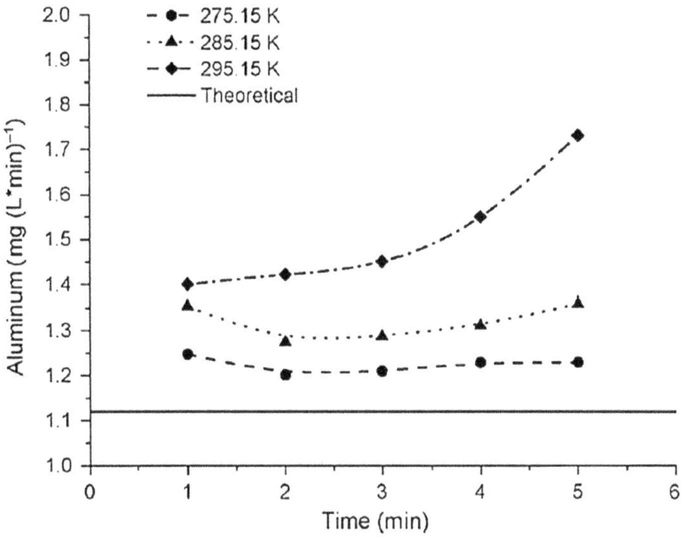

Figure 4.3 Dissolving rate of the aluminum (Al) electrodes in different water temperatures. Theoretical values are calculated according to Faraday's law for anodic dissolving reaction (Vepsäläinen et al., 2009).

Several researchers have considered EC for surface water treatment and natural organic matter (NOM) removal from waters (Vepsäläinen et al., 2009; Yildiz et al., 2007; Dubrawski and Mohseni, 2013a). Vepsäläinen et al. (2009) studied the effect of initial pH and temperature on EC treatment and found that high NOM removal was obtained using an Al electrode at a low pH of 4.3 and a temperature of 22.25 °C. The dissolving rate of the electrodes in different water temperatures is shown in Figure 4.3. Temperature had significant effect on the dissolving speed of the aluminum, but the impact of temperature on NOM removal was minor compared to that of initial pH (Vepsäläinen et al., 2009). Yildiz et al. (2007) studied the influence of different supporting electrolyte solutions on NOM removal, and Na_2SO_4 was observed to be the most favorable electrolyte type. Dubrawski and Mohseni (2013a) applied an iron EC reactor to drinking water treatment. Significant EC design variables were examined: current density, coagulant, and charge-loading rate. They observed a correlation between increased current density and decreased current efficiency, and optimal NOM removal was found at a current density of 10 mA cm^{-2} (Dubrawski and Mohseni, 2013a). Dubrawski et al. (2013) also compared zinc, iron, and aluminum electrodes for dissolved organic carbon (DOC) removal, finding that zinc electrodes could achieve better results than aluminum electrodes.

Mohora et al. (2012) studied the performance of a laboratory-scale EC/flotation reactor in removing NOM and arsenic from groundwater. Removal rates were rather high, satisfying the drinking water standards in both cases. Electrode passivation was reduced by adding NaCl to the electrolyte solution, although this did not improve the NOM removal rate significantly (Mohora et al., 2012). Ghernaout et al. (2009a) investigated the reduction of humic acids (HAs) by continuous electromagnetic (EM) treatment followed by EC. EM treatment seemed to increase the reactivity of HA macromolecules before their removal by EC, but alone in continuous mode, EM did not have any positive effect on the HA system. In another study by Ghernaout et al. (2010), EC in series using two electrocoagulators (BIEC or doubled EC) and BIEC assisted by an electromagnetic field (EMT-BIEC) proved efficient for HA removal. Ghernaout et al. (2008) also investigated EC for *Escherichia coli* removal in surface water treatment, rapidly achieving total removal of coliforms and algae. Aluminum electrodes were found to be slightly more efficient than ordinary steel or stainless-steel electrodes.

Kourdali et al. (2014) conducted a series of experimental assays to determine the optimal operating conditions for HA removal from surface waters. Under optimal conditions, up to 72% of HA could be removed. EC treatment mechanisms were proposed for HA removal from various waters (Ghernaout et al., 2009b, 2014). Based on current intensity as a function of applied voltage variation and the relevant literature, three mechanisms were proposed for acid, neutral, and alkaline pH (Ghernaout et al., 2009b).

Koparal et al. (2008) investigated the influence of initial pH on the removal of a high concentration of humic substances by EC. They observed the two possible effects of pH on EC systems: transformation of humic substances is related to the initial pH and gel layer formation on the surface of the electrode starts at higher humic substance concentrations. The authors prevented gel layer formation by adjusting the initial pH value of the wastewater to 5.0. Vepsäläinen et al. (2012) analyzed the effect of EC cell construction on NOM removal. Three types of cells were investigated: one that only had aluminum electrodes, one that had aluminum anodes and inert cathodes, and one that had inert anodes and aluminum cathodes. The reactions between NOM and aluminum were found to occur in the bulk water and not on the electrode surface. NOM destabilization mechanisms were similar to those for chemical coagulation; at low pH, double layer compression dominated, in contrast to adsorption and bridging at higher pH (Vepsäläinen et al., 2012).

El-Masry et al. (2004) successfully tested the ability of EC to purify raw surface water. EC efficiently reduced bacterial load, water turbidity, and chemical contaminants such as sulfates, nitrates, and phosphates. EC also removed 97.8% of HA from groundwater when the electrode interval, current density, and the initial pH value were optimized (Feng et al., 2007).

NOM was found to influence arsenic removal by the continuous-flow EC treatment of weakly mineralized waters (Pallier et al., 2011). Flocs generated during electroflocculation treatment were smaller than those generated during chemical coagulation/flocculation. The presence of HAs negatively influenced floc size because the more negative zeta potential involved influenced coagulation mechanisms. The As(V) removal rate was lower than that of chemical coagulation/flocculation, but during the electroflocculation process As(III) was oxidized to As(V), considerably improving removal rates regardless of the matrix conditions (Pallier et al., 2011).

Combined EC–EF treatment successfully removed suspended and organic matter from surface waters (Ricordel et al., 2010). In this process, coagulant ions Al^{3+} or Fe^{2+} are produced at the anode while the cathode reaction creates hydrogen bubbles that contribute to a flotation process. In contrast to chemical coagulation, less coagulant is needed; consequently, less sludge is formed using this technique.

Studies in which EC technology has been used for NOM removal are presented in Table 4.2.

The potential benefits of combining electric treatment with membrane technologies have been recognized in the last few years (Liu et al., 2012a,b,c). The fouling of membranes in a membrane bioreactor (MBR) can be reduced by EC to form bigger particles; an electric field can also successfully reduce fouling. Liu et al. (2012a) applied an electric field to a conductive membrane cathode and an iron anode effectively to reduce membrane fouling in an electro-MBR (eMBR). Electric currents of 1–5 mA could extend the membrane filtration cycle and reduce cleaning frequency. The other positive effects of EC included an increase in sludge activity and a decrease in both the extracellular polymeric substance (EPS) and the specific filtration resistances (Liu et al., 2012a).

Novel configurations of electrodes and membrane modules made it possible to achieve significant flux enhancement and fouling reduction using low voltage intensity (Liu et al., 2012b). The electrophoresis and/or electrostatic repulsion of sludge/EPS in an electric field decreased fouling. Liu et al. (2012c) also deposited electrically conductive polypyrrole membranes by coating them in the vapor phase. These membranes were used as cathodes to induce an electric field in a laboratory-scale MBR. EC in an eMBR further improved performance by reducing COD and total phosphorus in the effluent.

EC-membrane filtration (EC-MF) hybrid systems can drastically cut filtration times. EC pretreatment mitigated fouling by reducing the

Table 4.2 NOM removal by EC technology

Matrix	Pollutant	Electrodes	Removal efficiency	Current densities used (mA cm^{-2})	References
Synthetic water	Humic substances	Iron	92.69% for Na_2SO_4 electrolyte	30, 40 and 50V	Yildiz et al. (2007)
Synthetic water	Humic substance	Aluminum	96.84% for Na_2SO_4 electrolyte	20–50V	Yildiz et al. (2008)
Synthetic water	NOM isolate (DOC)	Iron	73%	2.43–26.8	Dubrawski and Mohseni (2013a)
Natural water	DOC	Iron	71%	1–25	Dubrawski and Mohseni (2013b)
Synthetic and natural waters	DOC	Iron, aluminum, and zinc	c. 70% for all electrodes by 1 min of electrolysis	2.43	Dubrawski et al. (2013)
Raw surface water	TOC	Aluminum	89%	10–30V	Mahvi et al. (2011)
Synthetic water	NOM	Iron and aluminum	80%	0.3A	Ben-Sasson et al. (2013)
Groundwater	DOC	Aluminum	71%	5.78	Mohora et al. (2012)
Natural water (lake water)	MS2 bacteriophage	Iron	1.0–1.5 log reduction	–	Tanneru and Chellam (2012)

Surface water	Algae	Aluminum, stainless steel, and ordinary steel	100%	0.25A	Ghernaout et al. (2008)
Synthetic water	Humic acid (HA)	Aluminum	96%	3.3	Ghernaout et al. (2009a)
Synthetic water	Humic substance	Aluminum	Almost 100%	50V	Koparal et al. (2008)
Surface water	Color	Aluminum and inert	>90%	0.48	Vepsäläinen et al. (2012)
Surface water	Turbidity	Aluminum	90%	0.6325	El-Masry et al. (2004)
Groundwater	HA	Aluminum	97.8%	4.76	Feng et al. (2007)
Surface water	Permanganate index	Aluminum	47%	1A	Ricordel et al. (2010)
Paper mill wastewater	Color	Iron	75%	0.4	Kliaugaite et al. (2013)
Paper mill wastewater	Color	Aluminum	91%	0.36–1.79	Vepsäläinen et al. (2007)
Paper mill wastewater	COD	Aluminum	>60%	10	Mansouri et al. (2012)

COD, chemical oxygen demand; DOC, dissolved organic carbon; NOM, natural organic matter; TOC, total organic carbon.

secondary effluent filtration period up to 36-fold (Ben-Sasson et al., 2011). Filtration energy was minimized by 90% when electroflocculation was applied in combination with microfiltration (Ben-Sasson and Adin, 2010a). The extent to which fouling was mitigated depended on fouling intensity, the dominant fouling mechanism, suspension pH, and electroflocculation operation time. Ben-Sasson and Adin (2010b) also investigated energy minimization and fouling mitigation by iron-based electroflocculation in dead-end microfiltration. They concluded that when internal fouling was the dominant mechanism, the iron hydroxide solids formed a layer that filtered out the primary particles. When external fouling was more significant, iron hydroxide solids also reduced the hydraulic resistance of the cake layer.

Using a novel anode material, titanium, Chen and Deng (2012) combined titanium-based EC with an ultrafiltration membrane process. This combination increased HA rejection, reduced membrane fouling, and decreased trans-membrane pressure. Hydrous TiO_2 flocs were formed, which had a large specific surface area and demonstrated high photocatalytic activity.

Ayu Sari and Chellam (2013) assessed the capability of four different advanced pretreatment processes to reduce the fouling of a thin-film composite nanofiltration (NF) membrane during surface water treatment. They observed that EC-MF was inefficient in removing the hydrophilic fraction of NOM, making this the major NF foulant. EC pretreatment significantly increased permeate fluxes during dead-end surface-water microfiltration (Gamage and Chellam, 2011; Gamage et al., 2012) by reducing both the cake mass and the cumulative hydraulic resistance. Also, as more aluminum was added, the floc zeta potential became less negative, indicating charge neutralization in destabilization (Gamage and Chellam, 2011).

Harif and Adin (2011) examined the first time size and structural evolution of flocs formed in during EC, using a static light scattering technique to elucidate kinetic aspects of the process. EC was found to be capable of generating a range of flocs, and their growth rates and structural characteristics were highly dependent on the operating conditions.

Membrane electrolysis and EC were used for the removal and recovery of humiclike substances from industrial secondary effluent (Kliaugaite et al., 2013). Membrane electrolysis was suitable for the decolorization of colored wastewater but was less efficient than EC treatment in reducing COD. The operational costs for EC were also lower than for membrane electrolyses. Thus, electrochemical methods proved an attractive alternative to other techniques, such as conventional coagulation and chemical oxidation (Kliaugaite et al., 2013).

4.3.2 Comparison of EC and Chemical Coagulation

Published comparative studies of chemical coagulation and EC are surprisingly few in number. Since these can be considered competing technologies, comparisons of treatment efficiencies are needed, but these studies are difficult to conduct due to the dynamic nature of the EC process.

Change of pH during EC does not significantly affect the aluminum species formed (Cañizares et al., 2009). The formation of monomeric and polymeric aluminum hydroxides was compared when aluminum was added as $AlCl_3$ or by EC. The results showed no significant differences in the speciation of aluminum obtained by these two methods. Under the same dynamic conditions, doses of aluminum, and steady-state pH, removal of wastewater pollutants was similar in both processes.

EC and chemical coagulation have been compared for the treatment of ultrafine quartz suspensions (Kiliç et al., 2009), the removal of Cr(VI) (Golder et al., 2007), and as a pretreatment before microfiltration membranes (Bagga et al., 2008; Zhu et al., 2005). Kiliç et al. (2009) used aluminum anodes and stainless-steel cathodes to compare chemical coagulation and EC and the effect of initial pH and aluminum dosage on the removal efficiency of a model pollutant (ultrafine quartz particles). The optimal pH for particle removal was 9 for EC and 6–9 for chemical coagulation. The broader pH range of the latter was probably a result of higher sulfate concentration because aluminum sulfate was used as a chemical coagulant. The effect of aluminum dosage on the destabilization of pollutants and optimal aluminum dosages was similar with both technologies. Both technologies removed turbidity equally well and the researchers suggested, based on the ζ-potential measurements, that destabilization mechanisms are similar with both technologies.

EC was more effective (42%) than chemical coagulation (approximately 11.5%) at removing Cr(VI) from aqueous solution (Golder et al., 2007). Some Cr(VI) was removed by deposition on the cathode. Alum and aluminum sulfate showed poor Cr(VI) removal, which was probably due to competitive adsorption of sulfate ions on the sludge matrix. Zhu et al. (2005) compared EC and chemical coagulation pretreatment for enhanced virus removal using microfiltration membranes. Both technologies improved virus removal, but EC outperformed chemical coagulation in removal efficiency, which was attributed to electrophoretic movement of viruses on the electrodes, although this has not been reliably verified by experimental studies. Rather different results were obtained from a comparison of chemical coagulation and EC as pretreatment for surface water microfiltration (Bagga et al., 2008). Chemical coagulation with ferric chloride was superior to EC with an iron anode and a stainless-steel cathode. Iron dissolved in Fe(II) form, which had higher solubility and poorer coagulation performance than Fe(III) in ferric chloride. NOM can prevent the oxidation of Fe(II) to the more effective Fe(III) form (Theis and Singer, 1974). Thus, the better performance of EC in research by Zhu et al. (2005) can be explained by the fact that they used synthetic freshwaters that did not include NOM.

These comparative studies seem to indicate that EC and chemical coagulation have similar destabilization mechanisms. Yet, a chemical coagulant is added with an anion, such as sulfate or chloride, which may have some influence on the efficiency of the process. In some cases side reactions, such as reduction on the cathode surface, may play an important role in the destabilization of pollutants. Iron is dissolved from iron electrodes in Fe(II) form, hence oxidation to the more effective Fe(III) form is a significant factor in high removal efficiency. Therefore, it would be more descriptive to compare Fe(II) coagulants, such as ferrous sulfate, with EC using iron electrodes.

4.3.3 EO in NOM Removal

Some work has been done on the removal of small model organic compounds (Scialdone et al., 2008, 2009a,b), although not many apply this specifically to NOM removal in drinking water treatment (Motheo and Pinhedo, 2000). One example is the removal of NOM by photoelectrocatalysis, where TiO_2 coated surfaces were employed in an electrochemical process (Selcuk et al., 2003, 2006; Fu et al., 2007).

Photoelectrocatalytic degradation of a model HA was compared to the photocatalytic destruction process (Selcuk et al., 2003). After 2 h of treatment, 96% of the color, 98% of the HA, and 85% of the TOC were removed in the photoanode cell, which was a TiO_2-coated electrode. Photoelectrocatalysis was much more effective than the photocatalytic method, especially at high pH values (Selcuk et al., 2003). Photoelectrocatalytic treatment can form bromate as a side reaction, however, when bromine is produced by this technique (Selcuk et al., 2006). Using response surface methodology, Fu et al. (2007) optimized the operating parameters (such as initial pH, potassium peroxodisulfate [$K_2S_2O_8$] and bias potential) of fulvic acid photoelectrocatalytic degradation. Regression analysis with an R2 value of 0.9754 showed that the experimental results closely correlated with the model predictions.

EO was effective in removing HA and algae from pond water. Liao et al. (2008) observed that the degradation followed first-order kinetics. Motheo and Pinhedo (2000) investigated EO of natural peat HA by DSA electrodes, demonstrating that electrolysis efficiency was highly dependent on electrode composition. Efficiency could be increased by changing the reactor design, using commercial electrodes, and exploring the semiconducting properties of these oxide mixtures (Motheo and Pinhedo, 2000).

Electrochemical oxidation of microcystin-LR in the presence of HA in a parallelepipedic electrolytic cell was studied by Tran and Drogui (2013). The pollutant was decomposed by an electrophilic attack by the oxidizing species, including hydroxyl radicals. Thus, the degradation was mainly attributed to direct anodic oxidation, although electrolyte composition influenced the treatment. The highest rates of color removal were recorded using sodium chloride, due to the generation of active chlorine, which is a powerful oxidizing agent.

Anglada et al. (2009) used BDD anodes to analyze the EO process of landfill leachate on a pilot-plant scale. They observed that organic matter and ammonia oxidation were highly influenced by the current density value, and the oxidation mechanism was changed when high current densities were applied. A later analysis (Anglada et al., 2011) showed that organic matter oxidation led to the formation of different disinfection by-products such as chloroform, dichloroacetonitrile, 1,2-dichloroethane and 1,1-dichloroacetone. COD and ammonia removals of 51% and 32%, respectively, could be achieved.

Reverse osmosis concentrate streams contain NOM, a major fraction of which are soluble humic substances. EO has been applied also in the treatment of these waters (Bagastyo et al., 2011; Radjenovic et al., 2011). Considerable differences between electrode materials were observed, and the main removal mechanism was recognized as indirect oxidation. Hazardous halogenated oxidation by-products were formed during the treatments, however, which may represent a significant barrier to the environmental application of EO treatment in NOM removal; thus, more detailed investigations are needed.

Studies of NOM removal by EO technology are shown in Table 4.3.

Table 4.3 Removal of NOM by EO technology

Matrix	Pollutant	Electrodes	Removal efficiency	Current densities used (mA cm^{-2})	References
Landfill leachate	Color	BDD	84%	18A	Anglada et al. (2011)
Reverse osmosis concentrate	DOC	$Ti/Ru_{0.7}Ir_{0.3}O_2$	25%	0.1–25	Radjenovic et al. (2011)
Reverse osmosis concentrate	DOC	Titanium coated with IrO_2–Ta_2O_5, RuO_2–IrO_2, Pt–IrO_2, PbO_2, SnO_2–Sb	<16%	10	Bagastyo et al. (2011)
Synthetic water	Color	Ti/BDD	81%	38	Tran and Drogui (2013)
Humic acid (HA) extracted from peat area	HA	DSA electrodes	100%	40	Motheo and Pinhedo (2000)
Pond water	HA	BDD	70%	17.5	Liao et al. (2008)
Synthetic water	TOC	DSA electrode	25%	20	Pinhedo et al. (2005)

DOC, dissolved organic carbon; BDD, boron-doped diamond; DSA, dimensional stable anode.

4.4 CONCLUSIONS

The advantages of EC technology over conventional coagulation include a reduction in the number of different chemicals needed and sludge formed, alkalinity consumption mitigation, and ease of adaptation to existing treatment units (Bagga et al., 2008). EC has removed not only hydrophobic, but also hydrophilic colloids (Vepsäläinen et al., 2009), which are not easily eliminated by conventional coagulation processes. Thus, EC is demonstrably effective for NOM removal, but the process conditions (e.g., pH, electric charge, and supporting electrolyte) must be determined carefully. Moreover, the EC mechanism is highly dependent on the chemistry of the aqueous medium, especially its conductivity (Yildiz et al., 2007, 2008; Vepsäläinen et al., 2009). Hence, it remains to be determined whether the benefits of this method are great enough for feasible application (Bagga et al., 2008).

EC has several advantages over other water treatment technologies (Mollah et al., 2001, 2004). Some of the important benefits for wastewater treatment are as follows:

- EC produces low amounts of sludge, and the sludge is dewatered easily.
- Flocs formed in EC are larger and more stable than those formed in chemical coagulation.
- Due to electrophoretic movement, the removal of DOC from water is better with EC than with chemical coagulation.
- Competing reactions, such as hydrogen and hypochlorite formation, can be beneficial in water treatment.
- EC can be run by green processes, such as solar or wind power and fuel cells.
- Adsorption of hydroxide on mineral surfaces is 100 times greater than precipitated hydroxides.
- No harmful anions, such as sulfate and chloride, are added to water in EC.

The disadvantages of EC in NOM removal during water treatment are the passivation of electrodes due to the presence of oxides and precipitation layers on the electrode surfaces, the dissolution of the electrodes, and the low conductivity of the surface waters (Mollah et al., 2001). High voltages can be avoided by reducing the distance between electrodes. Electrodes need to be replaced regularly as the electrode material dissolves, raising maintenance costs. Trihalomethane formation can occur in the treatment of waters with a high humic and fulvic acid content (Mollah et al., 2004).

Optimal DOC removal and low residual aluminum can be obtained simultaneously with EC, which may be a significant advantage of this method over chemical coagulation in surface water treatment. EC can produce high-quality water that is potable or suitable as fresh water for industrial applications. This technology seems capable of removing some toxic pollutants from wastewater and could be used as pretreatment prior to combination with some other purification technology. A thorough economic and ecological comparison of chemical coagulation and EC is recommended, however, because these methods seem to have similar pollutant destabilization mechanisms, metal consumption, and removal efficiencies in most applications. This investigation should measure operating and investment costs, including the coagulation, mixing, and separation phases of the treatment. Side reactions of EC, such as flotation due to hydrogen bubble production, should also be considered. These calculations would show what applications of EC are economically viable. If EC is more expensive than chemical coagulation, it may still have small-scale applications, for example in emergency water treatment systems or small communities, as it could be simpler to operate than chemical coagulation, especially if EF is used for particle separation. In addition to economic and ecological research, new technological innovations and investigations are also required to prevent electrode passivation and fouling, and to further optimize treatment parameters. Full-scale systems are likely to operate in continuous-flow mode, so it is proposed that bench-scale continuous-flow systems are used in future studies.

The electrochemical mineralization of organic compounds has interested several researchers. Nevertheless, BDD anodes have been found to be particularly effective for HA removal from aqueous solutions (Liao et al., 2008). Due to the complexity of humic substances, however, the formation of intermediates causes degradation. The observed disadvantages of using BDD anodes are the formation of disinfection by-products (DBPs), mainly perchlorate (Bergmann et al., 2009), and their high manufacturing costs.

Some traditional techniques such as chemical coagulation, adsorption, and advanced oxidation processes are still often more economical than EO for NOM removal. EO can provide almost complete mineralization of organic compounds, however, which makes it an attractive process in some applications. EO can be an effective pretreatment or posttreatment to guarantee the necessary water quality.

The efficiency of the EO process largely depends on cell configuration, electrode material, electrolyte composition, and other experimental parameters, such as current density or the temperature of the treated water. In future NOM treatment studies, it would be beneficial to measure different radical reactions at the surface of the anode and by-products formed after the treatment.

The discovery of an effective and stable, yet economical, electrode material would speed up the use of EO techniques for NOM removal. This would be further assisted by the development of some novel electric power source, such as solar energy. Combining treatment with current competitive techniques would enhance purification results. The economic viability of EO in the treatment of NOM-containing waters also needs to be calculated.

REFERENCES

Anglada, A., Urtiaga, A., Ortiz, I., 2009. Pilot scale performance of the electro-oxidation of landfill leachate at boron-doped diamond anodes. Environ. Sci. Technol. 43, 2035–2040.

Anglada, A., Urtiaga, A., Ortiz, I., Mantzavinos, D., Diamadopoulos, E., 2011. Boron-doped diamond anodic treatment of landfill leachate: evaluation of operating variables and formation of oxidation by-products. Water Res. 45, 828–838.

Awad, H.S., Abo Galwa, N., 2005. Electrochemical degradation of Acid Blue and Basic Brown dyes on Pb/PbO_2 electrode in the presence of different conductive electrolyte and effect of various operating factors. Chemosphere 61, 1327–1335.

Ayu Sari, M., Chellam, S., 2013. Surface water nanofiltration incorporating (electro) coagulation-microfiltration pretreatment: fouling control and membrane characterization. J. Membr. Sci. 437, 249–256.

Bagastyo, A.Y., Radjenovic, J., Mu, Y., Rozendal, R.A., Batstone, D.J., Rabaey, K., 2011. Electrochemical oxidation of reverse osmosis concentrate on mixed metal oxide (MMO) titanium coated electrodes. Water Res. 45, 4951–4959.

Bagga, A., Chellam, S., Clifford, D.A., 2008. Evaluation of iron chemical coagulation and electrocoagulation pretreatment for surface water microfiltration. J. Membr. Sci. 309, 82–93.

Ben-Sasson, M., Adin, A., 2010a. Fouling mechanisms and energy appraisal in microfiltration pretreated by aluminum-based electroflocculation. J. Membr. Sci. 352, 86–94.

Ben-Sasson, M., Adin, A., 2010b. Fouling mitigation by iron-based electroflocculation in microfiltration: mechanisms and energy minimization. Water Res. 44, 3973–3981.

Ben-Sasson, M., Lin, Y.M., Adin, A., 2011. Electrocoagulation-membrane filtration hybrid system for colloidal fouling mitigation of secondary-effluent. Sep. Purif. Technol. 82, 63–70.

Ben-Sasson, M., Calmano, W., Adin, A., 2009. Iron-oxidation processes in an electroflocculation (electrocoagulation) cell. J. Hazard. Mater. 171, 704–709.

Ben-Sasson, M., Zidon, Y., Calvo, R., Adin, A., 2013. Enhanced removal of natural organic matter by hybrid process of electrocoagulation and dead-end microfiltration. Chem. Eng. J. 232, 338–345.

Bergmann, M.E.H., Koparal, A.S., 2005. Studies on electrochemical disinfectant production using anodes containing RuO_2. J. Appl. Electrochem. 35, 1321–1329.

Bergmann, M.E.H., Rollin, J., Iourtchouk, T., 2009. The occurrence of perchlorate during drinking water electrolysis using BDD anodes. Electrochim. Acta 54, 2102–2107.

Cañizares, P., Jiménez, C., Martínez, F., Rodrigo, M.A., Sáez, C., 2009. The pH as a key parameter in the choice between coagulation and electrocoagulation for the treatment of wastewaters. J. Hazard. Mater. 163, 158–164.
Chatzisymeon, E., Xekoukoulotakis, N.P., Coz, A., Kalogerakis, N., Mantzavinos, D., 2006. Electrochemical treatment of textile dyes and dyehouse effluents. J. Hazard. Mater. B137, 998–1007.
Chen, G., 2004. Electrochemical technologies in wastewater treatment. Sep. Purif. Technol. 38, 11–41.
Chen, X., Chen, G., 2006. Anodic oxidation of Orange II on Ti/BDD electrode: variable effects. Sep. Purif. Technol. 48, 45–49.
Chen, X., Chen, G., 2011. Fabrication and application of Ti/BDD for wastewater treatment. In: Brillas, E., Martinez-Huitle, C.A. (Eds.), Synthetic Diamond Films: Preparation, Electrochemistry, Characterization, and Applications. John Wiley & Sons, New Jersey, USA, pp. 353–371.
Chen, X., Deng, H., 2012. Removal of humic acids from water by hybrid titanium-based electrocoagulation with ultrafiltration membrane processes. Desalination 300, 51–57.
Chen, X., Chen, G., Yue, P.L., 2000. Separation of pollutants from restaurant wastewater by electrocoagulation. Sep. Purif. Technol. 19, 65–76.
Chu, Y.Y., Qian, Y., Wang, W.J., Deng, X.L., 2012. A dual-cathode electro-Fenton oxidation coupled with anodic oxidation system used for 4-nitrophenol degradation. J. Hazard. Mater. 199–200, 179–185.
Comninellis, C., Kapalka, A., Malato, S., Parsons, S.A., Poulios, I., Mantzavinos, D., 2008. Advanced oxidation processes for water treatment: advances and trends for R&D. J. Chem. Technol. Biotechnol. 83, 769–776.
Dhaouadi, A., Monser, L., Adhoum, N., 2009. Anodic oxidation and electro-Fenton treatment of rotenone. Electrochim. Acta 54, 4473–4480.
Dubrawski, K.L., Mohseni, M., 2013a. Standardizing electrocoagulation reactor design: iron electrodes for NOM removal. Chemosphere 91, 55–60.
Dubrawski, K.L., Mohseni, M., 2013b. In-situ identification of iron electrocoagulation speciation and application for natural organic matter (NOM) removal. Water Res. 47, 5371–5380.
Dubrawski, K.L., Fauvel, M., Mohseni, M., 2013. Metal type and natural organic matter source for direct filtration electrocoagulation of drinking water. J. Hazard. Mater. 244–245, 135–141.
El-Ashtoukhy, E.-S.Z., Amin, N.K., Abdelwahab, O., 2009. Treatment of paper mill effluents in a batch-stirred electrochemical tank reactor. Chem. Eng. J. 146, 205–210.
El-Masry, M.H., Sadek, O.M., Mekhemer, W.K., 2004. Purification of raw surface water using electro-coagulation method. Water Air Soil Pollut. 158, 373–385.
Faouzi, A.M., Nasr, B., Abdellatif, G., 2007. Electrochemical degradation of anthraquinone dye Alizarin Red S by anodic oxidation on boron-doped diamond. Dyes Pigm. 73, 86–89.
Feng, Q., Li, X., Cheng, Y., Meng, L., Meng, Q., 2007. Removal of humic acid from groundwater by electrocoagulation. J. China Univ. Min. Technol. 17, 513–520.
Flox, C., Arias, C., Brillas, E., Savall, A., Groenen-Serrano, K., 2009. Electrochemical incineration of cresols: a comparative study between PbO_2 and boron-doped diamond anodes. Chemosphere 74, 1340–1347.
Fu, J., Zhao, Y., Wu, Q., 2007. Optimizing photoelectrocatalytic oxidation of fulvic acid response surface methodology. J. Hazard. Mater. 144, 499–505.
Gamage, N.P., Chellam, S., 2011. Aluminum electrocoagulation pretreatment reduces fouling during surface water microfiltration. J. Membr. Sci. 379, 97–105.
Gamage, N.P., Rimer, J.D., Chellam, S., 2012. Improvements in permeate flux by aluminum electroflotation pretreatment during microfiltration of surface water. J. Membr. Sci. 411–412, 45–53.

Ghernaout, D., Badis, A., Kellil, A., Ghernaout, B., 2008. Application of electrocoagulation in *Escherichia coli* culture and two surface waters. Desalination 219, 118–125.

Ghernaout, D., Ghernaout, B., Saiba, A., Boucherit, A., Kellil, A., 2009a. Removal of humic acids by continuous electromagnetic treatment followed by electrocoagulation in batch using aluminium electrodes. Desalination 239, 295–308.

Ghernaout, D., Irki, S., Boucherit, A., 2014. Removal of Cu^{2+} and Cd^{2+}, and humic acid and phenol by electrocoagulation using iron electrodes. Desalin. Water Treat. 52, 3256–3270.

Ghernaout, D., Ghernaout, B., Boucherit, A., Naceur, M.W., Khelifa, A., Kellil, A., 2009b. Study on mechanism of electrocoagulation with iron electrodes in idealized conditions and electrocoagulation of humic acids solution in batch using aluminium electrodes. Desalin. Water Treat. 8, 91–99.

Ghernaout, D., Mariche, A., Ghernaout, B., Kellil, A., 2010. Electromagnetic treatment doubled electrocoagulation of humic acid in continuous mode using response surface method for its optimization and application on two surface waters. Desalin. Water Treat. 22, 311–329.

Golder, A., Chanda, A., Samanta, A., Ray, S., 2007. Removal of Cr(VI) from aqueous solution: electrocoagulation vs chemical coagulation. Sep. Sci. Technol. 42, 2177–2193.

Gomes, J.A.G., Daida, P., Kesmez, M., Weir, M., Moreno, H., Parga, J.R., Irwin, G., McWhinney, H., Grady, T., Peterson, E., Cocke, D.L., 2007. Arsenic removal by electrocoagulation using combined Al–Fe electrode system and characterization of products. J. Hazard. Mater. 139, 220–231.

Gotsi, M., Kalogerakis, N., Psillakis, E., Samaras, P., Mantzavinos, D., 2005. Electrochemical oxidation of olive oil mill wastewaters. Water Res. 39, 4177–4187.

Guinea, E., Garrido, J.A., Rodriguez, R.M., Cabot, P.-L., Arias, C., Centellas, F., Brillas, E., 2010. Degradation of the fluoroquinolone enrofloxacin by electrochemical advanced oxidation processes based on hydrogen peroxide electrogeneration. Electrochim. Acta 55, 2101–2115.

Hamza, M., Ammar, S., Abdelhedi, R., 2011. Electrochemical oxidation of 1,3, 5-trimethoxybenzene in aqueous solutions at gold oxide and lead dioxide electrodes. Electrochim. Acta 56, 3785–3789.

Harif, T., Adin, A., 2011. Size and structure evolution of kaolin-$Al(OH)_3$ flocs in the electroflocculation process: a study using static light scattering. Water Res. 45, 6195–6206.

Holt, P.K., Barton, G.W., Mitchell, C.A., 2005. The future for electrocoagulation as a localized water treatment technology. Chemosphere 59, 355–367.

Jeong, J., Kim, C., Yoon, J., 2009. The effect of electrode material on the generation of oxidants and microbial inactivation in the electrochemical disinfection processes. Water Res. 43, 895–901.

Kabdaslı, I., Vardar, B., Arslan-Alaton, I., Tunay, O., 2009. Effect of dye auxiliaries on color and COD removal from simulated reactive dye bath effluent by electrocoagulation. Chem. Eng. J. 148, 89.

Kapalka, A., Foti, G., Comninellis, C., 2010. Basic principles of the electrochemical mineralization of organic pollutants for wastewater treatment. In: Comninellis, C., Chen, G. (Eds.), Electrochemistry for the Environment. Springer, New York, USA, pp. 1–23.

Kiliç, M.G., Hosten, Ç., Demirci, S., 2009. A parametric comparative study of electrocoagulation and coagulation using ultrafine quartz suspensions. J. Hazard. Mater. 171, 247–252.

Klavarioti, M., Mantzavinos, D., Kassinos, D., 2009. Removal of residual pharmaceuticals from aqueous systems by advanced oxidation processes. Environ. Int. 35, 402–417.

Kliaugaite, D., Yasadi, K., Euverink, G.-J., Bijmans, M.F.M., Racys, V., 2013. Electrochemical removal and recovery of humic-like substances from wastewater. Sep. Purif. Technol. 108, 37–44.

Kobya, M., Gebologlu, U., Ulu, F., Oncel, S., Demirbas, E., 2011. Removal of arsenic from drinking water by the electrocoagulation using Fe and Al electrodes. Electrochim. Acta 56, 5060–5070.
Kobya, M., Hiz, H., Senturk, E., Aydiner, C., Demirbas, E., 2006. Treatment of potato chips manufacturing wastewater by electrocoagulation. Desalination 190, 2201–2211.
Koparal, A.S., Yildiz, Y.Ş., Keskinler, B., Demircioglu, N., 2008. Effect of initial pH on the removal of humic substances from wastewater by electrocoagulation. Sep. Purif. Technol. 59, 175–182.
Kourdali, S., Badis, A., Saiba, A., Boucherit, A., Boutoumi, H., 2014. Humic acid removal by electrocoagulation using aluminium sacrificial anode under influencing operational parameters. Desalin. Water Treat. 52, 5442–5453.
Kraft, A., Stadelmann, M., Blaschke, M., 2003. Anodic oxidation with doped diamond electrodes: a new advanced oxidation process. J. Hazard. Mater. B103, 247–261.
Kraft, A., Stadelmann, M., Blaschke, M., Kreysig, D., Sandt, B., Schröder, F., Rennau, J., 1999. Electrochemical water disinfection part I: hypochlorite production from very dilute chloride solutions. J. Appl. Electrochem. 29, 861–868.
Liao, A.A., Spitzer, M., Motheo, A.J., Bertazzoli, R., 2008. Electrocombustion of humic acid and removal of algae from aqueous solutions. J. Appl. Electrochem. 38, 721–727.
Liu, J., Liu, L., Gao, B., Yang, F., 2012a. Cathode membrane fouling reduction and sludge property in membrane bioreactor integrating electrocoagulation and electrostatic repulsion. Sep. Purif. Technol. 100, 44–50.
Liu, L., Liu, J., Gao, B., Yang, F., 2012b. Minute electric field reduced membrane fouling and improved performance of membrane bioreactor. Sep. Purif. Technol. 86, 106–112.
Liu, L., Liu, J., Gao, B., Yang, F., Chellam, S., 2012c. Fouling reductions in a membrane bioreactor using an intermittent electric field and cathodic membrane modified by vapor phase polymerized pyrrole. J. Membr. Sci. 394–395, 202–208.
Ma, H., Wang, B., Wang, Y., 2007. Application of molybdenum and phosphate modified kaolin in electrochemical treatment of paper mill wastewater. J. Hazard. Mater. 145, 417–423.
Mahvi, A.H., Malakootian, M., Heidari, M.R., 2011. Comparison of polyaluminum silicate chloride and electrocoagulation process, in natural organic matter removal from surface water in Ghochan, Iran. J. Water Chem. Technol. 33, 377–385.
Mansouri, K., Hannachi, A., Abdel-Wahab, A., Bensalah, N., 2012. Electrochemically dissolved aluminum coagulants for the removal of natural organic matter from synthetic and real industrial wastewaters. Ind. Eng. Chem. Res. 51, 2428–2437.
Martins, A.F., Wilde, M.L., Vasconcelos, T.G., Henriques, D.M., 2006. Nonylphenol polyethoxylate degradation by means of electrocoagulation and electrochemical Fenton. Sep. Purif. Technol. 50, 249–255.
Meas, Y., Ramirez, J.A., Villalon, M.A., Chapman, T.W., 2010. Industrial wastewaters treated by electrocoagulation. Electrochim. Acta 55, 8165–8171.
Migliorini, F.L., Braga, N.A., Alves, S.A., Lanza, M.R.V., Baldan, M.R., Ferreira, N.G., 2011. Anodic oxidation of wastewater containing the Reactive Orange 16 Dye using heavily boron-doped diamond electrodes. J. Hazard. Mater. 192, 1683–1689.
Miwa, D.W., Malpass, G.R.P., Machado, S.A.S., Motheo, A.J., 2006. Electrochemical degradation of carbaryl on oxide electrodes. Water Res. 40, 3281–3289.
Mohora, E., Roncevic, S., Dalmacija, B., Agbaba, J., Watson, M., Karlovic, E., Dalmacija, M., 2012. Removal of natural organic matter and arsenic from water by electrocoagulation/flotation continuous flow reactor. J. Hazard. Mater. 235–236, 257–264.
Mollah, M.Y.A., Morkovsky, P., Gomez, J.A.G., Kesmez, M., Parga, J., Cocke, D.L., 2004. Fundamentals, present and future perspectives of electrocoagulation. J. Hazard. Mater. B114, 199–210.

Mollah, M.Y.A., Schennach, R., Parga, J.R., Cocke, D.L., 2001. Electrocoagulation (EC) – science and applications. J. Hazard. Mater. B84, 29–41.
Montilla, F., Michaud, P.A., Morallon, E., Vazquez, J.L., Comninellis, C., 2002. Electrochemical oxidation of benzoic acid at boron-doped diamond electrodes. Electrochim. Acta 47, 3509–3513.
Moreno, H., Cocke, D., Gomes, J., Morkovsky, P., Parga, J., Peterson, E., Garcia, C., 2009. Electrochemical reactions for electrocoagulation using iron electrodes. Ind. Eng. Chem. Res. 48, 2275–2282.
Morgan, B., Lahav, O., 2007. The effect of pH on the kinetics of spontaneous Fe(II) oxidation by O_2 in aqueous solution – basic principles and a simple heuristic description. Chemosphere 68, 2080–2084.
Motheo, A.J., Pinhedo, L., 2000. Electrochemical degradation of humic acid. Sci. Total Environ. 256, 67–76.
Mouedhen, G., Feki, M., Wery, M.D.P., Ayedi, H.F., 2008. Behavior of aluminum electrodes in electrocoagulation process. J. Hazard. Mater. 150, 124–135.
Pallier, V., Feuillade-Cathalifaud, G., Serpaud, B., 2011. Influence of organic matter on arsenic removal by continuous flow electrocoagulation treatment of weakly mineralized waters. Chemosphere 83, 21–28.
Panizza, M., Cerisola, G., 2005. Application of diamond electrodes to electrochemical processes. Electrochim. Acta 51, 191–199.
Parga, J., Cocke, D., Valverde, V., Gomes, J., Kesmez, M., Moreno, H., Weir, M., Mencer, D., 2005. Characterization of electrocoagulation for removal of chromium and arsenic. Chem. Eng. Technol. 28, 605–612.
Patel, U.D., Suresh, S., 2008. Electrochemical treatment of pentachlorophenol in water and pulp bleaching effluent. Sep. Purif. Technol. 61, 115–122.
Patermarakis, G., Fountoukidis, E., 1990. Disinfection of water by electrochemical treatment. Water Res. 24, 1491–1496.
Picard, T., Cathalifaud-Feuillade, G., Mazet, M., Vandensteendam, C., 2000. Cathodic dissolution in the electrocoagulation process using aluminium electrodes. J. Environ. Monit. 2, 77–80.
Pinhedo, L., Pelegrini, R., Bertazzoli, R., Motheo, A.J., 2005. Photoelectrochemical degradation of humic acid on a $(TiO_2)_{0.7}(RuO_2)_{0.3}$ dimensionally stable anode. Appl. Catal. B Environ. 57, 75–81.
Polcaro, A.M., Vacca, A., Mascia, M., Palmas, S., Pompei, R., Laconi, S., 2007. Characterization of a stirred tank electrochemical cell for water disinfection processes. Electrochim. Acta 52, 2595–2602.
Qu, X., Gao, W.J., Han, M.N., Chen, A., Liao, B.Q., 2012. Integrated thermophilic submerged aerobic membrane bioreactor and electrochemical oxidation for pulp and paper effluent treatment – towards system closure. Bioresour. Technol. 116, 1–8.
Radjenovic, J., Bagastyo, A., Rozendal, R.A., Mu, Y., Keller, J., Rabaey, K., 2011. Electrochemical oxidation of trace organic contaminants in reverse osmosis concentrate using RuO_2/IrO_2-coated titanium anodes. Water Res. 45, 1579–1586.
Rajkumar, D., Kim, J.G., 2006. Oxidation of various reactive dyes with in situ electro-generated active chlorine for textile dyeing industry wastewater treatment. J. Hazard. Mater. B136, 203–212.
Rajkumar, D., Song, B.J., Kim, J.G., 2007. Electrochemical degradation of Reactive Blue 19 in chloride medium for the treatment of textile dyeing wastewater with identification of intermediate compounds. Dyes Pigm. 72, 1–7.
Ricordel, C., Darchen, A., Hadjiev, D., 2010. Electrocoagulation–electroflotation as a surface water treatment for industrial uses. Sep. Purif. Technol. 74, 342–347.
Rodrigo, M.A., Cañizares, P., Sanchez-Carretero, A., Saez, C., 2010. Use of conductive-diamond electrochemical oxidation for wastewater treatment. Catal. Today 151, 173–177.

Sakalis, A., Fytianos, K., Nickel, U., Voulgaropoulos, A., 2006. A comparative study of platinized titanium and niobe/synthetic diamond as anodes in the electrochemical treatment of textile wastewater. Chem. Eng. J. 119, 127–133.

Scialdone, O., Galia, A., Guarisco, C., Randazzo, S., Filardo, G., 2008. Electrochemical incineration of oxalic acid at boron doped diamond anodes: the role of operative parameters. Electrochim. Acta 53, 2095–2108.

Scialdone, O., Randazzo, S., Galia, A., Filardo, G., 2009a. Electrochemical oxidation of organics at metal oxide electrodes: the incineration of oxalic acid at IrO_2-TaO_2O_5. Electrochim. Acta 54, 1210–1217.

Scialdone, O., Randazzo, S., Galia, A., Silvestri, G., 2009b. Electrochemical oxidation of organics in water: role of operative parameters in the absence and in the presence of NaCl. Water Res. 43, 2260–2272.

Selcuk, H., Sarikaya, H.Z., Bekbolet, M., Anderson, M.A., 2006. Bromate formation on the non-porous TiO_2 photoanode in the photoelectrocatalytic system. Chemosphere 62, 715–721.

Selcuk, H., Sene, J.J., Anderson, M.A., 2003. Photoelectrocatalytic humic acid degradation kinetics and effect of pH, applied potential and inorganic ions. J. Chem. Technol. Biotechnol. 78, 979–984.

Särkkä, H., Kuhmonen, K., Vepsäläinen, M., Pulliainen, M., Selin, J., Rantala, P., Kukkamäki, E., Sillanpää, M., 2009. Electrochemical oxidation of sulphides in paper mill wastewater by using mixed oxide anodes. Environ. Technol. 30, 885–892.

Tanneru, C.T., Chellam, S., 2012. Mechanisms of virus control during iron electrocoagulation – microfiltration of surface water. Water Res. 46, 2111–2120.

Tavares, M.G., da Silva, L.V.A., Sales Solano, A.M., Tonholo, J., Martinez-Huitle, C.A., Zanta, C.L.P.S., 2012. Electrochemical oxidation of Methyl Red using $Ti/Ru_{0.3}Ti_{0.7}O_2$ and Ti/Pt anodes. Chem. Eng. J. 204–206, 141–150.

Theis, T.L., Singer, P.C., 1974. Complexation of iron(II) by organic matter and its effect on iron(II) oxygenation. Environ. Sci. Technol. 8, 569–573.

Tran, N., Drogui, P., 2013. Electrochemical removal of microcystin-LR from aqueous solution in the presence of natural organic pollutants. J. Environ. Manage. 114, 253–260.

Trompette, J.L., Vergnes, H., 2009. On the crucial influence of some supporting electrolytes during electrocoagulation in the presence of aluminum electrodes. J. Hazard. Mater. 163, 1282–1288.

Vepsäläinen, M., Ghiasvand, M., Selin, J., Pienimaa, J., Repo, E., Pulliainen, M., Sillanpää, M., 2009. Investigations of the effects of temperature and initial sample pH on natural organic matter (NOM) removal with electrocoagulation using response surface method (RSM). Sep. Purif. Technol. 69, 255–261.

Vepsäläinen, M., Pulliainen, M., Sillanpää, M., 2012. Effect of electrochemical cell structure on natural organic matter (NOM) removal from surface through electrocoagulation (EC). Sep. Purif. Technol. 99, 20–27.

Vepsäläinen, M., Selin, J., Pulliainen, M., Sillanpää, M., 2007. Combined electrocoagulation and chemical coagulation of paper mill mechanically cleaned water. J. Pulp Pap. Sci. 33, 233–239.

Vik, E.A., Carlson, D.A., Eikum, A.S., Gjessing, E.T., 1984. Electrocoagulation of potable water. Water Res. 18, 1355–1360.

Wang, B., Gu, L., Ma, H., 2007a. Electrochemical oxidation of pulp and paper making wastewater assisted by transition metal modified kaolin. J. Hazard. Mater. 143, 198–205.

Wang, B., Kong, W., Ma, H., 2007b. Electrochemical treatment of paper mill wastewater using three-dimensional electrodes with $Ti/Co/SnO_2$-Sb_2O_5 anode. J. Hazard. Mater. 146, 295–301.

Yildiz, Y.Ş., Koparal, A.S., Keskinler, B., 2008. Effect of initial pH and supporting electrolyte on the treatment of water containing high concentration of humic substances by electrocoagulation. Chem. Eng. J. 138, 63–72.

Yildiz, Y.Ş., Koparal, A.S., İrdemez, Ş., Keskinler, B., 2007. Electrocoagulation of synthetically prepared waters containing high concentration of NOM using iron cast electrodes. J. Hazard. Mater. B139, 373–380.
Zhao, X., Zhang, B., Liu, H., Chen, F., Li, A., Qu, J., 2012. Transformation characteristics of refractory pollutants in plugboard wastewater by an optimal electrocoagulation and electro-Fenton process. Chemosphere 87, 631–636.
Zhu, B., Clifford, D.A., Chellam, S., 2005. Comparison of electrocoagulation and chemical coagulation pretreatment for enhanced virus removal using microfiltration membranes. Water Res. 39, 3098–3108.
Zhu, X., Ni, J., Wei, J., Chen, P., 2011. Scale-up of B-doped diamond anode system for electrochemical oxidation of phenol simulated wastewater in batch mode. Electrochim. Acta 56, 9439–9447.

CHAPTER 5

Membranes

Mika Sillanpää[*], Sari Metsämuuronen[**], Mika Mänttäri[**]

[*]Lappeenranta University of Technology, LUT Faculty of Technology, LUT Chemtech, Laboratory of Green Chemistry, Sammonkatu 12, 50130 Mikkeli, Finland; [**]Lappeenranta University of Technology, LUT Faculty of Technology, LUT Chemtech, Skinnarilankatu 34, 53850 Lappeenranta, Finland

Abstract

Aquatic natural organic matter (NOM) is a heterogeneous mixture of biopolymers and their degradation products that cause harmful by-products during drinking water production. The great variability in NOM composition makes it difficult to remove from drinking water completely by conventional coagulation or by any single technique. This chapter reviews NOM removal by membranes—namely, micro-, ultra-, and nanofiltration, and also reverse osmosis.

Keywords: Fouling; Membrane technology; Microfiltration; Nanofiltration; Natural organic matter (NOM); Reverse osmosis; Ultrafiltration; Water treatment.

ABBREVIATIONS

AOC Assimilable organic carbon
BDOC Biodegradable dissolved organic carbon
DBP Disinfection by-product
DOC Dissolved organic carbon
HAAs Haloacetic acids
MF Microfiltration
MM Molar mass
NF Nanofiltration
NOM Natural organic matter
PES Polyethersulfone
PVDF Polyvinylidene fluoride
RO Reverse osmosis
SUVA Specific UV absorbance
THM Trihalomethane
THMFP Trihalomethane formation potential
TOC Total organic carbon
UF Ultrafiltration

5.1 INTRODUCTION

Since the late 1980s, membrane processes have increasingly been considered an effective means of removing turbidity, organics, microorganisms, and disinfection by-product (DBP) precursors from drinking water in compliance with the more stringent regulations. This is the result of significant improvements in membrane technology, design, and materials during the past three decades, making these drinking water production processes more environmentally friendly.

It is estimated that, today, the overall plant capacity for microfiltration (MF) and ultrafiltration (UF) is around 20 million m^3 day^{-1}, with 60% for drinking water production (Schrotter and Bozkaya-Schrotter, 2010), and UF applications represent about 74% of the total capacity (Fane et al., 2008).

Membrane processes use a semipermeable film called a *membrane* to separate particles and solutes from water. Most membrane processes use pressure differences across the membrane as a driving force. These can be divided into pressure and vacuum systems (Figure 5.1). In pressure systems, membranes are installed in modules, which are pressurized from the retentate side of the membrane. In a dead-end filtration feed, raw water permeates through the membrane, and solutes larger than the membrane pore size are retained, whereas in a cross-flow configuration, the majority of the feed flows tangentially across the membrane surface. In vacuum systems, membrane elements are submerged in open tanks.

The membranes reject uncharged macromolecules mainly by size exclusion, and thus the rejection is determined by the membrane pore size and

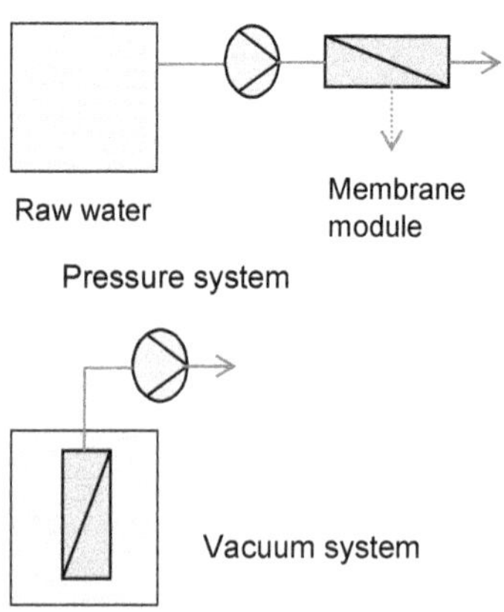

Figure 5.1 Schematic representation of pressure and vacuum membrane configurations.

pore size distribution compared with the size, size distribution, and shape of the natural organic matter (NOM) molecules. The molar mass (MM) of NOM is in the colloidal size range, from a few 100 Da (Daltons equals grams per mole) to more than 100 kDa (Ericsson and Trägårdh, 1997). The size distribution of NOM varies between different water sources, but generally a very high proportion of small molecules, less than 1 kDa, have been observed (Fan et al., 2001; Kim and Yu, 2005; Zularisam et al., 2007b; Zularisam et al., 2009) (Figure 5.2). Thus, the average MM is quite low. Weight-averaged MM values of 833–1031 Da (Spain (de la Rubia et al., 2008)), 1254–1493 Da (Australia (Fabris et al., 2008)), 1442–1812 Da (Norway (Fabris et al., 2008)), and 2190 Da (Georgia, USA (Chin et al., 1994)) have been reported for surface waters. The smallest NOM fractions consist of hydrophilic compounds, organic acids, and building blocks, which are degradation products of humic substances (Huber et al., 2011). Polysaccharides, proteins, and amino sugars are the highest MM components in NOM, whereas the most aromatic compounds (i.e., lignin and tannin derivatives) are abundant in the intermediate–high MM fractions of NOM (Lankes et al., 2008).

In addition to size exclusion, membrane rejection depends on solute–solute and solute–membrane interactions, and hydrodynamic conditions.

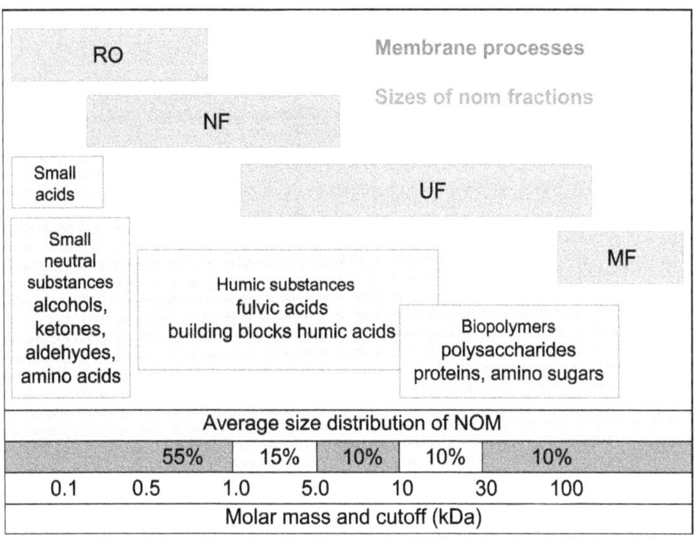

Figure 5.2 Cutoffs of different membrane processes and the sizes of natural organic matter (NOM) fractions. MF, microfiltration; NF, nanofiltration; RO, reverse osmosis; UF, ultrafiltration. *Data from Fan et al. (2001), Zularisam et al. (2007a), Huber et al. (2011), Thorsen (2004), Kennedy et al. (2005), Kaewsuk and Seo (2011), and Jeong et al. (2013).*

Rejection can be greater than what is expected according to the nominal cutoff value of the membrane if there is a repulsive electrostatic force between the membrane and solutes. Therefore, the properties of both the membrane and NOM are important factors in NOM removal, including acidic group content and hydrophobic/hydrophilic character. Other physicochemical effects that affect rejection when charge repulsion is used are pH, ionic strength, and divalent cation—especially Ca^{2+}—content. This chapter reviews NOM removal by membrane technologies and is an update of an earlier publication (Metsämuuronen et al., 2014).

5.2 MICROFILTRATION

MF membranes were first used for color, turbidity, and pathogen removal in municipal drinking water plants during the late 1980s (Ericsson et al., 1997; Jacangelo et al., 1997; Thorsen, 1998; Laîné et al., 2000). The pore sizes of MF membranes (0.1–5 μm or >100 kDa) are much larger than the size of most NOM fractions; thus, MF alone is not effective in the removal of NOM substances other than polysaccharides (Kennedy et al., 2005). MF membranes are able to remove less than 10% of NOM, whereas turbidity removal is generally more than 97%.

The advantage of MF is that its large pore size enables operation at relatively low transmembrane pressures. MF is used widely in variety of surface water purification processes and membrane bioreactors, and as a pretreatment method in seawater desalination plants (Hilal et al., 2006). Submerged membrane bioreactors are used mainly for wastewater treatment, but also increasingly for freshwater polishing (Sutherland, 2012). MF is usually coupled with other methods, either at the beginning of the treatment chain to remove particular and microbial materials or after coagulation (Sanchez et al., 2013; Schrotter and Bozkaya-Schrotter, 2010). Precoagulation can enhance NOM removal of MF membranes several fold and simultaneously mitigates membrane fouling (Li et al., 2011; Liu et al., 2011; Ben-Sasson et al., 2013).

Flat plate, hollow fiber, tubular, and spiral wound membranes are used (Sutherland, 2012), but hollow fiber is the most used MF and UF configuration in drinking water plants (Schrotter and Bozkaya-Schrotter, 2010; Rachwal and Judd, 2006; Kajitvichyanukul et al., 2008; Fane et al., 2008). Many industrial plants use inside-out hollow fibers (Remize et al., 2010). The MF and UF membranes are generally made of cellulose acetate, polysulfone, polyethersulfone (PES), and polyvinylidene fluoride (PVDF) polymers (Schrotter and Bozkaya-Schrotter, 2010). The hydrophilic regenerated cellulose and cellulose acetate membranes are only slightly charged, so the primary mechanism of NOM removal with these membranes is size exclusion. The other membrane materials exhibit greater surface charge densities and hydrophobicities based on their composition. Membrane surface charge is caused also by the adsorption of anions from water. Hence, the membrane surface charge depends on the concentration of ions in the solution (Bodzek et al., 2011).

5.3 ULTRAFILTRATION

The pore size of UF membranes is 10–100 nm and their cutoff value is 1–100 kDa, which matches the size of the higher MM humic substances and biopolymers. In general, NOM removal increases as the membrane pore size decreases (Amy and Cho, 1999; Cho et al., 2000a; Schäfer et al., 2000; Syafei et al., 2008; Ates et al., 2009). When considering surface water treatment by UF membranes, loose UF membranes (cutoff, ≥60 kDa) have shown dissolved organic carbon (DOC) removal ranging from 14% to 49% and medium membranes (cutoff, 10–60 kDa) ranging from 20% to 75% (Table 5.1). The pore sizes of tight UF membranes (cutoff, <10 kDa) are in the range of the larger NOM molecules, so their DOC removal is already quite high (38–87%).

Yet, it is not straightforward to predict NOM removal on the basis of the membrane cutoff value. Apart from size exclusion, electrostatic repulsion is also an important separation mechanism in UF. The lowest NOM removal rates have been obtained with uncharged membranes, whereas with charged UF membranes, much greater removal rates have been observed than would be expected on the basis of the membrane cutoff value and the size of the surface water NOM (Cho et al., 1999, 2000a,b; Yoon et al., 2005; Lee et al., 2005).

Lee et al. (2005) filtered surface waters with a negatively charged 8-kDa cutoff membrane and noticed that the MM of NOM reduced from feed water values of 1210 and 2360 Da to 620 and 900 Da in the permeates, respectively. In another study, an 8-kDa cutoff membrane was observed to remove almost completely the MM fraction larger than 2 kDa (Cho et al., 2000b).

Hydrophobic interactions and electrostatic exclusion between the hydrophobic and charged membrane surface and NOM molecules can increase NOM removal significantly. Yoon et al. (2005) reported NOM removal with hydrophobic PES membranes eight times greater than expected, considering the average MM of NOM and membrane pore sizes. These results clearly show that much smaller molecules are rejected than the membrane cutoff value presumes.

Table 5.1 Studies concerning NOM removal by MF and UF membranes from natural waters and surrogate waters containing humic substances

Solution	Membrane, material, cutoff/pore diameter, module type, TMP	Variable	DOC removal (%)	Quality of permeate; content and/or removal (%)	References
6.3% NOM, Australia	Different membranes, stirred batch cell, lab scale, 100–300 kPa	GVWP, mod; PVDF, 220 nm	17		Schäfer et al. (2000)
		PLHK, RC, 100 kDa	9		
		PLTK, 30 kDa	10		
		PLGC, RC, 10 kDa	51		
		PLCC, 5 kDa	74		
		PLBC, 3 kDa	77		
		PLAC, 1 kDa	86		
Humic acid, 15 mg L^{-1}	MF, polycarbonate track etched, 0.1 μm	pH 6		TOC, 20%	Ben-Sasson et al. (2013)
		pH 7		TOC, 33%	
		pH 8		TOC, 14%	
	UF, UE50, PES, 100 kDa	pH 6		TOC, 59%	
		pH 7		TOC, 49%	
		pH 8		TOC, 56%	
Surface water; DOC, 1.3 mg L^{-1}	PVDF, 100 nm, hollow fiber	Submerged	35		Chae et al. (2009)
		Pressurized	35		
Surface water; DOC, 2.2 mg L^{-1}; SUVA, 2.9; THMFP, 75 μg L^{-1}; HAAFP, 80 μg L^{-1}	VS-70-10, PE textile, 100 nm, cross-flow		18	DOC, 1.8 mg L^{-1}; THMFP, 55 μg L^{-1}, 8%; HAAFP, 48 μg L^{-1}, 4%	Siddiqui et al. (2000)

Surface water; DOC, 5.3 mg L^{-1}; SUVA, 1.8	ZeeWeed 1000, 20 nm, full scale		66	DOC, 1.9 mg L^{-1}; THMFP, 54%; HAAFP, 77%	Lamsal et al. (2012)
Different NOM fractions; DOC, 12.5 mg L^{-1}	PLHK, RC, 100 kDa/18.2 nm, stirred batch cell, lab scale, 100 kPa	IHSS humic acid	~45		Aoustin et al. (2001)
		IHSS fulvic acid	~10		
		6.3% NOM, Australia	~10		
		NOM–humic acid, Australia	10		
		NOM–fulvic acid, Australia	<5		
		NOM–hydrolysate, Australia	>10		
	PLGC, RC, 10 kDa/5.2 nm, stirred batch cell, lab scale, 300 kPa	IHSS humic acid	75		
		IHSS fulvic acid	45		
		6.3% NOM, Australia	~50		
		NOM–humic acid, Australia	51		
		NOM–fulvic acid, Australia	41		
		NOM–hydrolysate, Australia	19		
Moorland water; TOC, 9.8 mg L^{-1}	PSu, 100 kDa, flat sheet, bench scale, 100 kPa		18	UV_{254}, 22%	Bessiere et al. (2009)

(Continued)

Table 5.1 Studies concerning NOM removal by MF and UF membranes from natural waters and surrogate waters containing humic substances—cont'd

Solution	Membrane, material, cutoff/pore diameter, module type, TMP	Variable	DOC removal (%)	Quality of permeate; content and/or removal (%)	References
Humic acid (Aldrich), 2 mg L^{-1}; DOC, 8.7 mg L^{-1}; different pHs	Different flat sheet membranes or charges, stirred cell, lab scale, 69 kPa	RC, 100 kDa; neutral; pH 3.5		UV_{254}, 55%	Shao et al. (2011)
		RC, 100 kDa; neutral; pH 7.5		UV_{254}, 59–97%	
		RC, 100 kDa; charged; pH 3.5		UV_{254}, 79%	
		RC, 100 kDa; charged; pH 7.5		UV_{254}, 92–98%	
		PES, 100 kDa; zeta, −12.3 mV		UV_{254}, 66%	
Humic acid (Aldrich); DOC, 5 mg L^{-1}; different size fractions	PSu, 100 kDa, hydrophobic, hollow fiber, cross-flow	MM, 6.5–22.6 kDa; DOC, 10.9 mg L^{-1}	96	THMFP, <20 μg L^{-1}, 95%	Lin et al. (1999)
		MM, 2.2–6.5 kDa; DOC, 7.0 mg L^{-1}	74	THMFP, ~50 μg L^{-1}	
		MM, 0.65–2.2 kDa; DOC, 7.5 mg L^{-1}	64	THMFP, ~60 μg L^{-1}	
		MM, 180–650 Da; DOC, 5.8 mg L^{-1}	14	THMFP, ~480 μg L^{-1}, 10%	

Humic acid (Aldrich); DOC, 5 mg L^{-1}; different pHs	Different membranes, hollow fiber, cross-flow, 150–170 kPa	PSu, 100 kDa, hydrophobic, pH 4	60		Lin et al. (2001)
		PSu, 100 kDa, hydrophobic, pH 7	~25		
		PSu, 10 kDa, hydrophobic, pH 7	~80		
		RCA, 10 kDa; hydrophilic; pH 7	~80		
6.3% NOM, Australia	Different membranes, stirred batch cell, lab scale, 100–300 kPa	PLHK, RC, 100 kDa	9		Schäfer et al. (2000)
		PLTK, 30 kDa	10		
		PLGC, RC, 10 kDa	51		
		PLCC, 5 kDa	74		
		PLBC, 3 kDa	77		
		PLAC, 1 kDa	86		
Surface (SW) and reservoir (RW) waters	PSu, 68 kDa, hydrophobic, submerged hollow fiber	SW; DOC, 6.7 mg L^{-1}; SUVA, 1.3	~17	UV_{254}, ~26%	Zularisam et al. (2007a)
		RW; DOC, 8.0 mg L^{-1}; SUVA, 2.8	~38	UV_{254} 44–50%	
Surface water; DOC, 6.8 mg L^{-1}; SUVA, 1.3; different fractions	MRUTM55, PSu, 68 kDa, hollow fiber	Hydrophobic fraction	49		Zularisam et al. (2007a)
		Transphilic fraction	31		
		Hydrophilic fraction	14		
	MRUTM66, CA, 50 kDa, hollow fiber	Hydrophobic fraction	34		
		Transphilic fraction	27		
		Hydrophilic fraction	22		

(Continued)

Table 5.1 Studies concerning NOM removal by MF and UF membranes from natural waters and surrogate waters containing humic substances—cont'd

Solution	Membrane, material, cutoff/pore diameter, module type, TMP	Variable	DOC removal (%)	Quality of permeate; content and/or removal (%)	References
Surface water; DOC, 2.2 mg L^{-1}; SUVA, 2.9; THMFP, 75 μg L^{-1}; HAAFP, 80 μg L^{-1}	192-PT3, TFC, 60 kDa, spiral wound		27	DOC, 1.6 mg L^{-1}; THMFP, 32 μg L^{-1}, 47%; HAAFP, 36 μg L^{-1}, 28%	Siddiqui et al. (2000)
Humic acid (Aldrich); DOC, 8.7–9.4 mg L^{-1}; pH 7	Different pore size ceramic membranes, flat sheet, cross-flow	50 kDa 15 kDa 1 kDa	~61 ~83 ~84		Syafei et al. (2008)
Surface water; DOC, 11.7 mg L^{-1}; SUVA, 2.0	YMJWSP3001, PVDF, 30 kDa, negative charge		11–17		Byun et al. (2011)
Humic acid (Aldrich); DOC, 10 mg L^{-1}	KERMBMU1, ceramic, 15 kDa/3.54 nm, single tube, bench scale	pH 2.4; zeta, −2.9 mV; pI, 1 mmol pH 7.9; zeta, −15.6 mV; pI, 1 mmol pI, 1 mmol L^{-1}; pH 7 pI, 152 mmol L^{-1}; pH 7	59 99 98 74	UV_{254} 85%	de la Rubia et al. (2006)

Surface water; DOC, 3.1 mg L^{-1}; SUVA, 2.9	Different flat-sheet membranes, dead-end stirred cell, bench scale	NTR7410, S-PES, 20 kDa, hydrophobic	~75		Yoon et al. (2005)
		PM10, PES, 10 kDa, hydrophobic	~70		
		10PESUF40, PES, 10 kDa hydrophobic	~75		
		GM, S-PES, 8 kDa, hydrophilic	~86		
Surface water; TOC, 2.3 mg L^{-1}; SUVA, 1.7; THMFP, 70 μg L^{-1}	Different flat-sheet membranes, stirred cell, lab scale	NTR-7410, S-PSu, 20 kDa	20	UV_{254}, 25%	Lee and Lee (2007)
		GR90, PSu, 10 kDa	40	UV_{254}, 50%	
		ETNA01A, HPC, 1 kDa	44	UV_{254}, 47%	
		HEKLA01A, amine + DIC, 1 kDa	48	UV_{254}, 53%	
Surface water; DOC, 4.8 ± 2.2 mg L^{-1}	PW-4040F, 10 kDa, spiral wound, pilot plant		29	DOC, 3.4 ± 1.9 mg L^{-1}; UV_{254}, 40%	Mierzwa et al. (2012)

(Continued)

Table 5.1 Studies concerning NOM removal by MF and UF membranes from natural waters and surrogate waters containing humic substances—cont'd

Solution	Membrane, material, cutoff/pore diameter, module type, TMP	Variable	DOC removal (%)	Quality of permeate; content and/or removal (%)	References
Surface water; DOC, 2.0 mg L^{-1}; SUVA, 2.4	Different flat-sheet membranes, cross-flow, bench scale	PM10, PES, 10 kDa	0		Cho et al. (1999)
		GM, PA TFC, 8 kDa, negative charge	38		
		YM3, RC, 3 kDa	0		
Agricultural drain; DOC, 47.8 mg L^{-1}; SUVA, 3.7		PM10, PES, 10 kDa	37		Cho et al. (1999)
		GM, PA TFC, 8 kDa, negative charge	56	UV_{254}, 60%	
		YM3, RC, 3 kDa	51		
Surface water; DOC, 3.1 mg L^{-1}; SUVA, 3.0		NTR-7410, S-PES, 20 kDa; zeta, −22.6 mV	32	UV_{254}, 49%	Amy and Cho (1999) and Cho et al. (2000a)
		10PESUF40, PES, 10 kDa; zeta, −12.8 mV	32	UV_{254}, 49%	
		GM, PA TFC, 8 kDa; zeta, −17.0 mV	57	UV_{254}, 68%	
Groundwater; DOC, 6.8 mg L^{-1}; SUVA, 5.7		NTR7410, S-PES, 20 kDa; zeta, −22.6 mV	74	UV_{254}, 86%	Amy and Cho (1999) and Cho et al. (2000a)
		10PESUF40, PES, 10 kDa; zeta, −12.8 mV	85	UV_{254}, 88%	
		GM, PA TFC, 8 kDa; zeta, −17.0 mV	87	UV_{254}, 93%	

Surface water; DOC, 3.07 mg L^{-1}; SUVA, 2.7	GM, PA TFC, 8 kDa; zeta, −20.0 mV; cross-flow; bench scale	Colloidal fraction; SUVA, 1.8; pH 4	60		Kim et al. (2011)
		Colloidal fraction; SUVA, 1.8; pH 10	27		
		Hydrophobic fraction; SUVA, 3.0; pH 4	30		
		Hydrophobic fraction; SUVA, 3.0; pH 7	40		
		Hydrophobic fraction; SUVA, 3.0; pH 10	15		
		Transphilic fraction; SUVA, 2.6; pH 4	63		
Surface water; DOC, 3.9 mg L^{-1}; SUVA, 4.5	GM, PA TFC, 8 kDa, negative charge, flat sheet, cross-flow, bench scale		72		Cho et al. (1999)
Different surface water and groundwater	GM, PA TFC, 8 kDa, flat sheet, tangential cross-flow, bench scale	DOC, 2.0 mg L^{-1}; SUVA, 2.4	38	UV_{254}, 64%	Amy and Cho (1999)
		DOC, 3.9 mg L^{-1}; SUVA, 4.4	60	UV_{254}, 84%	
		DOC, 9.8 mg L^{-1}; SUVA, 4.9	85	UV_{254}, 93%	
		DOC, 6.8 mg L^{-1}; SUVA, 5.7	87	UV_{254}, 93%	
Different surface waters	GM, PA TFC, 8 kDa, flat sheet, cross-flow, bench scale	DOC, 9.8 mg L^{-1}; SUVA, 4.9	84		Cho et al. (2000b)
		DOC, 47.80 mg L^{-1}; SUVA, 3.7	60		

(Continued)

Table 5.1 Studies concerning NOM removal by MF and UF membranes from natural waters and surrogate waters containing humic substances—cont'd

Solution	Membrane, material, cutoff/pore diameter, module type, TMP	Variable	DOC removal (%)	Quality of permeate; content and/or removal (%)	References
Surface water; DOC, 3.8 mg L^{-1}; SUVA, 2.3	Different membranes, flat sheet, cross-flow, bench scale	GM, PA TFC, 8 kDa	~39	UV_{254}, ~42%; HAAFP, 57%	Lee and Cho (2004)
		CeRAM28, T-8000, TiO_2, 8 kDa	~40	UV_{254}, ~42%; HAAFP, 63%	
		CeRAM25, T-1000, TiO_2, 1 kDa	~73	UV_{254}, ~72%; HAAFP, 78%	
Reservoir water; DOC, 4.0 mg L^{-1}; UV_{280}/DOC, 2.0	Ceramic, 4 nm, single-channel tubular, lab scale	TMP, 400 kPa	55	UV_{280}, 72%	Harman et al. (2010)
		TMP, 1200 kPa	73	UV_{280}, 83%	
Different surface waters	YM3, RC, 3 kDa, flat sheet, cross-flow, bench scale	DOC, 2.0 mg L^{-1}; SUVA, 2.4	0		Cho et al. (2000b)
		DOC, 3.9 mg L^{-1}; SUVA, 4.5	55		
		DOC, 9.8 mg L^{-1}; SUVA, 4.9	47		
		DOC, 47.80 mg L^{-1}; SUVA, 3.7	51		

Natural water; DOC, 3.4 mg L^{-1}; SUVA, 2.5; HMM, ~12 kDa; LMM, ~1.8 kDa	Different flat-sheet membranes, cross-flow, lab scale	PT, PES, 5 kDa	61	UV_{254}, 74%; HMM, 90%; LMM, 7%; THMs, 60 μg L^{-1}, 63%; HAAs, 34 μg L^{-1}, 38%	Ates et al. (2009)
		GH, PA TFC/S-PES, 1 kDa	85	UV_{254}, 90%; HMM, 92%; LMM, 11%; THMs, 22 μg L^{-1}, 82%; HAAs, 21 μg L^{-1}, 60%	
Surface water; TOC, 3.6–5.4 mg L^{-1}	GK-4040F, 3.5 kDa, TFC, spiral wound, pilot plant		85	DOC, 0.3–1.2 mg L^{-1}	Mierzwa et al. (2008)
Surface water; TOC, 3.2 mg L^{-1}	ETNA10PP, 1 kDa, flat sheet, cross-flow, lab scale, 400 kPa		68	UV_{254}, 68%	Costa and de Pinho (2006a)
Surface waters from 10 different sources; DOC, 1.4–7.1 mg L^{-1}	Different flat-sheet membranes, stirred cell, lab scale	YM1, RCA, 1 kDa	22–68	UV_{254}, 63–78%	de la Rubia et al. (2008)
		YC05, CA, 500 Da	57–93	UV_{254}, 72–93%	

CA, cellulose acetate; DIC, diisocyanate; DOC, dissolved organic carbon; HAA, haloacetic acid; HAAFP, haloacetic acid formation potential; HMM, high molar mass; HPC, hydroxypropyl cellulose; IHSS, international humic substance society; LMM, low molar mass; MF, microfiltration; NOM, natural organic matter; PA, polyamide; PE, polyester; PES, polyethersulfone; PSu, polysulfone; PVDF, polyvinylidene fluoride; RC, regenerated cellulose; RCA, regenerated cellulose acetate; S-PES, sulfonated polyethersulfone; S-PSu, sulfonated polysulfone; TFC, thin-film composite; THM, trihalomethane; THMFP, trihalomethane formation potential; TMP, transmembrane pressure; TOC, total organic carbon; UF, ultrafiltration; zeta, zeta potential; SUVA, specific ultraviolet absorbance. SUVA values are in liters per meter per milligram.

5.3.1 Effect of NOM Characteristics

Some of the important properties of NOM that enhance its removal by membranes include high MM, negative charge, and a hydrophobic and aromatic structure. All these characteristics are associated with the hydrophobic NOM fraction. This is consistent with the fact that the studies performed with humic acids show greater DOC removal than those performed with surface waters or NOM solutions (Table 5.1). Lamsal et al. (2012) noted an overall DOC removal of 66%, whereas the removal of hydrophobic acids was as high as 93% in river water UF.

Several researchers have applied a ratio of UV_{254} to DOC (specific ultraviolet absorbance (SUVA)) in the identification of NOM removal and reported that the majority of DOC removed from raw waters by UF was aromatic or hydrophobic compounds (humic and fulvic acids (de la Rubia et al., 2008; Amy and Cho, 1999; Cho et al., 1999, 2000a,b; Ates et al., 2009)). The high SUVA value corresponds to high hydrophobicity and to a high concentration of humic substances (Edzwald and Tobiason, 1999). Amy and Cho (1999) reported that DOC rejection increased from 38% to 87% when SUVA increased from 2.4 to $5.7\,\mathrm{L\,m^{-1}\,mg^{-1}}$. In conclusion, NOM rejection increases with an increase in the proportion of humic substances in NOM, and thus with its SUVA value.

Hydrophobic NOM fractions have a significantly greater charge density than hydrophilic NOM. Acidic contents of 5.1, 3.6, and $1.0\,\mathrm{meq\,g^{-1}}$ DOC were noted for the humic, fulvic, and hydrophilic acid fractions, respectively (Sharp et al., 2006). Furthermore, zeta potential values of about −25 and −13 mV have been observed at pH greater than 4 for the humic and fulvic acid fractions, respectively (Sharp et al., 2006). Hence, humic and fulvic acids are removed more favorably by charge repulsion than by hydrophilic acids.

This corresponds to the results of Zularisam et al. (2007a,b). They isolated surface water into different NOM fractions and investigated their behavior in UF equipped with a 68-kDa polysulfone membrane. The hydrophobic fraction showed the greatest DOC removal (45%) of all the fractions. This phenomenon was presumed to occur as a result of charge repulsion. DOC removal of hydrophilic NOM was low, only 14%, as was expected because of its relatively low MM, aliphatic molecular structure, and lesser hydrophobic interactions compared with hydrophobic NOM.

The charge of the hydrophobic fraction of NOM is associated largely with the carboxylic and phenolic groups (Edwards et al., 1996; van Loon and Duffy, 2005). The former have pKa values in the range of 2.5–5, whereas the phenolic hydrogens have pKa values around 9 or 10 (van Loon and

Duffy, 2005). As a result, humic and fulvic acids are negatively charged as a result of the ionization of carboxylic groups in the pH range of natural waters. The usual concentration of carboxyl groups in humic substances ranges from 2 to 6 mmol g^{-1} (Leenheer et al., 1995).

5.3.2 Effect of Solution Composition

The importance of pH in membrane filtration performance has been demonstrated in various research studies, but mainly in relation to membrane fouling rather than NOM removal. Yet, the degree of DOC removal is influenced strongly by the feed solution pH and ionic content because they affect all removal mechanisms by changing the surface properties of both the membrane and NOM molecules.

The volumetric size of humic and fulvic acids changes according to the solution conditions. Low concentrations, relatively low ionic strength, and high pH renders a large, flexible, and linear shape, which changes to a small, rigid, and spherical shape at high humic concentration, high ionic strength, and low pH (Ghosh and Schnitzer, 1980).

NOM removal has been shown to increase with pH (de la Rubia et al., 2008; Shao et al., 2011; Hong and Elimelech, 1997), although greater removal at lower pH has also been reported (Lin et al., 2001; Dong et al., 2006). Removal is expected to be greater at higher pH because, in this case, humic molecules have a more linear configuration and a larger radius than at low pH, and because the ionization and charge repulsion between the membrane and the molecules are also greater. Thus, the effect of pH on NOM removal is much greater for charged membranes than for neutral membranes.

A very high removal of humic acid (99%) has been observed at pH 6.9 and pH 7.9 compared with 59% at pH 2.4 (de la Rubia et al., 2006). This was the result of the high charge density of the ceramic membrane used; its zeta potentials were −2.9 and −15.6 mV at a pH of 2.4 and 7.9, respectively.

Lin et al. (2001) observed greater DOC removal (~60%) at pH 4 than at pH 7 (~25%), but this was the result of the fouling of a hydrophobic polysulfone membrane at pH 4. In addition, the polysulfone membrane may have a positive net charge at this pH, so charge attraction—rather than repulsion—occurs between the membrane and the molecules. Charge attraction and hydrophobic interactions might lead to fouling of the membrane and, as a consequence, to greater retention.

Removal mechanisms by hydrophobic interactions are more obvious at low pH because the protonation of the phenolic and carboxylic hydroxyl groups of humic and fulvic acids leads to lower water solubility and greater hydrophobicity (Jucker and Clark, 1994). This phenomenon is more pronounced for humic acids than for fulvic acids because they have more phenolic groups and relatively fewer carboxylic groups, and hence are relatively more hydrophobic than fulvic acids (Jucker and Clark, 1994).

High ionic strength decreases the thickness of the double layer around the molecules and of the membrane pore walls (shielding of electrostatic charges) and, as a result, decreases charge repulsion. Furthermore, an increase in ionic strength would decrease intra- and intermolecular repulsion of the humic acid molecules, leading to more coiled structure. As a consequence, NOM removal decreases (Cho et al., 1999; Shao et al., 2011; Lin et al., 2001; de la Rubia et al., 2006). For example, de la Rubia et al. (2006) noted that increasing ionic strength from 0.001 to 0.152 mol KCl L^{-1} decreased the average DOC reduction from 98% to 74%.

Calcium ions do more than just affect ionic strength; they can bind to the acidic functional groups of NOM and the membrane surface, significantly neutralizing the charges of both (Lee et al., 2005; Shao et al., 2011; Hong and Elimelech, 1997). Therefore, NOM rejection decreases when calcium ions are in the solution, but at greater concentrations it can also enhance NOM rejection by causing NOM aggregation (Cho et al., 1999; Shao et al., 2011; Schäfer et al., 2001).

DOC concentration of raw water seems to be less significant to NOM removal than the aromatic content of NOM. This can be seen clearly in the results of Cho et al. (2000b). They filtered two different surface waters containing 3.9 DOC L^{-1} and 47.8 mg DOC L^{-1} with SUVA values of 4.5 and 3.7 L mg^{-1} m^{-1}, respectively. Despite the differences between these two waters, they reported that NOM rejection was similar for both (60%), whereas the greater DOC rejection, 84%, was achieved with surface water characterized by 9.8 mg DOC L^{-1} and SUVA of 4.9 L mg^{-1} m^{-1} (third surface water).

5.3.3 Effect of Hydrodynamic Conditions

Hydrodynamic operation conditions such as transmembrane pressure, permeate flux, cross-flow velocity, aeration, filtration mode, and membrane module configuration can all affect NOM removal via changes in the concentration and transport of NOM molecules at the membrane interface.

Amy and Cho (1999) used a ratio of permeate flux (J) to a diffusional back-transport coefficient (k) to describe the hydrodynamic conditions at the membrane interface. The greater J/k value corresponds to greater concentration polarization, and the greater cross-flow velocity corresponds to the greater back-transport velocity of the retained solutes and lower concentration polarization. Particle back-transport velocity depends also on its size; it is lowest for particles with a diameter of approximately 0.2 μm (Lahoussine-Turcaud et al., 1990). Smaller than this, as in the case of dissolved NOM, back-transport is dominated by Brownian diffusion, and the smaller the particle diameter, the greater its particle back-transport.

Furthermore, solution chemistry and NOM-membrane interactions have an impact on NOM diffusivity (Cheryan, 1998; Clark and Lucas, 1998; Lee et al., 2004b; Wang et al., 2001).

Lee et al. (2005) reported that NOM removal decreased with increasing J/k values, and transmission of the greater MM compounds was also found to increase. Cho et al. (2000a) did not notice significant differences in DOC rejection between the J/k values 2 and 1 with the 8-kDa polyamide membrane, whereas with the 10-kDa PES membrane, NOM rejection decreased and flux decline increased with increasing J/k values. Yet, the J/k value had no effect on the rejection of NOM containing more fulvic/humic acids (Cho et al., 2000a).

5.3.4 NOM Fractions and DBPs

Some researchers have isolated different NOM fractions to understand their removal and, especially, DBP precursors more completely. The proportions of the hydrophobic, transphilic, and hydrophilic fractions in raw waters have been observed to vary in the ranges of 38–79%, 11–31%, and 9–38%, respectively (Zularisam et al., 2007a; Kennedy et al., 2005; Yoon et al., 2005; Lee et al., 2004b).

Yee et al. (2009) investigated the DBP formation potentials of the different NOM fractions. They found that the hydrophobic and hydrophilic bases were the most reactive fractions to form trihalomethanes (THMs), but still they contributed the least to the total THM formation potential because of their low proportion (14% of DOC) in the raw water. The greatest THM formation potential was associated with the major fractions of NOM: hydrophobic acids (40% of DOC) and hydrophilic neutrals (22% of DOC).

Lin et al. (1999) reported that a 100-kDa UF membrane was able to remove 95% of DOC and THM formation potential in the greatest MM fraction of humic acid. The concentration of THM formation potential in the permeate was reduced to less than the European (EU) regulation limit of 100 $\mu g L^{-1}$ (The Council of the European Union, 1998). Yet, only 10% of THM formation potential from the smallest MM fraction was reduced, and its concentration in the permeate was still ~480 $\mu g L^{-1}$.

In relation to their relatively high cutoff values, some UF membranes have been shown to be quite effective in the removal of DBP precursors

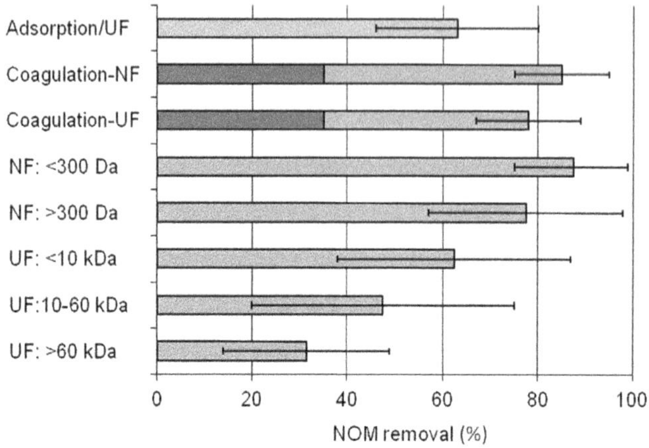

Figure 5.3 Natural organic matter (NOM) removal from raw water applying different membrane technology methods. NF, nanofiltration; UF, ultrafiltration.

from surface waters. Removals of DOC, trihalomethane formation potentials, and haloacetic acid (HAA) with formation potentials of 27–85%, 47–82%, and 28–63%, respectively, have been observed in lab-scale experiments with 1–60-kDa membranes (Table 5.1). The corresponding removals of 66, 54, and 77% have been observed in full-scale plants operating with 20-nm pore size membranes (Lamsal et al., 2012).

In contrast, the removal of assimilable organic carbon (AOC) with UF membranes has not been successful. Hem and Efraimsen (2001) noted that AOC is related mainly to NOM with an MM less than 1 kDa, which corresponded to 16–38% of the total organic carbon in the tested waters. AOC removal rates of less than 30 and 49% have been observed with 10-kDa and 1-kDa membranes, respectively. Hence, reduction in biofilm formation may be difficult without combining UF with a tight nanofiltration (NF)/reverse osmosis (RO) or biological treatment.

Tight UF membranes can be used for the removal of greater fractions of NOM from raw water, including high MM DBP precursors, whereas these should be combined with other processes to remove low and medium MM compounds (Figure 5.3).

5.4 NANOFILTRATION

NF was developed in the United States at the end of 1970s for the treatment of salt-containing groundwater and was used to soften surface waters during the 1980s (Schäfer et al., 2005). Since the 1990s, NF has been used to remove DBP precursors and micropollutants (Bodzek et al., 2011). NF membrane cutoff values are in the range of 100–500 Da and are able to remove solution components with diameters of about 1–3 nm, but at a greater transmembrane pressure as UF. In general, NF operates at a transmembrane pressure range of 1–3 MPa (Bodzek et al., 2011), but operation at 0.8 MPa pressure is common in several full-scale plants (Lopes et al., 2013).

Size exclusion and charge repulsion are the major removal mechanisms for humic substances in NF (Ates et al., 2009; Cho et al., 1999; Hong and Elimelech, 1997; Nyström et al., 1996; Visvanathan et al., 1998; Schäfer et al., 1998; Ribau Teixeira and Rosa, 2006). The charge effect becomes less important when the molecules are larger than the membrane pores (van der Bruggen et al., 1999). Braeken et al. (2005) noticed that hydrophobicity is the most significant parameter determining the retention of molecules with an MM less than the membrane cutoff value.

The transport of NOM through NF membrane pores is influenced either by convection or diffusion, as in RO, depending on the hydrodynamic conditions and electrostatic interactions between the membrane surface and NOM molecules. Moreover, different mechanisms influencing the removal of small organic compounds have been suggested. The polarity and differences in diffusion rates in a nonporous structure may influence the rejection of small organic compounds in NF (van der Bruggen et al., 1999; van der Bruggen and Vandecasteele, 2003). The molecules with high dipole moments are reported to show a lower rejection than molecules with a lower dipole moment, although the sizes of the molecules have been approximately the same (van der Bruggen et al., 1999). This indicates that a lower fraction of polar molecules is removed compared with nonpolar molecules of equivalent size. The transport and rejection mechanisms by NF membranes are still partly unclear, however.

Almost total NOM rejection (>90%) has been achieved in lab- and bench-scale studies using NF membranes with cutoff values in the range of 200–400 Da (de la Rubia et al., 2008; Cho et al., 1999; Yoon et al., 2005; Costa and de Pinho, 2006a; Meylan et al., 2007) (Table 5.2). Large NOM compounds like polysaccharides and humic substances are clearly retained by NF membranes, whereas low-MM organic compounds permeate through the membrane (Schäfer et al., 2000). Meylan et al. (2007) noted 98–99% DOC removal with 300-Da NF membranes and an RO membrane. The high-MM polysaccharides and humic substances were almost completely rejected, charged low-MM organic compounds had 97%

Table 5.2 Studies concerning NOM removal from natural waters and surrogate waters containing humic substances with nanofiltration (NF) and RO membranes

Solution	Membrane, material, cutoff/pore diameter, module type, TMP	Variable	DOC removal (%)	Quality of permeate, content and/or removal (%)	References
Surface water; TOC, 2.3 mg L^{-1}; SUVA, 1.7; THMFP, 70 μg L^{-1}	Different flat-sheet membranes, stirred cell, lab scale	NTR-7450, S-PSu, 1 kDa	38	UV_{254}, 67%	Lee and Lee (2007)
		MPF-34, unknown, 200 Da	75	UV_{254}, 85%	
		NF-45, PPiA, 400 Da	73	UV_{254}, 86%	
		NTR-7250, PPiA, 300–400 Da	60	UV_{254}, 81%	
		CA865, CA; NaCl, 60%	47	UV_{254}, 52%	
		NTR-729HF, PVA, 700 Da	82	UV_{254}, 87%	
6.3% NOM, Australia	Different membranes, stirred batch cell, lab scale, 500 kPa	TFC-SR, <180 Da/<0.64 nm	69		Schäfer et al. (2000)
		TFC-S, <180 Da/<0.64 nm	95		
		TFC-ULP, <180 Da/<0.64 nm	95		
Aquatic NOM; DOC, 10 mg L^{-1}	Different flat-sheet membranes, cross-flow, bench scale	NF-PES-10, PES, 1 kDa	58		Kilduff et al. (2000)
		NTR-7450, S-PSu, 1 kDa	83		

Surface water; DOC, 2.5 mg L^{-1}; SUVA, 1.5	Different flat-sheet submerged membranes, lab scale	NE70, S-PES, 350 Da	80		Kaewsuk and Seo (2011)
		NE90, *m*-phenylene-diamine, 550 Da	88		
Surface waters from 10 different sources; DOC, 1.4–7.1 mg L^{-1}	Different flat-sheet membranes, stirred cell, lab scale	NF270, PPi based, 150–430 Da	69–99	UV_{254}, 83–93%	de la Rubia et al. (2008)
		NF90, PA TFC, 200–300 Da	70–99	UV_{254}, 76–94%	
		NFT50, PPi, 150 Da	35–100	UV_{254}, 59–95%	
Surface water; DOC, 7.7–9.4 mg L^{-1}	2 membranes, pilot plant	NF45, PA TFC, 400 Da	>87	DOC, <1 mg L^{-1}	Ericsson et al. (1997)
		Desal DL5, TFC	>87	DOC, <1 mg L^{-1}	
Surface water; TOC, 3.5–3.6 mg L^{-1}	2 membranes, pilot plant	NF270, ~400 Da		TOC, >71%, <1 mg L^{-1}	Lopes et al. (2013)
		Desal DK, ~250 Da		TOC, >72%, <1 mg L^{-1}	
Surface water after sedimentation; TOC, 3.2–3.6 mg L^{-1}	NF270, ~400 Da	Recovery rate, 91%		TOC, >72.5%, <1 mg L^{-1}	Lopes et al. (2013)
		Recovery rate, 95%		TOC, >69%, <1 mg L^{-1}	
Different surface waters	NF45, PA TFC, 400 Da, flat sheet, cross-flow, bench scale	DOC, 2.0 mg L^{-1}; SUVA, 2.4	85		Cho et al. (1999)
		DOC, 47.8 mg L^{-1}; SUVA, 3.7	98		

(Continued)

Table 5.2 Studies concerning NOM removal from natural waters and surrogate waters containing humic substances with nanofiltration (NF) and RO membranes—cont'd

Solution	Membrane, material, cutoff/pore diameter, module type, TMP	Variable	DOC removal (%)	Quality of permeate, content and/or removal (%)	References
Surface water; DOC, 1.1 mg L^{-1}; AOC, 103 μg L^{-1}	Different flat-sheet membranes, cross-flow, lab scale, 2000 kPa	HL, ~300 Da	98	DOC, 68 μg L^{-1}; AOC, 81 μg L^{-1}	Meylan et al. (2007)
		NF270, PA TFC, ~300 Da	98	DOC, 55 μg L^{-1}; AOC, 84 μg L^{-1}	
		XLE, NaCl, 98.6%	99	DOC, 42 μg L^{-1}; AOC, 19 μg L^{-1}	
Surface water; DOC, 3.2 mg L^{-1}; AOC, 229 μg L^{-1}		HL, ~300 Da	99	DOC, 1.2 mg L^{-1}; AOC, 153 μg L^{-1}	Meylan et al. (2007)
		NF270, PA TFC, ~300 Da	99	DOC, 129 μg L^{-1}; AOC, 92 μg L^{-1}	
		XLE; NaCl, 98.6%	99	DOC, 110 μg L^{-1}; AOC, 31 μg L^{-1}	
Surface water; DOC, 11.7 mg L^{-1}; SUVA, 2.0	NF90, aromatic PA, 200–400 Da		92–96		Byun et al. (2011)

Surface water; DOC, $mg\,L^{-1}$; SUVA, 2.7	HL, PA TFC, 150–300 Da, zeta, −9.9 mV, cross-flow, bench scale	Colloidal fraction; pH 4, 7, and 10	75–80		Kim et al. (2011)
		Hydrophobic fraction; pH 4	70		
		Hydrophobic fraction; pH 7	80		
		Hydrophobic fraction; pH 10	85		
		Transphilic fraction; pH 4, 7, and 10	~80		
Surface water; DOC, $2.2\,mg\,L^{-1}$; SUVA, 2.9; THMFP, $75\,\mu g\,L^{-1}$; HAAFP, $80\,\mu g\,L^{-1}$	NF70, NF45, and Desal 5; TFC; 100–400 Da; all spiral wound	Averages of all 4 membranes	88	DOC, $0.27\,mg\,L^{-1}$; THMFP, $3\,\mu g\,L^{-1}$, 96%; HAAFP, $5\,\mu g\,L^{-1}$, 94%	Siddiqui et al. (2000)
Surface water; TOC, $3.2\,mg\,L^{-1}$	NF200-400B, flat sheet, cross-flow, lab scale, 1500 kPa		>90	UV_{254}, >97%	Costa and de Pinho (2006a)
Surface water; DOC, $3.8\,mg\,L^{-1}$	ESNA, PA TFC, 250 Da, flat sheet, cross-flow, bench scale		~78	UV_{254}, ~82%; HAAFP, 79%	Lee and Cho (2004)

(Continued)

Table 5.2 Studies concerning NOM removal from natural waters and surrogate waters containing humic substances with nanofiltration (NF) and RO membranes—cont'd

Solution	Membrane, material, cutoff/pore diameter, module type, TMP	Variable	DOC removal (%)	Quality of permeate, content and/or removal (%)	References
Surface water after UF; DOC, 1.9 $mg L^{-1}$; SUVA, 3.2	PRO-100NF, 98% salt rejection, full scale		83	DOC, 0.32 $mg L^{-1}$	Lamsal et al. (2012)
Natural water; DOC, 3.4 $mg L^{-1}$; SUVA, 2.5; HMM, ~12 kDa; LMM, ~1.8 kDa	DK and DL PPiA TFC, 98 and 96% $MgSO_4$, flat sheet, cross-flow, lab scale	Averages of the membranes	85	UV_{254}, 90%; HMM, 98%; LMM, 30%; AOC, 82 $\mu g L^{-1}$, 76%; THMs, 8 $\mu g L^{-1}$, 95%; HAAs, 6 $\mu g L^{-1}$, 90%	Ates et al. (2009)
Groundwater	NF200B, PPi, 200 Da, spiral wound, pilot scale		>96	UV_{254}, >96%; THMFP, 95%	Gorenflo et al. (2003)

Surface water and groundwater	ESNA, PA, 200 Da, flat sheet, cross-flow, bench scale	DOC, 3.1 mg L^{-1}; SUVA, 2.9	76	UV_{254}, 96%	Amy and Cho (1999)
		DOC, 6.8 mg L^{-1}; SUVA, 5.7	88	UV_{254}, 98%	
Surface and groundwater	ESNA, PA TFC, 200 Da, hydrophobic, flat sheet, stirred cell, bench scale, 396 kPa	DOC, 3.1 mg L^{-1}; SUVA, 2.9	95		Yoon et al. (2005)
		DOC, 6.8 mg L^{-1}; SUVA, 5.6	~95		
Surface water	AFC30, 200 Da, tubular, cross-flow, pilot plant			TOC, 94%; UV_{254}, 100%	Orecki et al. (2004)
Seawater after DAF and UF; DOC, 1.22 mg L^{-1}	TFC RO		92	DOC, 0.097 mg L^{-1}	Simon et al. (2013)
Surface water after UF; DOC, 2.8 mg L^{-1}; HAAFP, 22.6 μg L^{-1}; THMFP, 149 μg L^{-1}	4040-UHY-ESPA, ultralow-pressure RO		93	HAAFP, <1.0 μg L^{-1}, >96%; THMFP, <6.8 μg L^{-1}, 95%	Kruithof et al. (1998)

AOC, assimilable organic carbon; CA, cellulose acetate; DAF, dissolved air flotation; DOC, dissolved organic carbon; HAA, haloacetic acid; HAAFP, haloacetic acid formation potential; HMM, high molar mass; LMM, low molar mass; NOM, natural organic matter; PA, polyamide; PES, polyethersulfone; PPi, polypiperazine; PPiA, polypiperazine amide; PVA, polyvinyl alcohol; RO, reverse osmosis; S-PES, sulfonated polyethersulfone; S-PSu, sulfonated polysulfone; TFC, thin-film composite; THM, trihalomethane; THMFP, trihalomethane formation potential; TMP, transmembrane pressure; TOC, total organic carbon; UF, ultrafiltration; SUVA, specific ultraviolet absorbance. SUVA values are in liters per meter per milligram.

rejection, and low-MM neutrals and hydrophobic organic compounds had rejections of 94 and 88%, respectively.

In general, the hydrophilic fraction of NOM is removed less easily than the hydrophobic fraction (Siddiqui et al., 2000; Ribau Teixeira and Rosa, 2006). Escobar et al. (2000) observed that AOC removal by NF membranes is strongly dependent on charge repulsion, and that biodegradable dissolved organic carbon (BDOC) removal was determined by both size exclusion and charge repulsion. In a full-scale drinking water treatment plant, they found NF a very effective means of reducing BDOC but not AOC.

Unlike UF permeates, several researchers have reported that NF permeates consist of NOM that appears to have an MM larger than the cutoff value of the membrane. Lee et al. (2005) noted 93–96% DOC rejections with a negatively charged 250-Da membrane, but still the apparent MMs of the permeated NOMs were about twice the membrane cutoff. Furthermore, the proportions of BDOC were the same—10 and 20% of NOM—as in the two raw waters filtered.

Ates et al. (2009) noted an 85% DOC rejection with tight NF membranes for natural water. The high MM fraction (average MM, ~12 kDa) was rejected by 98%, but the low MM fraction (average MM, 1.8 kDa) was rejected by only 30%. Because of the broad range of molecular sizes found in NOM, a membrane with a low cutoff is needed for complete removal of small organic compounds.

5.4.1 Effect of Solution Composition

It has been reported that NOM removal by NF membranes is not influenced significantly by factors such as DOC levels, humic content as identified by SUVA, pH, and ionic strength (Cho et al., 1999; Siddiqui et al., 2000; Glucina et al., 2000; Ventresque et al., 2000). Yet, NOM removal increases with water recovery rate (Costa and de Pinho, 2006b) (i.e., with increasing concentration). As a result, a stronger concentration polarization layer may form on the membrane surface (Visvanathan et al., 1998; Jarusutthirak et al., 2007).

The effect of concentration polarization is more pronounced when the raw water has a high NOM concentration, especially when the rejection of a membrane is high. This layer may become so dense that it prevents permeability and thus enhances rejection, as shown by Jarusutthirak et al. (2007). They found that an increase in NOM concentration increased the NOM and salt rejection and permeate flux decline by enhancing accumulation on the membrane surface.

Ribau Teixeira and Rosa (2006) noted that NF performance was influenced by the pH and calcium hardness, rather than by the type of NOM. In their study, DOC rejection increased with recovery rate and pH, whereas the UV_{254} rejection was almost complete (90–100%) at all pH values and recovery rates. They also found that the rejection was reduced at the membrane isoelectric point. This is a result of the lower membrane surface and pore charges, which result in a lower charge repulsion between the membrane and NOM, and larger pores resulting from the lower electrical double layer (Al-Amoudi, 2010). The membrane becomes negatively charged above its isoelectric point, and with the increase in pH, the charge repulsion between the membrane surface and NOM molecules increases. This is in accordance with the results of Hong and Elimelech (1997), who found that humic acid is better removed at pH 8 than at pH 4.

In contrast, Küchler and Miekeley (1994) observed that the retention of humic acid decreased with increasing pH in the pH range of 4.3–9.0, but that of fulvic acid increased. In their study, humic and fulvic acid retention with the 1-kDa membrane varied in the range of 80–90% and 60–70%, respectively. The differences between humic and fulvic acids may be explained by their different pKa and solubility values and sizes. Visvanathan et al. (1998) found that rejection increases with increasing pH up to a certain level, but above this level the solubility of the humic substances increases so that they can again pass through the membrane easily, thereby decreasing rejection. Optimum rejection was obtained at near-neutral conditions, between pH 7.5 and 8.5.

The salt concentration of raw water may have a greater influence in NF than in UF as a result of partial salt rejection and, hence, increasing salt concentration on the membrane surface. As a consequence, osmotic pressure increases and, because it opposes the transmembrane pressure, the permeate flux decreases.

Jarusutthirak et al. (2007) reported that increased ionic strength ranging from 0.004 to 0.1 M led to a greater flux decline, possibly as a result of increasing osmotic pressure. The same effect was observed when the solution pH was increased from 4 to 10. They also reported that a solution with high ionic strength (0.05 M) showed greater NOM rejection than that of low ionic strength (0.01 M), which is contradictory to the observation of Hong and Elimelech (1997), who noted that both permeate flux and humic acid rejection decreased drastically when NF was carried out at low pH and high ionic strength.

The presence of calcium and other divalent ions reduces the rejection capacity by shielding the interchain charges of the humic molecules (Visvanathan et al., 1998; Ribau Teixeira and Rosa, 2006). Another possible explanation proffered by Visvanathan et al. (1998) is the Donnan effect: calcium and magnesium cations tend to pass through the negatively charged membrane and, to maintain electroneutrality on both sides of the membrane, the negatively charged humic acid passes through the membrane, thus reducing rejection. A tight NF membrane (NF200B) with a high passage of calcium, more than 55% as $CaCl_2$, demonstrated stable rejections (>95%) of DOC, AOC, and THM formation potential at water recovery rates from 6% to 85%, although the DOC content in the feed water increased up to 20 mg L^{-1} (Gorenflo et al., 2003).

5.4.2 Effect of Hydrodynamic Conditions

Studies of how permeate flux affects NOM rejection achieved contradictory results. On the one hand, Lee and Lee (2007) observed that the greater the flux, the lower the DOC rejection when they filtered surface water with a range of NF membranes using small stirred-cell equipment. Thus, they concluded that it is difficult to achieve high permeate flux and high NOM rejection simultaneously. On the other hand, several researchers have found that an increase in the transmembrane pressure produces a greater flux but does not have a significant effect on rejection (Visvanathan et al., 1998; Chang et al., 2009; Uyak et al., 2008). The rejection of the low MM fraction may increase with an increase in the transmembrane pressure, however (Ates et al., 2009).

Alborzfar et al. (1998) found that high cross-flow velocity and low transmembrane pressure reduced the flux decline of tight NF membranes, but NOM removal remained at 93%. Lee et al. (2004a,b) observed that NOM removal improved with an increasing cross-flow rate, which was attributed to an increase in the back diffusive transport away from the membrane surface and, hence, to a reduced-concentration polarization layer. Furthermore, as the concentration at the membrane surface decreased, the driving force for the NOM diffusive transport through the membrane decreased. Although J/k values are not available, it seems that a low J/k value is favorable for NOM removal.

High permeate flux may result in severe fouling resulting from high solute transport toward the membrane and more compressed fouling layers (Cho et al., 1999; Hong and Elimelech, 1997). Thus, although greater operating pressures allow greater initial permeate flux, the following rapid decline in flux may counteract this advantage. Although NF of surface waters generally achieves a high removal of organics and salts, high flux decline is common as a result of colloidal, organic, and biological fouling and scaling. The most practical way to mitigate NF fouling is to treat raw water before NF. Various appropriate pretreatment processes have been tested, including coagulation, adsorption, ozonation, chlorination, dissolved air flotation, and MF/UF treatment (Lee and Lee, 2007).

5.4.3 DBP Removal

NF and tight UF membranes with a cutoff ranging from 150 to 1000 Da have proved to be effective in DBP formation potential removal without any chemical treatment processes (de la Rubia et al., 2008; Ates et al., 2009; Siddiqui et al., 2000; Glucina et al., 2000).

THM formation potential removal is supposed to be proportional to NOM removal (Lee et al., 2005). Yet, THM formation potential per DOC can be even greater in the NF permeate than in raw water (de la Rubia et al., 2008). Siddiqui et al. (2000) observed average reductions of 90, 97, and 94% in DOC, THM, and HAA formation potentials, respectively, with four 100–300-Da spiral-wound NF membranes. The THM formation potential in permeates was well below the EU regulation limit of $100\,\mu g\,L^{-1}$ (The Council of the European Union, 1998). The 500-Da membrane was less effective, providing THM and HAA formation potential reductions of 79 and 72%, respectively. In their study, prefiltration by a 60-kDa UF membrane removed DOC, THM, and HAA formation potentials by ~30, 50, and 32%, respectively, whereas MF was virtually ineffective in precursor removal.

More than 90% removal in THM and HAA formation potentials were also observed by other researchers with the 100–300-Da membranes (Ates et al., 2009; Glucina et al., 2000). Removal of 85, 82, and 60% of DOC, THMs, and HAAs, respectively, was noticed with the 1-kDa UF membrane (Ates et al., 2009). In general, THM formation potential was slightly better removed than HAA formation potential, suggesting that lower MM and/or hydrophilic DOC may be more responsible for HAA formation.

5.5 REVERSE OSMOSIS

RO has been used mainly for the production of drinking water from seawater by desalination. It is being applied increasingly to drinking water production from high-hardness and NOM-rich surface water and groundwater. RO plants generally operate with polyamide-based hollow fiber and spiral-wound membranes (Schrotter and Bozkaya-Schrotter, 2010).

In general, NOM concentration in seawater is low (DOC, $<3\,mg\,L^{-1}$) (Jeong et al., 2013) and it contains more hydrophilic compounds than surface waters. Hydrophilic, transphilic, and hydrophobic fractions of 60, 26, and 14% of DOC, respectively, have been reported in raw seawater (Jeong et al., 2013).

Currently, new desalination plants are being designed to use MF or UF systems upstream from the RO membranes for the removal of turbidity; nonsoluble, colloidal organics; and pathogens such as *Giardia* and *Cryptosporidium* from raw seawater. Yet, they are not able to remove significant amounts of dissolved organics and marine microorganisms, which typically are responsible for RO membrane biofouling (Voutchkov, 2010).

Jeong et al. (2013) studied submerged MF with coagulation and with coagulation and adsorption as a pretreatment to seawater RO. A DOC removal of 23% was reported for MF pretreatment with coagulant and 48% for MF pretreatment with coagulant and adsorbent. The hydrophilic fraction was removed by 19 and 38%, respectively. Both pretreatments were able to remove more than 900-Da MM compounds and reduce flux decline from 49% to 24–26%.

The flotation UF pretreatment method reduced low-MM neutrals and biopolymers by 33–40% and 18–19%, respectively. RO membranes acted as an almost complete barrier for humic compounds, building blocks, and low-MM neutrals, with rejections of 97–100% (Simon et al., 2013).

Apart from seawater desalination, RO is often used to remove THMs, HAAs, and other halogenated hydrocarbons from water (Bodzek et al., 2011). A DOC removal of 99% (Meylan et al., 2007), and THM and HAA formation potential removals of more than 95% (Kruithof et al., 1998) from surface waters have been reported.

The main challenges in NF and RO plants are the concentration of scale-forming minerals and organic materials that decrease recovery rates, and treatment and disposal of membrane concentrates. Concentrate treatment may be necessary before its final disposal to a wastewater treatment plant or surface waters, or via deep-well injection. Comstock et al. (2011) investigated treatment of NF/RO concentrate by precipitative softening with lime, ferric sulfate coagulation, and anion exchange using MIEX-Cl resin. Both coagulation and anion exchange were able to reduce DOC concentration to the raw water level, whereas coagulation was not recommended for treating the membrane concentrate because the method is sensitive to alkalinity and pH, and does not remove inorganic chemicals. Anion exchange removed organic and inorganic chemicals, except calcium, which was removed by lime softening only.

5.6 MEMBRANE FOULING

The major problem in drinking water production by membrane processes is the decrease of flux and yield over time resulting from concentration polarization and fouling. Periodic cleaning and renewal of the membranes are needed, which considerably increase the price of produced water (Guizard and Amblard, 2009). Therefore, a number of methods for the removal of matter accumulated on the membrane surface have been introduced to prevent fouling and to lengthen operation time between cleanings.

Concentration polarization depends on the selectivity of a membrane. When particles or solutes are retained by the membrane, they accumulate in a boundary layer adjacent to the membrane surface. This leads to a reduction in solvent flow through the membrane, which results in a reduction in the effective transmembrane pressure resulting from an osmotic pressure difference between the filtrate and retentate sides of the membrane surface. This phenomenon is reversible and it can be reduced by simply releasing the transmembrane pressure or by using methods that increase particle back-transport away from the membrane surface (Fane and Chang, 2009).

Membrane fouling is a buildup of material on the membrane surface or within the membrane structure. Different fouling mechanisms, such as adsorption, pore blocking, pore constriction, or gel/cake layer formation, may occur, depending on the process (Figure 5.4).

The typical fouling mechanisms for MF and UF are pore constriction, pore blocking, and gel/cake formation. If particles are smaller than

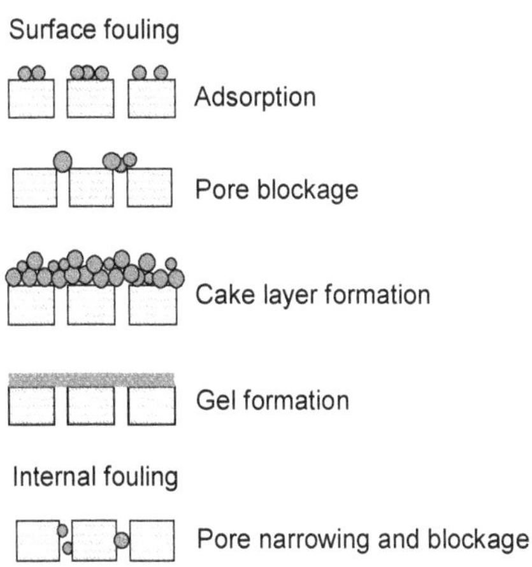

Figure 5.4 Major fouling mechanisms.

membrane pores, they can enter and stick to the pores. If particles are about the same size as membrane pores, deposition of the particles onto the membrane surface may cause pore blocking. In a membrane filtration process, fouling may be a combination of different mechanisms resulting from the distribution of membrane pore size and component sizes of the feed water. During the initial stages, membrane fouling is thought to be dominated by pore constriction and pore blocking (Fane and Chang, 2009).

Fouling is a very complex phenomenon, and it has to be investigated for every single application to avoid it. It depends on the composition of the feed solution (nature, concentration and interactions of components, pH, and ionic strength), membrane properties (hydrophobicity, charge, roughness, pore size, pore size distribution, and porosity), and operating conditions (temperature, transmembrane pressure, and cross-flow velocity) (Fane et al., 2008). Membrane-fouling mitigation may involve pretreatment of the raw water (e.g., coagulation, adsorption, oxidation, and prefiltration) and the development of fouling-resistant membranes (Zhou et al., 2010; Huang et al., 2011).

Fouling can be classified according to the type of fouling material: colloidal, organic, inorganic (scaling), and microbial/biological (biofouling). The fouling caused by organic macromolecules may overlap organic, colloidal, and microbial fouling. Only the fouling caused by NOM molecules is discussed here. Aquatic colloidal, organic molecules are classified into rigid biopolymers, humic and fulvic acids, and flexible biopolymers (Buffle et al., 1998; Tang et al., 2011). Rigid biopolymers consist mainly of large-MM polysaccharides with a long persistence length; flexible biopolymers consist of proteins and some low-molecular weight organic molecules (molecular weight, <1000 Da).

NOM is believed to be the major membrane foulant in drinking water treatment plants using surface water. A propensity for fouling depends on the composition of NOM. Of the two NOM types—allochthonous (terrestrially derived, mainly humic substances) and autochthonous or algal (microbially derived) NOMs—the autochthonous types have greater fouling potential (Amy, 2008). They consist of more extracellular and intracellular macromolecules and cellular debris. Surface water dominated by autochthonous NOM can cause serious fouling when algae bloom, especially in spring (Plottu et al., 2003).

Several authors observed that biopolymers cause more fouling than humic substances (Amy, 2008; Kennedy et al., 2008b; Lee et al., 2006; Jermann et al., 2008; Her et al., 2008; Zazouli et al., 2010). Jermann et al. (2008)

investigated the fouling of a 100-kDa PES membrane using alginate, which is a polysaccharide secreted by brown algae, and humic acid in a dead-end UF. They reported that most alginate fouling was reversible, whereas humic acid fouling was more irreversible. Zazouli et al. (2010) investigated the fouling effects of humic acid and alginate on 150–300-Da polyamide NF membranes. Increasing organic concentration or ionic concentration, or decreasing pH increased the rate and extent of flux reduction. Alginate exhibited greater flux decline than humic acid fouling under the same conditions.

Lee et al. (2006) investigated the fouling of 100-kDa membranes made of regenerated cellulose and PES, and 0.22-μm pore size MF membranes made of mixed cellulose ester and PVDF with different isolated fractions of surface waters dominated by both autochthonous and allochthonous NOM. They found that colloidal NOM and autochthonous NOM derived from algae, both containing polysaccharides and protein compounds, were responsible for significant MF membrane fouling. Hydrophobic fractions also fouled the MF membranes. The UF membranes were more fouling resistant, and the PES membrane was fouled significantly only by the NOM derived from algae. The hypothesized fouling mechanisms suggested by atomic force microscope (AFM) were (1) a largely hydraulically reversible surface (gel layer) fouling that was dominant for UF and (2) less hydraulically reversible pore blockage, which was dominant for MF.

Amy (2008) reported that the most severe foulants correspond to polysaccharide and protein macromolecules and organic colloids (peptidoglycans and amino sugars). It was hypothesized that polysaccharidelike and proteinlike foulants interact with a membrane whether by hydrogen bonding and dipole interactions, respectively, or, in a colloidal form, contributing to a cake/gel layer. Foulants larger than the pore size—for example, the high-MM polysaccharides, proteins, and organic colloids that cause only surface fouling—may form a gel layer that is largely hydraulically reversible, and hydraulic backwashing alone may be adequate to remove it (Amy, 2008).

5.6.1 Effects of pH, Ionic Strength and Divalent Ion Concentration

Many existing studies have shown that fouling by humic acid is promoted by low pH, high divalent ion concentration (Ca^{2+} and Mg^{2+}), high ionic strength, and high foulant concentration (Kim et al., 2011; Hong and Elimelech, 1997; Zazouli et al., 2010; Seidel and Elimelech, 2002). The adverse effect of low pH and high ionic strength can be attributed to the reduced electrostatic repulsion between humic acid molecules and the membrane surface, as well as between humic acid molecules. Calcium has a strong affinity for the carboxylic moieties of humic acid (Hong and Elimelech, 1997; Wang et al., 2001; Hering and Morel, 1988). Calcium ions can complex with humic acid and thus partially neutralize humic acid molecules. Shao et al. (2011) reported that the filtrate flux and the initial removal rate of humic acid decreased sharply as the calcium concentration increased from 0–0.5 mM; when the calcium concentration was increased further to 10 mM, both the flux and retention increased. The most likely explanation for these phenomena is electrostatic interactions and coagulation of humic acid, respectively.

Many polysaccharides can form an extensive gel in the presence of some multivalent metal ions. For example, calcium has a strong affinity to carboxylic groups (Hering and Morel, 1988) and can interact with some polysaccharides by crosslinking the biopolymers into a solidlike gel layer via metal–ligand complex formation (Wang and Waite, 2008; Wang et al., 2009).

Amy (2008) found that the addition of calcium (Ca^{2+}) to the hydrophobic NOM fraction did not increase fouling, because the resultant solute was still small enough to pass through the pores of 100-kDa UF and MF membranes. The negative impact of Ca^{2+} is clearer when the membrane pore size is smaller. Flux decline increased with the J/k value and, finally, a significant flux decline was observed. This result was attributed to greater concentration polarization and the aggregation of humic substances in the boundary layer. It was assumed that the addition of Ca^{2+} enhanced fouling by neutralizing the negative charge of the humic and fulvic acids, thus promoting aggregation.

Jermann et al. (2007) studied fouling of a 100-kDa PES membrane by humic acid and alginate with and without calcium. Calcium increased humic acid adsorption onto the membrane surface and in the pores, whereas alginate formed a cake layer, which was relatively reversible because of strong electrostatic repulsion. Only marginal alginate adsorption was noticed, and the addition of calcium let to complexation of alginate and gel formation.

Wang and Waite (2008) also observed that alginate gel formed a very compressible and porous cake layer (water content, >96%) on a 100-kDa UF membrane. In another study conducted with a membrane bioreactor, they noticed significant gel formation associated mainly with the polysaccharides through metal–ligand complexation provoked by multivalent metals (Ca^{2+}, Mg^{2+}) (Wang and Waite, 2009).

5.6.2 Biofouling

Microbial fouling is a result of the formation of biofilms on membrane surfaces. Wet membrane surfaces are ideal areas for microbial (bacterial, algal, or fungal) attachment and colonization, with a steady nutrient supply by convective flow (Chen et al., 2010). After attachment to the membrane, bacteria start to multiply and secrete extracellular polymeric substances such as heteropolysaccharides, proteins, and amino sugars, which form a viscous, slimy, hydrated gel (Kennedy et al., 2008a,b). This gel structure protects bacterial cells from hydraulic shearing and from chemical attacks by biocides such as chlorine (Kennedy et al., 2008a,b). In contrast to other foulants, microorganisms can repopulate soon after a cleaning procedure, if they are not removed completely from the system.

Microbial fouling is closely related to the characteristics of the feed water and is more severe in warm environments and at plants using surface waters that provide great potential for biological activity from numerous species of microorganisms. The growth of microorganisms is most pronounced in the temperature range of 30–45 °C (Koltuniewich and Drioli, 2008).

Microcystins, the most commonly occurring cyanotoxins in surface water reservoirs used for water supply, are the first to be adsorbed on the membrane surface as a result of their hydrophobic character (Ribau Teixeira and Sousa, 2013). The parts of the treatment chain most vulnerable to biofouling are UF and MF pretreatments when situated downstream of the cartridge filters or sand filters, which are potential breeding grounds for microorganisms (Chen et al., 2010).

5.7 CONCLUSIONS

All surface waters contain NOM, which is a heterogeneous mixture of biopolymers and their different-size degradation products. NOM causes problems in drinking water treatment and distribution systems, and is a health risk when it reacts with disinfectants. The great variability of NOM composition makes it difficult to remove it from drinking water completely, particularly because its concentration and quality are influenced by seasonal variation. The characteristics of NOM that influence membrane filtration most greatly are its size distribution and acidic functional group content. The MM of NOM is in the colloidal size range from a few 100 Da to more than 100 kDa, covering the pore sizes of NF, UF, and MF membranes. Although the size distribution of NOM varies among different water sources, in general very high proportions of small molecules (<1 kDa) have been observed.

During disinfection, NOM can react with chlorine to form chlorinated DBPs, such as THMs and HAAs, which are recognized human carcinogens. The formation of DBPs in treated water depends on the raw water quality, especially the concentration of NOM. Hydrophobic acids and hydrophilic neutrals are the major precursors to the formation of DBPs and, therefore, are the fractions that should be removed primarily to minimize DBP formation. Biodegradable dissolved organic matter that causes bacterial growth and biofilm formation in drinking water distribution systems, measured as biodegradable DOC or AOC, also needs to be minimized. AOC is observed to be related mainly to NOM with an MM less than 1 kDa.

Molecules larger than the membrane pores are rejected by size exclusion; smaller compounds may be rejected by charge repulsion if they contain negatively charged groups and if a negatively charged membrane is used. Hence, NOM removal increases with decreasing membrane pore size and with increasing charge density on the membrane surface and in the pores. These conditions are favorable at higher pH and low ionic strength.

MF membranes have large pores and, hence, are unable to reject most NOM compounds unless they are coagulated into bigger flocs. The advantages of these membranes are that they remove suspended solids and colloidal materials, such as pathogens and algae, more economically than tighter membranes. Negatively charged UF membranes are able to remove larger hydrophobic NOM fractions, even more than would be expected on the basis of the membrane cutoff value and NOM size. Yet, NOM contains plenty of small hydrophilic compounds with a low acidic content that are difficult to remove by size or charge exclusion through UF membranes. Hence, UF is ineffective for removing AOC and for preventing the bacterial regrowth potential in drinking water distribution systems. Therefore, to achieve the required water quality, MF and UF need to be combined with other processes such as coagulation, adsorption, oxidation/biological activated carbon (BAC), or tighter membranes (NF or RO).

NF membranes are able to remove compounds from macromolecular size to multivalent ions, but at higher transmembrane pressure than UF. Almost complete NOM rejection was achieved by NF membranes with cutoff values in the range of 100–400 Da. DOC and THM formation potential removals increase with decreasing pore size in UF and NF as follows: 60-kDa UF, ≥30%; 1-kDa UF, >75%; 500-Da NF, >80%; and 100–300 Da NF, ≥90%. HAA formation potential removal has been found to be slightly lower. Yet, AOC removal is incomplete, so the disinfection process is still necessary. NF membranes are very susceptible to fouling in surface water treatment and, in general, require extensive pretreatment to control colloidal, organic, and biological fouling and scaling. Pretreatment steps may include MF/UF, conventional treatment, or slow sand or dual-media filtration.

Membrane technology has a great potential to shorten and simplify the long treatment chains of physical–chemical and biological unit processes needed to satisfy stringent water quality criteria. Yet, disinfection is still needed because membranes are not able to remove AOC compounds economically if high quality water is required.

REFERENCES

Al-Amoudi, A.S., 2010. Factors affecting natural organic matter (NOM) and scaling fouling in NF membranes: a review. Desalination 259, 1–10.

Alborzfar, M., Jonsson, G., Grøn, C., 1998. Removal of natural organic matter from two types of humic ground waters by nanofiltration. Water Res. 32, 2983–2994.

Amy, G., 2008. Fundamental understanding of organic matter fouling of membranes. Desalination 231, 44–51.

Amy, G., Cho, J., 1999. Interactions between natural organic matter (NOM) and membranes: rejection and fouling. In: Ødegaard, H. (Ed.), International IAWQ-IWSA Joint Specialist Conference on Removal of Humic Substances from Water. June 24–26, 1999, p. 141.

Aoustin, E., Schäfer, A.I., Fane, A.G., Waite, T.D., 2001. Ultrafiltration of natural organic matter. Sep. Purif. Technol. 22–23, 63–78.

Ates, N., Yilmaz, L., Kitis, M., Yetis, U., 2009. Removal of disinfection by-product precursors by UF and NF membranes in low-SUVA waters. J. Membr. Sci. 328, 104–112.

Ben-Sasson, M., Zidon, Y., Calvo, R., Adin, A., 2013. Enhanced removal of natural organic matter by hybrid process of electrocoagulation and dead-end microfiltration. Chem. Eng. J. 232, 338–345.

Bessiere, Y., Jefferson, B., Goslan, E., Bacchin, P., 2009. Effect of hydrophilic/hydrophobic fractions of natural organic matter on irreversible fouling of membranes. Desalination 249, 182–187.

Bodzek, M., Konieczny, K., Kwiecińska, A., 2011. Application of membrane processes in drinking water treatment – state of art. Desalin. Water Treat. 35, 164–184.

Braeken, L., Ramaekers, R., Zhang, Y., Maes, G., Bruggen, B.V.d., Vandecasteele, C., 2005. Influence of hydrophobicity on retention in nanofiltration of aqueous solutions containing organic compounds. J. Membr. Sci. 252, 195–203.

van der Bruggen, B., Schaep, J., Wilms, D., Vandecasteele, C., 1999. Influence of molecular size, polarity and charge on the retention of organic molecules by nanofiltration. J. Membr. Sci. 156, 29–41.

van der Bruggen, B., Vandecasteele, C., 2003. Removal of pollutants from surface water and groundwater by nanofiltration: overview of possible applications in the drinking water industry. Environ. Pollut. 122, 435–445.

Buffle, J., Wilkinson, K.J., Stoll, S., Filella, M., Zhang, J., 1998. A generalized description of aquatic colloidal interactions: the three-colloidal component approach. Environ. Sci. Technol. 32, 2887–2899.

Byun, S., Taurozzi, J.S., Alpatova, A.L., Wang, F., Tarabara, V.V., 2011. Performance of polymeric membranes treating ozonated surface water: effect of ozone dosage. Sep. Purif. Technol. 81, 270–278.
Chae, S., Yamamura, H., Choi, B., Watanabe, Y., 2009. Fouling characteristics of pressurized and submerged PVDF (polyvinylidene fluoride) microfiltration membranes in a pilot-scale drinking water treatment system under low and high turbidity conditions. Desalination 244, 215–226.
Chang, E.E., Chen, Y., Lin, Y., Chiang, P., 2009. Reduction of natural organic matter by nanofiltration process. Chemosphere 76, 1265–1272.
Chen, V., Mansouri, J., Charlton, T., 2010. Biofouling in membrane systems. In: Peinemann, K., Pereira Nunes, S. (Eds.), Membranes for Water Treatment. Wiley-VCH Verlag, Weinheim, pp. 25–49.
Cheryan, M., 1998. Ultrafiltration and Microfiltration Handbook. Technomic Publishing, Lancaster, USA.
Chin, Y., Aiken, G., O'Loughlin, E., 1994. Molecular weight, polydispersity, and spectroscopic properties of aquatic humic substances. Environ. Sci. Technol. 28, 1853–1858.
Cho, J., Amy, G., Pellegrino, J., 2000a. Membrane filtration of natural organic matter: comparison of flux decline, NOM rejection, and foulants during filtration with three UF membranes. Desalination 127, 283–298.
Cho, J., Amy, G., Pellegrino, J., 2000b. Membrane filtration of natural organic matter: factors and mechanisms affecting rejection and flux decline with charged ultrafiltration (UF) membrane. J. Membr. Sci. 164, 89–110.
Cho, J., Amy, G., Pellegrino, J., 1999. Membrane filtration of natural organic matter: initial comparison of rejection and flux decline characteristics with ultrafiltration and nanofiltration membranes. Water Res. 33, 2517–2526.
Clark, M.M., Lucas, P., 1998. Diffusion and partitioning of humic acid in a porous ultrafiltration membrane. J. Membr. Sci. 143, 13–25.
Comstock, S.E.H., Boyer, T.H., Graf, K.C., 2011. Treatment of nanofiltration and reverse osmosis concentrates: comparison of precipitative softening, coagulation, and anion exchange. Water Res. 45, 4855–4865.
Costa, A.R., de Pinho, M.N., 2006a. Comparison of the performance of ultrafiltration and nanofiltration in surface water treatment. Desalination 199, 73–75.
Costa, A.R., de Pinho, M.N., 2006b. Performance and cost estimation of nanofiltration for surface water treatment in drinking water production. Desalination 196, 55–65.
Dong, B.Z., Chen, Y., Gao, N.Y., Fan, J.C., 2006. Effect of pH on UF membrane fouling. Desalination 195, 201–208.
Edwards, M., Benjamin, M.M., Ryan, J.N., 1996. Role of organic acidity in sorption of natural organic matter (NOM) to oxide surfaces. Colloids Surf. A Physicochem. Eng. Aspects 107, 297–307.
Edzwald, J.K., Tobiason, J.E., 1999. Enhanced coagulation: US requirements and a broader view. Water Sci. Technol. 40, 63–70.
Ericsson, B., Hallberg, M., Wachenfeldt, J., 1997. Nanofiltration of highly colored raw water for drinking water production. Desalination 108, 129–141.
Ericsson, B., Trägårdh, G., 1997. Treatment of surface water rich in humus — membrane filtration vs. conventional treatment. Desalination 108, 117–128.
Escobar, I.C., Hong, S., Randall, A.A., 2000. Removal of assimilable organic carbon and biodegradable dissolved organic carbon by reverse osmosis and nanofiltration membranes. J. Membr. Sci. 175, 1–17.
Fabris, R., Chow, C.W.K., Drikas, M., Eikebrokk, B., 2008. Comparison of NOM character in selected Australian and Norwegian drinking waters. Water Res. 42, 4188–4196.
Fan, L., Harris, J.L., Roddick, F.A., Booker, N.A., 2001. Influence of the characteristics of natural organic matter on the fouling of microfiltration membranes. Water Res. 35, 4455–4463.

Fane, A.G., Chang, S., 2009. Techniques to enhance performance of membrane process. In: Paddy, A.K., Rizvi, S.S.H., Sastre, A.M. (Eds.), Handbook of Membrane Separations, Chemical, Pharmaceutical, Food, and Biotechnological Applications. CRC Press, Taylor & Francis Group, Boca Raton, USA, pp. 193–232.
Fane, A.G., Wang, R., Jia, Y., 2008. Membrane technology: past, present and future. In: Wang, L.K., Chen, J.P., Hung, Y., Shammas, N.K. (Eds.), Handbook of Environmental Engineering. Membrane and Desalination Technologies, vol. 13. Springer Science & Business Media, New York, pp. 1–45.
Ghosh, K., Schnitzer, M., 1980. Macromolecular structures of humic substances. Soil Sci. 129, 266–276.
Glucina, K., Alvarez, A., Laîné, J.M., 2000. Assessment of an integrated membrane system for surface water treatment. Desalination 132, 73–82.
Gorenflo, A., Velázquez-Padrón, D., Frimmel, F.H., 2003. Nanofiltration of a German groundwater of high hardness and NOM content: performance and costs. Desalination 151, 253–265.
Guizard, C., Amblard, P., 2009. Current status and prospects for ceramic membrane applications. In: Paddy, A.K., Rizvi, S.S.H., Sastre, A.M. (Eds.), Handbook of Membrane Separations, Chemical, Pharmaceutical, Food, and Biotechnological Applications. CRC Press, Taylor & Francis Group, Boca Raton, USA, pp. 139–179.
Harman, B.I., Koseoglu, H., Yigit, N.O., Beyhan, M., Kitis, M., 2010. The use of iron oxide-coated ceramic membranes in removing natural organic matter and phenol from waters. Desalination 261, 27–33.
Hem, L.J., Efraimsen, H., 2001. Assimilable organic carbon in molecular weight fractions of natural organic matter. Water Res. 35, 1106–1110.
Her, N., Amy, G., Chung, J., Yoon, J., Yoon, Y., 2008. Characterizing dissolved organic matter and evaluating associated nanofiltration membrane fouling. Chemosphere 70, 495–502.
Hering, J.G., Morel, F.M.M., 1988. Humic acid complexation of calcium and copper. Environ. Sci. Technol. 22, 1234–1237.
Hilal, N., Al-Abri, M., Al-Hinai, H., 2006. Enhanced membrane pre-treatment processes using macromolecular adsorption and coagulation in desalination plants: a review. Sep. Sci. Technol. 41, 403–453.
Hong, S., Elimelech, M., 1997. Chemical and physical aspects of natural organic matter (NOM) fouling of nanofiltration membranes. J. Membr. Sci. 132, 159–181.
Huang, G., Meng, F., Zheng, X., Wang, Y., Wang, Z., Liu, H., Jekel, M., 2011. Biodegradation behavior of natural organic matter (NOM) in a biological aerated filter (BAF) as a pre-treatment for ultrafiltration (UF) of river water. Appl. Microbiol. Biotechnol. 90, 1795–1803.
Huber, S.A., Balz, A., Abert, M., Pronk, W., 2011. Characterization of aquatic humic and non-humic matter with size-exclusion chromatography – organic carbon detection – organic nitrogen detection (LC-OCD-OND). Water Res. 45, 879–885.
Jacangelo, J.G., Rhodes Trussell, R., Watson, M., 1997. Role of membrane technology in drinking water treatment in the United States. Desalination 113, 119–127.
Jarusutthirak, C., Mattaraj, S., Jiraratananon, R., 2007. Factors affecting nanofiltration performances in natural organic matter rejection and flux decline. Sep. Purif. Technol. 58, 68–75.
Jeong, S., Kim, S., Min Kim, C., Vigneswaran, S., Vinh Nguyen, T., Shon, H., Kandasamy, J., Kim, I.S., 2013. A detailed organic matter characterization of pretreated seawater using low pressure microfiltration hybrid systems. J. Membr. Sci. 428, 290–300.
Jermann, D., Pronk, W., Kägi, R., Halbeisen, M., Boller, M., 2008. Influence of interactions between NOM and particles on UF fouling mechanisms. Water Res. 42, 3870–3878.
Jermann, D., Pronk, W., Meylan, S., Boller, M., 2007. Interplay of different NOM fouling mechanisms during ultrafiltration for drinking water production. Water Res. 41, 1713–1722.

Jucker, C., Clark, M.M., 1994. Adsorption of aquatic humic substances on hydrophobic ultrafiltration membranes. J. Membr. Sci. 97, 37–52.
Kaewsuk, J., Seo, G.T., 2011. Verification of NOM removal in MIEX-NF system for advanced water treatment. Sep. Purif. Technol. 80, 11–19.
Kajitvichyanukul, P., Shammas, N.K., Hung, Y.-., Wang, L.K., Ananpattarachai, J., 2008. Portable water biotechnology, membrane filtration and biofiltration. In: Wang, L.K., Chen, J.P., Hung, Y., Shammas, N.K. (Eds.), Handbook of Environmental Engineering. Membrane and Desalination Technologies, vol. 13. Springer Science & Business Media, New York, pp. 477–523.
Kennedy, M.D., Kamanyi, J., Salinas Rodríguez, S.G., Lee, N.H., Schippers, J.C., Amy, G., 2008a. Water treatment by microfiltration and ultrafiltration. In: Li, N.N., Fane, A.G., Ho, W.S.W., Matsuura, T. (Eds.), Advanced Membrane Technology and Applications. John Wiley & Sons, Hoboken, USA, pp. 131–170.
Kennedy, M.D., Chun, H.K., Quintanilla Yangali, V.A., Heijman, B.G.J., Schippers, J.C., 2005. Natural organic matter (NOM) fouling of ultrafiltration membranes: fractionation of NOM in surface water and characterization by LC-OCD. Desalination 178, 73–83.
Kennedy, M.D., Kamanyi, J., Heijman, B.G.J., Amy, G., 2008b. Colloidal organic matter fouling of UF membranes: role of NOM composition & size. Desalination 220, 200–213.
Kilduff, J.E., Mattaraj, S., Pieracci, J.P., Belfort, G., 2000. Photochemical modification of poly(ether sulfone) and sulfonated poly(sulfone) nanofiltration membranes for control of fouling by natural organic matter. Desalination 132, 133–142.
Kim, D.H., Shon, H.K., Sharma, G., Cho, J., 2011. Charge effect of natural organic matter for ultrafiltration and nanofiltration membranes. J. Ind. Eng. Chem. 17, 109–113.
Kim, M.H., Yu, M.J., 2005. Characterization of NOM in the Han River and evaluation of treatability using UF–NF membrane. Environ. Res. 97, 116–123.
Koltuniewich, A.B., Drioli, E., 2008. Membranes in Clean Technologies, Theory and Practice. Wiley-VCH Verlag, Weinheim.
Kruithof, J.C., Schippers, J.C., Kamp, P.C., Folmer, H.C., Hofman, J.A.M.H., 1998. Integrated multi-objective membrane systems for surface water treatment: pretreatment of reverse osmosis by conventional treatment and ultrafiltration. Desalination 117, 37–48.
Küchler, I.L., Miekeley, N., 1994. Ultrafiltration of humic compounds through low molecular mass cut-off level membranes. Sci. Total Environ. 154, 23–28.
Lahoussine-Turcaud, V., Wiesner, M.R., Bottero, J., 1990. Fouling in tangential-flow ultrafiltration: the effect of colloid size and coagulation pretreatment. J. Membr. Sci. 52, 173–190.
Laîné, J.M., Vial, D., Moulart, P., 2000. Status after 10 years of operation — overview of UF technology today. Desalination 131, 17–25.
Lamsal, R., Montreuil, K.R., Kent, F.C., Walsh, M.E., Gagnon, G.A., 2012. Characterization and removal of natural organic matter by an integrated membrane system. Desalination 303, 12–16.
Lankes, U., Lüdemann, H.D., Frimmel, F.H., 2008. Search for basic relationships between "molecular size" and "chemical structure" of aquatic natural organic matter—answers from ^{13}C and ^{15}N CPMAS NMR spectroscopy. Water Res. 42, 1051–1060.
Lee, N., Amy, G., Croué, J., 2006. Low-pressure membrane (MF/UF) fouling associated with allochthonous versus autochthonous natural organic matter. Water Res. 40, 2357–2368.
Lee, N., Amy, G., Croué, J., Buisson, H., 2004a. Identification and understanding of fouling in low-pressure membrane (MF/UF) filtration by natural organic matter (NOM). Water Res. 38, 4511–4523.
Lee, S., Lee, C., 2007. Effect of membrane properties and pretreatment on flux and NOM rejection in surface water nanofiltration. Sep. Purif. Technol. 56, 1–8.
Lee, S., Amy, G., Cho, J., 2004b. Applicability of sherwood correlations for natural organic matter (NOM) transport in nanofiltration (NF) membranes. J. Membr. Sci. 240, 49–65.

Lee, S., Cho, J., 2004. Comparison of ceramic and polymeric membranes for natural organic matter (NOM) removal. Desalination 160, 223–232.
Lee, S., Kwon, B., Sun, M., Cho, J., 2005. Characterizations of NOM included in NF and UF membrane permeates. Desalination 173, 131–142.
Leenheer, J.A., Wershaw, R.L., Reddy, M.M., 1995. Strong-acid, carboxyl-group structures in fulvic acid from the Suwannee River, Georgia. 1. Minor structures. Environ. Sci. Technol. 29, 393–398.
Li, Y., Wang, J., Zhang, W., Zhang, X., Chen, C., 2011. Effects of coagulation on submerged ultrafiltration membrane fouling caused by particles and natural organic matter (NOM). Chin. Sci. Bull. 56, 584–590.
Lin, C., Huang, Y., Hao, O.J., 1999. Ultrafiltration processes for removing humic substances: effect of molecular weight fractions and PAC treatment. Water Res. 33, 1252–1264.
Lin, C., Liu, S., Hao, O.J., 2001. Effect of functional groups of humic substances on uf performance. Water Res. 35, 2395–2402.
Liu, T., Chen, Z., Yu, W., Shen, J., Gregory, J., 2011. Effect of two-stage coagulant addition on coagulation–ultrafiltration process for treatment of humic-rich water. Water Res. 45, 4260–4268.
van Loon, G.W., Duffy, S.J., 2005. Environmental Chemistry: A Global Perspective, second ed. Oxford University Press, USA.
Lopes, M.P., Matos, C.T., Pereira, V.J., Benoliel, M.J., Valério, M.E., Bucha, L.B., Rodrigues, A., Penetra, A.I., Ferreira, E., Cardoso, V.V., Reis, M.A.M., Crespo, J.G., 2013. Production of drinking water using a multi-barrier approach integrating nanofiltration: a pilot scale study. Sep. Purif. Technol. 119, 112–122.
Metsämuuronen, S., Sillanpää, M., Bhatnagar, A., Mänttäri, M., 2014. Natural organic matter removal from drinking water by membrane technology. Sep. Purif. Rev. 43, 1–61.
Meylan, S., Hammes, F., Traber, J., Salhi, E., von Gunten, U., Pronk, W., 2007. Permeability of low molecular weight organics through nanofiltration membranes. Water Res. 41, 3968–3976.
Mierzwa, J.C., Hespanhol, I., da Silva, M.C.C., Rodrigues, L.D.B., Giorgi, C.F., 2008. Direct drinking water treatment by spiral-wound ultrafiltration membranes. Desalination 230, 41–50.
Mierzwa, J.C., da Silva, M.C.C., Veras, L.R.V., Subtil, E.L., Rodrigues, R., Li, T., Landenberger, K.R., 2012. Enhancing spiral-wound ultrafiltration performance for direct drinking water treatment through operational procedures improvement: a feasible option for the Sao Paulo Metropolitan Region. Desalination 307, 68–75.
Nyström, M., Ruohomäki, K., Kaipia, L., 1996. Humic acid as a fouling agent in filtration. Desalination 106, 79–87.
Orecki, A., Tomaszewska, M., Karakulski, K., Morawski, A.W., 2004. Surface water treatment by the nanofiltration method. Desalination 162, 47–54.
Plottu, A., Houssais, B., Democrate, C., Gatel, D., Cavard, J., 2003. Autopsies of membranes fouled on Mery-sur-Oise pilot units: many lessons for the behaviour of the water treatment plant. Desalination 157, 367. 367.
Rachwal, T., Judd, S., 2006. A synopsis of membrane technologies in UK municipal potable water treatment: history, status and prospects. Water Environ. J. 20, 110–113.
Remize, P.J., Guigui, C., Cabassud, C., 2010. Evaluation of backwash efficiency, definition of remaining fouling and characterization of its contribution in irreversible fouling: case of drinking water production by air-assisted ultra-filtration. J. Membr. Sci. 355, 104–111.
Ribau Teixeira, M., Rosa, M.J., 2006. The impact of the water background inorganic matrix on the natural organic matter removal by nanofiltration. J. Membr. Sci. 279, 513–520.
Ribau Teixeira, M., Sousa, V.S., 2013. Fouling of nanofiltration membrane: effects of NOM molecular weight and microcystins. Desalination 315, 149–155.

de la Rubia, Á., Rodríguez, M., Prats, D., 2006. pH, ionic strength and flow velocity effects on the NOM filtration with TiO_2/ZrO_2 membranes. Sep. Purif. Technol. 52, 325–331.
de la Rubia, Á., Rodríguez, M., León, V.M., Prats, D., 2008. Removal of natural organic matter and THM formation potential by ultra- and nanofiltration of surface water. Water Res. 42, 714–722.
Sanchez, N.P., Skeriotis, A.T., Miller, C.M., 2013. Assessment of dissolved organic matter fluorescence PARAFAC components before and after coagulation–filtration in a full scale water treatment plant. Water Res. 47, 1679–1690.
Schäfer, A.I., Fane, A.G., Waite, T.D., 2005. Nanofiltration – Principles and Applications. Elsevier Advanced Technology, Kidlington, UK.
Schäfer, A.I., Fane, A.G., Waite, T.D., 2001. Cost factors and chemical pretreatment effects in the membrane filtration of waters containing natural organic matter. Water Res. 35, 1509–1517.
Schäfer, A.I., Fane, A.G., Waite, T.D., 2000. Fouling effects on rejection in the membrane filtration of natural waters. Desalination 131, 215–224.
Schäfer, A.I., Fane, A.G., Waite, T.D., 1998. Nanofiltration of natural organic matter: removal, fouling and the influence of multivalent ions. Desalination 118, 109–122.
Schrotter, J., Bozkaya-Schrotter, B., 2010. Current and emerging membrane processes for water treatment. In: Peinemann, K., Pereira Nunes, S. (Eds.), Membranes for Water Treatment. Wiley-VCH Verlag, Weinheim, pp. 53–91.
Seidel, A., Elimelech, M., 2002. Coupling between chemical and physical interactions in natural organic matter (NOM) fouling of nanofiltration membranes: implications for fouling control. J. Membr. Sci. 203, 245–255.
Shao, J., Hou, J., Song, H., 2011. Comparison of humic acid rejection and flux decline during filtration with negatively charged and uncharged ultrafiltration membranes. Water Res. 45, 473–482.
Sharp, E.L., Parsons, S.A., Jefferson, B., 2006. The impact of seasonal variations in DOC arising from a moorland peat catchment on coagulation with iron and aluminium salts. Environ. Pollut. 140, 436–443.
Siddiqui, M., Amy, G., Ryan, J., Odem, W., 2000. Membranes for the control of natural organic matter from surface waters. Water Res. 34, 3355–3370.
Simon, F.X., Penru, Y., Guastalli, A.R., Esplugas, S., Llorens, J., Baig, S., 2013. NOM characterization by LC-OCD in a SWRO desalination line. Desalin. Water Treat. 51, 1776–1780.
Sutherland, K., 2012. Drinking and pure water: filtration improvements progress global clean water provision. Filtr. Sep. 49, 12–16.
Syafei, A.D., Lin, C., Wu, C., 2008. Removal of natural organic matter by ultrafiltration with TiO_2-coated membrane under UV irradiation. J. Colloid Interface Sci. 323, 112–119.
Tang, C.Y., Chong, T.H., Fane, A.G., 2011. Colloidal interactions and fouling of NF and RO membranes: a review. Adv. Colloid Interface Sci. 164, 126–143.
The Council of the European Union, 1998. Council directive 98/83/EC. Off. J. Eur. Communities L 330, 32–54.
Thorsen, T., 2004. Concentration polarization by natural organic matter (NOM) in NF and UF. J. Membr. Sci. 233, 79–91.
Thorsen, T., 1998. Flux and recovery influences on treatment efficiency in ultrafiltration of coloured drinking water. Desalination 117, 131–138.
Uyak, V., Koyuncu, I., Oktem, I., Cakmakci, M., Toroz, I., 2008. Removal of trihalomethanes from drinking water by nanofiltration membranes. J. Hazard. Mater. 152, 789–794.
Ventresque, C., Gisclon, V., Bablon, G., Chagneau, G., 2000. An outstanding feat of modern technology: the Mery-sur-Oise nanofiltration treatment plant (340,000 m^3/d). Desalination 131, 1–16.
Visvanathan, C., Marsono, B.D., Basu, B., 1998. Removal of THMP by nanofiltration: effects of interference parameters. Water Res. 32, 3527–3538.

Voutchkov, N., 2010. Considerations for selection of seawater filtration pretreatment system. Desalination 261, 354–364.
Wang, X., Waite, T.D., 2009. Role of gelling soluble and colloidal microbial products in membrane fouling. Environ. Sci. Technol. 43, 9341–9347.
Wang, X., Waite, T.D., 2008. Gel layer formation and hollow fiber membrane filterability of polysaccharide dispersions. J. Membr. Sci. 322, 204–213.
Wang, Y.F., Shu, L., Jegatheesan, V., Gao, B.Y., 2009. Coagulation and nano-filtration: a hybrid system for the removal of lower molecular weight organic compounds (LMWOC). Desalin. Water Treat. 11, 23–31.
Wang, Y., Combe, C., Clark, M.M., 2001. The effects of pH and calcium on the diffusion coefficient of humic acid. J. Membr. Sci. 183, 49–60.
Yee, L.F., Abdullah, P.M., Abdullah, A., Ishak, B., Abidin, K.N.Z., 2009. Hydrophobicity characteristics of natural organic matter and the formation of THM. Malays. J. Anal. Sci. 13, 94–99.
Yoon, Y., Amy, G., Cho, J., Her, N., 2005. Effects of retained natural organic matter (NOM) on NOM rejection and membrane flux decline with nanofiltration and ultrafiltration. Desalination 173, 209–221.
Zazouli, M.A., Nasseri, S., Ulbricht, M., 2010. Fouling effects of humic and alginic acids in nanofiltration and influence of solution composition. Desalination 250, 688–692.
Zhou, J., Li, W., Gu, J., Yu, H., Tang, Z., Wei, X., 2010. Development of a novel RAFT-UV grafting technique to modify polypropylene membrane used for NOM removal. Sep. Purif. Technol. 71, 233–240.
Zularisam, A.W., Ismail, A.F., Salim, M.R., Sakinah, M., Hiroaki, O., 2007a. Fabrication, fouling and foulant analyses of asymmetric polysulfone (PSF) ultrafiltration membrane fouled with natural organic matter (NOM) source waters. J. Membr. Sci. 299, 97–113.
Zularisam, A.W., Ismail, A.F., Salim, M.R., Sakinah, M., Matsuura, T., 2009. Application of coagulation–ultrafiltration hybrid process for drinking water treatment: optimization of operating conditions using experimental design. Sep. Purif. Technol. 65, 193–210.
Zularisam, A.W., Ismail, A.F., Salim, M.R., Sakinah, M., Ozaki, H., 2007b. The effects of natural organic matter (NOM) fractions on fouling characteristics and flux recovery of ultrafiltration membranes. Desalination 212, 191–208.

CHAPTER 6

NOM Removal by Advanced Oxidation Processes

Mika Sillanpää*, Anu Matilainen**

*Lappeenranta University of Technology, LUT Faculty of Technology, LUT Chemtech, Laboratory of Green Chemistry, Sammonkatu 12, 50130 Mikkeli, Finland; **Finnish Safety and Chemicals Agency, Kalevantie 2, 33100 Tampere, Finland

Abstract

Efficient methods for the removal of natural organic matter (NOM) in waters have emerged. Among these are advanced oxidation processes (AOPs). These processes include O_3/H_2O_2, O_3/UV, UV/H_2O_2, TiO_2/UV, H_2O_2/catalyst, Fenton and photo-Fenton processes, and ultrasound. This chapter presents an overview of the recent research studies dealing with AOP methods for the removal of NOM and related compounds from drinking water.

Keywords: Advanced oxidation processes (AOP)s; Fenton; Natural organic matter (NOM); Ozone; Photocatalysis; Ultrasound; UV light; Water treatment.

ABBREVIATIONS

AC Activated carbon
ALD Atomic layer deposition
AOPs Advanced oxidation processes
BAC Biologically activated carbon
COD Chemical oxygen demand
DBP Disinfection by-product
DBPFP Disinfection by-product formation potential
DOC Dissolved organic carbon
FP Formation potential
HA Humic acids
HAAs Haloacetic acids
HMM High molar mass
LMM Low molecular mass
MM Molar mass
MWD Molecular weight distribution
NEU (hydrophilic) neutral
NOM Natural organic matter
SUVA Specific UV absorbance
THM Trihalomethane
THMFP Trihalomethane formation potential

TOC Total organic carbon
US Ultrasonic
VUV Vacuum UV

6.1 INTRODUCTION

The natural organic matter (NOM) removal results of advanced oxidation processes (AOPs) are always study specific because they depend on the water characteristics, such as the amount of organic matter. Therefore, characterization of the NOM in water necessarily precludes the design and optimization of any AOP treatment. The organic characteristics of the treated water must also be determined to assess their influence on downstream processes. Furthermore, individual disinfection by-product (DBP) measurements and disinfection by-product formation potential (DBPFP) measurements are important when assessing the suitability of an AOP treatment.

AOP research is well established. Various combinations of oxidants, radiation, and catalysts have been developed for the removal of NOM and organic pollutants, e.g., O_3/H_2O_2, UV/H_2O_2, UV/O_3, UV/TiO_2, Fe^{2+}/H_2O_2, $Fe^{2+}/H_2O_2 + hv$, vacuum ultraviolet (VUV) radiation or ionizing radiation (Fujishima, 1971; Glaze et al., 1987, 1992; Legrini et al., 1993; Frimmel, 1994; Nagata et al., 1996; Fukushima et al., 2001; Thomson et al., 2002; Suty et al., 2004). These processes involve the generation of highly reactive radical intermediates, especially the $^{\cdot}OH$ radical (Glaze et al., 1987). The appeal of AOPs is their ability to achieve complete oxidation or mineralization of organic contaminants at near ambient temperature and pressure (Matilainen and Sillanpää, 2010).

The $^{\cdot}OH$ radical is one of the most powerful known oxidants. The reaction rate of a compound in $^{\cdot}OH$ radical mediated oxidation is usually several orders of magnitude higher than that achieved by molecular ozone under the same conditions. The reaction rate constants between $^{\cdot}OH$ radicals and organic species are in the range of 10^8–$10^{10}\,M^{-1}\,s^{-1}$ (Parsons and Byrne, 2004). $^{\cdot}OH$ radicals are highly nonselective oxidants, enabling a very large number of reactions. Once free radical reaction has been initiated, whether by photolysis, ozone, hydrogen peroxide, or heat, a series of simple reactions will ensue. The reactions of $^{\cdot}OH$ radicals with NOM proceeds in three different ways: (1) by the addition of $^{\cdot}OH$ radicals to double bonds, (2) by H-atom abstraction, which yields carbon-centered radicals, and (3) by the $^{\cdot}OH$ radical gaining an electron from an organic substituent. The carbon-centered radicals then react very rapidly with oxygen to form organic peroxyl radicals. The mutual reactions of peroxyl radicals can lead to the production of ketones or aldehydes and/or carbon dioxide (Kleiser and Frimmel, 2000). The rate of oxidation depends on radical, oxygen, and pollutant concentrations. Other factors affecting radical formation are pH, temperature, the presence of ions, the type of pollutant, and the presence of scavengers such as bicarbonate ions (Parsons and Byrne, 2004).

Various AOP investigations have found that NOM decreases more when measured as UV_{254} than as dissolved organic carbon (DOC), for example, in UV/H_2O_2 (Goslan et al., 2006; Sanly et al., 2007; Moncayo-Lasso et al., 2008), TiO_2/UV (Uyguner et al., 2007b; Huang et al., 2008; Liu et al., 2008a,b, 2010), and Fenton processes (Sanly et al., 2007; Moncayo-Lasso et al., 2008, 2009; Nie et al., 2010). The greater drop in UV_{254} values suggests that the chromophores in the NOM macromolecules, which mostly consist of high molar mass (HMM) aromatic rings, are rapidly broken down into lower MM by-products with no UV absorbance. These by-products are thus less susceptible to attack by $^{\cdot}OH$ radicals, and therefore are not mineralized completely (Sanly et al., 2007). The intermediate fragments usually undergo a series of complicated reactions prior to mineralization. Many researchers have observed the preferential degradation of hydrophobic compounds (Zhang and Jian, 2006; Katsumata et al., 2008; Liu et al., 2010). Fractionation studies have revealed that although the fraction of NOM with very hydrophobic acids has been reduced, the corresponding increase happens in the hydrophilic charged and/or hydrophilic neutral (NEU) NOM fractions (Buchanan et al., 2006; Sanly et al., 2007; Liu et al., 2008b, 2010). However, this hydrophilic NEU fraction has not been found to be very reactive in forming, e.g., trihalomethanes (THMs) (Buchanan et al., 2006; Liu et al., 2008b, 2010).

It has been observed that HMM compounds are removed more readily, and the removal of the smaller compounds occurs only after the large molecules have been degraded. Nevertheless, a small amount of organic matter with low molecular mass (LMM) has remained after treatment in several instances (Liu et al., 2008a,b, 2010; Tercero Espinoza and Frimmel, 2009; Tercero Espinoza et al., 2009b). Also, the shift in molecular weight distribution (MWD) toward lower MM compounds during oxidation has been noted in many cases, e.g., on UV/H_2O_2 (Goslan et al., 2006; Wang et al., 2006; Sanly et al., 2007; Sarathy and Mohseni, 2007), TiO_2/UV (Liu et al., 2008a,b, 2010; Tercero Espinoza and Frimmel, 2009; Tercero Espinoza et al., 2009b), and Fenton processes (Goslan et al., 2006; Sanly et al., 2007; Katsumata et al., 2008; Nie et al., 2010). It has been noted that incomplete oxidation increases DBPFP under VUV (Buchanan et al., 2006; Bond et al., 2009), TiO_2/UV

(Liu et al., 2008a,b), and UV/H_2O_2 treatments (Toor and Mohseni, 2007). Moreover, during some oxidation processes, the DBP speciation has been found to shift to more brominated DBPs, especially in VUV (Buchanan et al., 2006) and UV/H_2O_2 (Wang et al., 2006) treatments, but contradictory results have also been reported (Tercero Espinoza and Frimmel, 2008). The formation of brominated DBPs is demonstrably dependent on the characteristics of water and the design parameters, e.g., a higher concentration of TiO_2 has been observed to produce more bromoform (Tercero Espinoza et al., 2009a). In addition, UV/H_2O_2 treatment was found to increase the formation potential of nitroso-DBP (Zhao et al., 2008).

A vast amount of research has been conducted on the removal of organic and inorganic micropollutants (Hu et al., 2008; Li et al., 2008; Swaim et al., 2008), including pharmaceuticals (Benotti et al., 2009), endocrine-disrupting chemicals (Chen et al., 2007; Benotti et al., 2009) or pesticides (Rastogi et al., 2009), and the inactivation of microorganisms (Gerrity et al., 2008; Lanao et al., 2008; Navalon et al., 2009). Reviews of such AOP efficiency research include Pera-Titus et al. (2004), Ikehata and El-Din (2005), and Klavarioti et al. (2009). Yet, many of these AOP NOM studies focus on improving the removal of DBP precursors. An overview of the recent research on NOM removal by AOPs was presented by Matilainen and Sillanpää (2010). This chapter is an update of that earlier work.

6.2 AOPS IN NOM REMOVAL

6.2.1 Ozone Based Applications

Ozone has long been used for disinfection, taste and odor management, and color removal in water treatment facilities (Langlais et al., 1991). Ozone reacts with NOM by an electrophilic addition to double bonds. This reaction is very selective. In addition to the direct reaction of ozone with NOM, a nonselective and fast reaction occurs with the $^{\bullet}OH$ radicals formed when ozone decomposes in water. The $^{\bullet}OH$ radical formation potential is much lower in ozonation than in AOPs. The oxidation potentials for ozone and $^{\bullet}OH$ radicals are 2.42 and 2.86 eV, respectively (Kleiser and Frimmel, 2000). There are several solutions to promote the $^{\bullet}OH$ radical production in ozonation processes. The simplest and most cost-efficient way to convert a conventional ozonation process into an AOP is the addition of H_2O_2 (Duguet et al., 1985; Glaze et al., 1992; von Gunten, 2003). Another alternative is the combination of ozone with UV irradiation or the combination of all three (Glaze et al., 1982, 1987; Legrini et al., 1993).

Ozone-AOP systems have recently been tested for the disinfection and detoxification of pollutants in different types of waters (Agustina et al., 2005; Ikehata and El-Din, 2005; Lanao et al., 2008). Ozonation has been used for NOM removal in a number of cases. Molnar et al. (2013) observed recently that ozone treatment results in partial oxidation of the fulvic acid fraction, increasing the amount of hydrophilic NOM. This finding is in agreement with Yan et al. (2007) and Treguer et al. (2010), who reported that an elevated ozone dose makes the NOM more hydrophilic and lower in molecular weight, and decreases DOC content in the hydrophobic fraction. Adding H_2O_2 into the ozonation process enhances the reduction of DOC compared to the ozone alone according to Irabelli et al. (2008). This research demonstrated that a postozonation O_3/H_2O_2 system achieves better removal of DOC than a preozonation O_3/H_2O_2 system. Yet, total THM concentrations were higher in the O_3/H_2O_2 samples than in ozonated samples in both cases (Irabelli et al., 2008). Similarly, Mosteo et al. (2009) did not observe any significant gains in THM reduction after adding H_2O_2 or TiO_2 to the ozone treatment. Tubić et al. (2011) concluded that a higher ratio of H_2O_2 (2:1 H_2O_2:O_3) is less favorable for NOM removal than a lower one (1:2 H_2O_2:O_3). Yet, the simultaneous addition of H_2O_2 and TiO_2 to ozonation resulted in lower THM production. The influence of ozonation on NOM fraction properties and weight distribution has been the subject of several recent studies (Bose and Reckhow, 2007; Lamsal et al., 2011; Chen et al., 2011; Molnar et al., 2013; Molnar et al., 2012b).

Combined O_3-UV treatment results in significant mineralization of DOC, reduction of trihalomethane formation potential (THMFP), and haloacetic acid formation potential (HAAFP), according to Chin and Bérubé (2005). In addition, photocatalytic ozonation, $O_3/UV/TiO_2$, has been used to remove recalcitrant organic compounds from water (Giri et al., 2008) and appears to degrade NOM much more efficiently than sole ozonation or photocatalysis (Zhang and Jian, 2006). Furthermore, O_3 alone was found to be less efficient than O_3/TiO_2 in NOM oxidation (Molnar et al., 2012a). Ozone based AOPs are summarized in Table 6.1.

6.3 UV LIGHT BASED APPLICATIONS

UV photolysis is widely used for disinfection purposes. It involves the interaction of artificial or natural light with the target molecules and compounds, which then induce a series of photochemical reactions (Gjessing and Källqvist, 1991; Frimmel, 1994, 1998). Direct photolysis is dependent on the ability of the compounds to absorb the emitted light. Its efficiency is

Table 6.1 Ozone based AOPs

Target compounds	Matrix	AOP features	Parameters monitored	Summary of results	References
O_3/H_2O_2, O_3/UV and O_3/H_2O_2/UV					
THM precursors	Synthetic humic solution (HA and FA)	H_2O_2/O_3 ratios 0.088, 0.032, 0.04	TOC, COD	The best NOM removal was achieved with 0.04 H_2O_2/O_3 at 20 min, removing 78% and 50% of TOC and COD, respectively. COD removal was enhanced with time. The best result achieved was with H_2O_2/O_3 0.088 at 30 min.	Alsheyab and Muñoz (2006)
Probe compound pCBA	Two raw waters and simulated groundwater	$H_2O_2:O_3$ ratios 1:2 and 1:1		Addition of peroxide to the ozone process resulted in faster formation of $^{\cdot}OH$ radicals, but did not appear to significantly increase the ultimate $^{\cdot}OH$ radical exposure.	Rosenfeldt et al. (2006)

THM precursors	Raw water	H_2O_2:O_3 ratios in the range 0.0–0.35	DOC, THMs	The removal of DOC from raw water by peroxone was more effective than by ozone alone (61% and 53%, respectively), with the postozonation peroxone system achieving the greatest removal. Yet, THM concentration was higher in the peroxone samples.	Irabelli et al. (2008)
	Synthetic HA solution (Fluka HA)	Ozone 3 mg O_3 L^{-1}, H_2O_2 1.5 mg L^{-1}	DOC, THMFP, THMs	The O_3/H_2O_2 combination did not seem to improve THM reduction, since the scavenger effect of H_2O_2 was produced by the working pH 8.	Mosteo et al. (2009)

(Continued)

Table 6.1 Ozone based AOPs—cont'd

Target compounds	Matrix	AOP features	Parameters monitored	Summary of results	References
HAAs (DCAA and TCAA)	DCAA and TCAA solutions	Ozone 0.3 mg min^{-1}/LP UV at 254 nm with addition of H_2O_2 2.5 mg L^{-1}	HAAs decomposition	Within 30 min reaction time, TCAA was decomposed >50% and DCAA >90%. Humic acids (conc. 1.2 mgTOC L^{-1}) decreased the decomposition rates of DCAA and TCAA by 41% and 24%, respectively. O_3/H_2O_2/UV was not remarkably more efficient than O_3/UV in HAA decomposition.	Wang et al. (2009)
Bromate	Simulated natural waters	LP UV at 254 nm	Bromate formation	Bromate formation and HA removal by ozone via O_3/UV with ozone generation by UV lamp and silent electric discharge. The concentrations of bromate formed were lower with O_3/UV than with ozone alone, once HA concentrations reached a certain level.	Zhao et al. (2013)

O_3/TiO_2 and $O_3/H_2O_2/TiO_2$					
THM precursors	Synthetic HA solution (Fluka HA)	Ozone $3\,mg\,O_3\,L^{-1}$, suspended Degussa TiO_2 $1\,mg\,L^{-1}$ with H_2O_2 $1.5\,mg\,L^{-1}$	DOC, THMFP, THMs	O_3/TiO_2 did not improve THM reduction, due to the working pH 8, since at higher pH an inhibition takes place. Simultaneous addition of H_2O_2 and TiO_2 produced a synergic effect: Lower production of THMs.	Mosteo et al. (2009)
	Groundwater	O_3 and O_3/TiO_2	DOC	O_3/TiO_2 catalyzed process resulted in significantly more effective reduction. The nature of residual NOM is key to DBP formation. Ozonation leads to partial oxidation of the fulvic acid fraction and increases hydrophilic NOM fractions.	Molnar et al. (2012b); Molnar et al. (2013)
O_3/catalyst					
DBP precursors	Natural river water after coagulation and sand filtration	O_3/several metal oxides	DOC, THMFP	Catalytic ozonation with metal oxides significantly reduced THMFP compared to ozonation.	Zhang and Jian (2006)

COD, chemical oxygen demand; pCBA, parachlorobenzoic acid; DBP, disinfection by-products; DCAA, dichloroacetic acid; DOC, dissolved organic carbon; FA, fulvic acid; HAA, haloacetic acid; HA, humic acid; LP, low pressure; NOM, natural organic matter; TCAA, trichloroacetic acid; THM, trihalomethane; THMFP, trihalomethane formation potential; TOC, total organic carbon.

enhanced when irradiation is combined with hydrogen peroxide, whose photocatalytic dissociation yields ˙OH radicals (Legrini et al., 1993; Rosenfeldt et al., 2006), thus facilitating the degradation processes of different contaminants in water and water disinfection (Alkan et al., 2007; Chen et al., 2007; Kruithof et al., 2007; Hu et al., 2008; Li et al., 2008; Swaim et al., 2008). Under VUV radiation, water molecules dissociate into hydrogen atoms, hydroxyl radicals, and hydrated electrons (Kutschera et al., 2009). The degradation of target compounds with UV/H_2O_2 and VUV occurs several times more efficiently than with direct UV photolysis, as observed by Dobrovic et al. (2007), Bond et al. (2009) and Imoberdorf and Mohseni (2011).

Although it has been proved that UV photolysis alone changes the chemical and biological properties of NOM, e.g., MWD (Frimmel, 1998), NOM reduction has been observed only with rather high UV doses (Goslan et al., 2006; Huang et al., 2008). The combination of UV irradiation at 254 nm and H_2O_2 treatment (with an oxidation potential of 1.8 eV) has been shown to enhance NOM reduction and promote ˙OH radical formation (Goslan et al., 2006; Toor and Mohseni, 2007).

UV/H_2O_2 treatment combines direct photolysis and advanced oxidation through indirect photolysis (Pereira et al., 2007; Sindelar et al., 2014). Differences in the spectrum of medium-pressure and low-pressure lamps can affect chemical bonds in various ways, generating different degradation products (Metz et al., 2011). It has been concluded, however (Sarathy and Mohseni, 2007), that NOM is not removed under the fluence and H_2O_2 concentrations typically applied in drinking water treatment applications (the standard UV dose for disinfection is about 400 mJ cm^{-2}; Chin and Bérubé, 2005), although a recent study observed changes in NOM structure with economically feasible UV fluences and H_2O_2 concentrations (Sarathy et al., 2011). Yet, other research demonstrated >60% reduction of NOM with a UV dose of 540 mJ cm^{-2} and a H_2O_2 dose of 6 mg L^{-1} (Kruithof et al., 2007). UV/H_2O_2 treatment has been found to decrease molecular size (Sarathy and Mohseni, 2009), create more biodegradable decomposition products (Thomson et al., 2004; Sarathy et al., 2011), reduce NOM aromaticity (Thomson et al., 2004; Kleiser and Frimmel, 2000; Sarathy et al., 2011), and decrease hydrophobicity (Sarathy and Mohseni, 2010; Lin and Wang, 2011). Moreover, significant changes in NOM structural characteristics have been noted even at lower fluences (Wang et al., 2006; Sanly et al., 2007; Sarathy and Mohseni, 2007). DOC degradation has been rather low, however (Sanly et al., 2007). Taken together, these results indicate that a combination of high UV fluence and high peroxide concentration is required to generate enough $^{\bullet}OH$ radicals to significantly reduce NOM and DBPFP (Tuhkanen, 2004; Sarathy and Mohseni, 2007; Toor and Mohseni, 2007). A pilot investigation into seasonal variations concluded that some NOM mineralization (2–7%) occurred. It is important to note that proper pretreatment before a UV/H_2O_2 process increases the NOM mineralization rate (Sarathy and Mohseni, 2010). Suitable pretreatment methods include coagulation, sand filtration, and membrane technology, which increase UV transmittance and remove OH scavengers.

The UV fluence and oxidant dose needed to reduce UV_{254} and DOC in water containing high levels of NOM was tested for three oxidants, namely hydrogen peroxide, percarbonate, and perborate. Sindelar et al. (2014) concluded that, although H_2O_2 is the best chemical oxidant to use in combination with H_2O_2, percarbonate and perborate are rather comparable in efficiency and offer other potential benefits over H_2O_2. Advances in UV light efficiency may make UV/H_2O_2 AOPs feasible for natural water applications in the future, however (Sindelar et al., 2014).

It is noteworthy that excess H_2O_2 can play the role of $^{\bullet}OH$ radical scavenger, thus making the process less effective (Tuhkanen, 2004; Wang et al., 2006; Rosenfeldt and Linden, 2007; Song et al., 2008). Wang et al. (2006) observed that H_2O_2 at a concentration of 0.0082 M could absorb most of the UVC energy (200–280 nm), and dissociate into $^{\bullet}OH$ radicals. They suggested that the optimal H_2O_2 concentration for HA degradation would be between 0.0032 and 0.0163 M. For surface waters with high DOC, even higher concentrations (~0.032 M) would be necessary.

AOPs can also be used as a pretreatment to enhance the biodegradability of NOM if the resulting intermediates are easily degradable in a further biological treatment. The more biodegradable compounds are readily utilized as substrates by microorganisms in biologically activated carbon (BAC) filtration (Toor and Mohseni, 2007). In combined UV/H_2O_2 AOP-BAC treatment, significant reductions of DBPs, TOC, and UV_{254} have been observed compared to AOP treatment without BAC (Toor and Mohseni, 2007). Similarly, photocatalysis combined with activated carbon (AC) filtration has been found to greatly improve removal efficiencies (measured as $color_{436}$ and UV_{254}) compared to photocatalysis results without AC. A combination of ozonation, photocatalytic oxidation, and AC adsorption obtained complete decolorization (Uyguner et al., 2007b).

In VUV treatment, highly oxidizing hydroxyl radicals are generated by the photolysis of water at wavelengths lower than 190 nm (Gonzalez et al., 2004).

VUV oxidation has been found highly efficient in NOM reduction (Thomson et al., 2002; Imoberdorf and Mohseni, 2011). NOM surrogates were degraded very efficiently when measured as DOC with high fluent dose (47–48 J cm^{-2}, Bond et al., 2009). At lower doses (21 J cm^{-2}), however, incomplete mineralization led to an increase in HAA levels, which was particularly dramatic in the case of two amino acids (Bond et al., 2009). These findings are in accordance with results obtained by others concerning THMFP (Buchanan et al., 2006) and UV/H_2O_2 treatment (Toor and Mohseni, 2007). Raw water NOM was found to decline with increased VUV dosage (0–128 J cm^{-2}), resulting in total THMFP reduction (Buchanan et al., 2006). Despite this, and although bromoform accounted for only a small fraction of the total THMFP, the speciation of DBPs shifted to more brominated compounds, and the formation of bromoform was observed in the hydrophilic fraction of NOM (Buchanan et al., 2006). The UV/H_2O_2 treatment has also been noted to increase the formation potential of nitroso-DBPs (Zhao et al., 2008). Furthermore, inorganic compounds also play a role in the radiation propagation during VUV process.

The formation of undesirable by-products presents a challenge to the application of UV in general and VUV in particular. The main by-product is nitrite, which is subject to European Union regulation (EU Drinking Water Standard, maximum concentration 0.5 mg L^{-1}). The formation of nitrite can be avoided by the addition of ozone when using UV/VUV (Kutschera et al., 2009). Nitrite formation during the irradiation of natural water has not been observed in all cases. One investigation found, however, that rapid formation of nitrite occurs in nitrate-spiked water. The concentration significantly decreases after VUV radiation of 25 J cm^{-2} (Buchanan et al., 2006). UV light based AOPs are summarized in Table 6.2.

6.4 FENTON PROCESSES

The conventional Fenton process involves the use of an oxidizing agent (usually H_2O_2) and a catalyst (metal salt or oxide, usually iron). Fenton's reagent (H_2O_2 with Fe(II)/Fe(III) ions) in water produces $^{\bullet}OH$ radicals and peroxy radicals (Voelker and Sulzberger, 1996; Park and Yoon, 2007). The photo-Fenton process also involves irradiation with sunlight or an artificial light source, which leads to the production of additional $^{\bullet}OH$ radicals, thus increasing the rate of organic compound or contaminant degradation (Zepp et al., 1992; Fukushima et al., 2001; Moncayo-Lasso et al., 2009). A considerable body of research has been conducted on the application of

Table 6.2 UV light based AOPs

Target compounds	Matrix	AOP features	Parameters monitored	Summary of results	References
UV/H_2O_2					
NOM	Raw water	LP UV at 254 nm	DOC, UV_{254}, HPSEC	At a UVC dose of 22 J cm^{-1}, reduction of NOM as UV_{254} and DOC were 94% and 78%, respectively. DOC removal was increased with increasing UVC dosage up to 37 J cm^{-1} while UV_{254} reduction did not change significantly at doses higher than 22 J cm^{-1}.	Goslan et al. (2006)
Probe compound pCBA	Two raw waters and simulated groundwater	LP UV and MP UV/H_2O_2 2, 10, and 50 mg L^{-1}	UV_{254}, specific compound	H_2O_2 was the most energy-intensive portion of the process. The MP UV systems were consistently less efficient for $^{\cdot}OH$ radical production than LP UV systems.	Rosenfeldt et al. (2006)

(Continued)

Table 6.2 UV light based AOPs—cont'd

Target compounds	Matrix	AOP features	Parameters monitored	Summary of results	References
NOM	Raw water and synthetic HA (Aldrich HA)	HP UV/H_2O_2 0–32.4 mM	NPDOC, UV_{254}, IC, FTIR, THMFP	At an irradiation time of 30 min, 90% of NPDOC was mineralized (0.01% H_2O_2), THMFP was reduced, although a shift from chlorine to bromine THMs was noted.	Wang et al. (2006)
HA	Synthetic HA solution (Fluka HA)	UVA with maximum centered at 365 nm	DOC, UV_{254}, HPSEC, fractionation	UVA/H_2O_2 treatment exhibited preferential degradation of organic compounds with HMM (1.5–10 kDa). There was a significant shift of average MWD to the lower range, although only 33% of DOC was removed after 1 h irradiation.	Sanly et al. (2007)

NOM	Raw surface water	LP UV	TOC, UV_{254}, HPSEC	Under AOP conditions typically applied in drinking water treatment applications (500 mJ cm^{-2}, 4 mg L^{-1} H_2O_2), NOM was not mineralized but was partially oxidized, resulting in significant reduction in aromaticity.	Sarathy and Mohseni (2007)
DBP precursors		LP UV at 254 nm	TOC, UV_{254}, UV_{203}, ADOC and BDOC, AOC, DBPFP	AOP was only effective at reducing DBPs at UV fluences >1000 mJ cm^{-1} and initial H_2O_2 concentrations of about or greater than 23 mg L^{-1}. Combined AOP-BAC treatment showed significant reductions.	Toor and Mohseni (2007)

(Continued)

Table 6.2 UV light based AOPs—cont'd

Target compounds	Matrix	AOP features	Parameters monitored	Summary of results	References
NOM and synthetic organic compounds	Groundwater and NOM-free water	LP UV at 254 nm	UV_{254} and UV_{310}, TOC, color, specific compounds	NOM decomposition kinetics was substantially enhanced with oxidation over time. Progressive decrease in $^{\cdot}OH$ radical scavengers and UV adsorbing species may increase NOM decomposition rates.	Song et al. (2008)
DBP precursors (especially nitrosamines)	Source waters from several locations	MP UV	TOC, UV_{254}, color	AOP oxidation treatment may be a source of nitroso-DBP precursors.	Zhao et al. (2008)

Model compounds (analogues of LMM and hydrophilic NOM)		N-lamp 254 nm	DOC, BDOC, HAAFP, specific compounds	DOC removal was 91% with a high fluence of 47–48 J cm^{-2} and 78% with UVC dose of 21 J cm^{-2}. Oxidation degraded both hydrophilic and hydrophobic surrogates. Partial biodegradation increased HAAFP, especially DCAAFP.	Bond et al. (2009)

(Continued)

Table 6.2 UV light based AOPs—cont'd

Target compounds	Matrix	AOP features	Parameters monitored	Summary of results	References
NOM	Surface water	LP UV at 254 nm	BDOC, TOC, SUVA	Partial mineralization of NOM occurred. This led to a reduction in aromatic structures and the degree of substitution of aromatic species, and to BDOC formation, including formaldehyde and acetaldehyde.	Sarathy and Mohseni (2010, 2011)
DBP precursors		LP UV and MP UV	SUVA, TOC, THM	Minor mineralization of NOM.	Metz et al. (2011)

VUV					
NOM	Raw surface water	VUV at 254 and 185 nm	DOC, BDOC, fractionation, THMFP, HAAFP	THMFP increased at VUV doses <40 J cm^{-2} but reduced higher fluences. HAAFP reduced by VUV irradiation. Concentrations of individual THM and HAA decreased with increasing radiation, but bromoform increased linearly.	Buchanan et al. (2006)
		LP UV at 254 and 185 nm	UV_{254}	NOM was reduced by 42–46%. VUV photodegradation was over 10 times faster than basic UV.	Dobrovic et al. (2007)

(Continued)

Table 6.2 UV light based AOPs—cont'd

Target compounds	Matrix	AOP features	Parameters monitored	Summary of results	References
LMM and hydrophilic NOM surrogates	Model compounds in solution referring to LMM and hydrophilic NOM	H-lamp 254 and 185 nm	DOC, BDOC, HAAFP, specific compounds	DOC removal was 97% with a UVC dose of 47–48 J cm^{-2} and 58% with 21 J cm^{-2}. Oxidation degraded both hydrophilic and hydrophobic surrogates. Partial oxidation and biodegradation increased HAAFP, especially DCAAFP.	Bond et al. (2009)
Taste and odor compounds (geosmin and 2-methylisoborneol (MIB))	Ultrapure and raw water	LP UV at 254 and 185 nm	DOC, UV_{254}, specific compound	In raw water, DOC and SUVA decreased by 6% and 18%, respectively. NOM acted as a scavenger for geosmin and MIB reduction, although they were oxidized quite well. Nitrite formed as a by-product.	Kutschera et al. (2009)

NOM	Surface water	VUV Hg lamp	TOC, HPSEC	VUV was much more efficient than UV/H_2O_2. Carbonates and bicarbonates were important in radiation propagation.	Imoberdorf and Mohseni (2011)

ADOC, adsorbable dissolved organic carbon; AOC, assimilable organic carbon; BDOC, biodegradable dissolved organic carbon; pCBA, parachlorobenzoic acid; DCAA, dichloroacetic acid; DOC, dissolved organic carbon; FTIR, Fourier transform infrared spectroscopy; HA, humic acid; HAAFP, haloacetic acid formation potential; HPSEC, high pressure size exclusion chromatography; IC, ion chromatography; LMM, low molecular mass; LP, low pressure; MP, medium pressure; MWD, molecular weight distribution; NOM, natural organic matter; NPDOC, nonpurgeable dissolved organic carbon; THM, trihalomethane; THMFP, trihalomethane formation potential; TOC, total organic carbon.

electro-Fenton and photoelectro-Fenton processes, especially in wastewater and contaminated soil treatment (Jiang and Zhang, 2007). Compared to the conventional methods, the electro-Fenton process produces H_2O_2 electrochemically via oxygen reduction on the cathode surface. Further reactions occur between H_2O_2 and Fe^{2+}, as in the conventional technique.

With UV/Fe(III) treatment, HMM compounds were formed as an agglomeration of the organic molecules by Fe, with a significant level of LMM compounds remaining even after a long irradiation time (Sanly et al., 2007). Thus, the UV/Fe(III) process was ineffective unless H_2O_2 was added, resulting in more efficient NOM removal across the whole range of MM. This, in turn, suggests that the UV/Fe(III)/H_2O_2 system has the potential to remove organics with a broad range of MMs (Sanly et al., 2007; Moncayo-Lasso et al., 2009). Even though high NOM removal rates were detected, the remaining organic compounds appear fairly reactive to chlorine, thus resulting in low reduction of THMFP. The presence of iron may have been responsible for the elevated THMFP values (Sanly et al., 2007). Nevertheless, efficient DBPFP reduction has been detected with an irradiation time of over 90 min (Nie et al., 2010). While the optimal pH for Fenton processes has been confirmed at about pH 4 in several cases (Park and Yoon, 2007; Sanly et al., 2007; Moncayo-Lasso et al., 2008), 254 nm absorption decay and TOC reduction is also most effective in the pH range of 3–7 (Goslan et al., 2006). This is more feasible and cost effective for full-scale treatment applications, since no pH adjustment would be necessary.

Solar light irradiation has gained advocates for environmental and economic reasons (Malato et al., 2007, 2009; Moncayo-Lasso et al., 2008). Solar light has been used in photo-Fenton processes, e.g., to inactivate bacteria, by Moncayo-Lasso et al. (2009), who demonstrated that NOM can be partially degraded at natural water pH by solar photo-Fenton. A significant limitation to the application of Fenton and photo-Fenton processes has been the low optimal pH (<3), which most often requires acidification and neutralization steps. Fenton based AOPs are summarized in Table 6.3.

6.5 HETEROGENEOUS PHOTOCATALYSIS AND CATALYTIC OXIDATION

Heterogeneous photocatalysis employs several different semiconductors such as TiO_2, ZnO, ZnS, WO_3, and $SrTiO_3$ (Hoffmann et al., 1995; Pirkanniemi and Sillanpää, 2002). TiO_2 particles are the most commonly used photocatalysts, due to their high photocatalytic activity and

Table 6.3 Fenton based AOPs

Target compounds	Matrix	AOP features	Parameters monitored	Summary of results	References
Fenton and photo-Fenton					
NOM	Raw water	Fenton's reagent/LP UV at 254 nm	DOC, UV_{254}, HPSEC	Efficient NOM removal at pH 3–7. DOC reduction after 20 min irradiation 82 (UV_{254}) and 96% (DOC) by Fenton reaction (FR), and after 1 min at 31 (UV_{254}) and 88% (DOC) by photo-FR (PFR), respectively. The DOC reduction was much faster for PFR. UV_{254} reduction was faster for FR. H_2O_2 and Fe^{2+} interfered with the UV_{254} measurement.	Goslan et al. (2006)

(Continued)

Table 6.3 Fenton based AOPs—cont'd

Target compounds	Matrix	AOP features	Parameters monitored	Summary of results	References
HA	Synthetic HA solution (Aldrich HA)	Fenton's reagent/ LP UV, mineral particles	DOC, UV-HA, THMFP, resin fractionation, oxidation byproducts	Modified-Fe-ions improved the H_2O_2 degradation rate in Fenton and photo-Fenton processes. Adding glass improved THM removal and decomposition rate of H_2O_2.	Park and Yoon (2007)
	Synthetic HA solution (Fluka HA)	Fenton's reagent/ UVA with maximum centered at 365 nm	DOC, UV_{254}, HPSEC, fractionation	More than 80% and 90% DOC and UV_{254} removal, respectively, over 2.5 h. Addition of H_2O_2 enhanced removal kinetics. The CHA and NEU fractions increased and were hardest to remove.	Sanly et al. (2007)
NOM	Natural source water	Fenton's reagent	DOC, UV_{254}, fractionation, HPSEC	Potential method for combined oxidation of any formed DBPs, poor removal of recalcitrant NOM.	Jarvis et al. (2008)

HA	Synthetic HA solution (Aldrich HA)	Fenton's reagent/UV	TOC, UV_{254}, HPSEC, HA concentration	Absorbance of HA disappeared after 8 h irradiation and TOC reduced 80% after 15 h. HMM compounds were preferentially removed.	Katsumata et al. (2008)
NOM, and especially DBP precursors	Natural river water before and after sand filtration, river water spiked with NOM model compounds	Fe^{3+} (1.0 mg L^{-1})/H_2O_2/sunlight	TOC, specific compounds, UV_{254}	Light/H_2O_2: UV_{254} reduction higher (c. 68%, 80%) than TOC reduction (c. 55%, 62%), before and after filtration, respectively. Addition of 1 mg L^{-1} Fe^{3+} significantly increased UV_{254} and TOC removal (70–90%). Prechlorination filtration led to almost total NOM removal, and remaining TOC was suitable to prevent THM formation.	Moncayo-Lasso et al. (2008)

(Continued)

Table 6.3 Fenton based AOPs—cont'd

Target compounds	Matrix	AOP features	Parameters monitored	Summary of results	References
Fenton and photo-Fenton					
Organic probes (benzoate and methanol)	Synthetic solutions	Fenton's reagent/ $PW_{12}O_{40}^{3-}$	Specific compound	Novel homogenous Fentonlike system capable of oxidizing recalcitrant organic compounds at neutral pH.	Lee and Sedlak (2009)
NOM and *Escherichia coli*	Natural river water after sand filtration	Fenton's reagent/ sunlight	TOC, *E. coli* inactivation	Addition of low conc. of Fe^{3+} in the presence of H_2O_2 at near-neutral pH using CPC solar reactor led to simultaneous decrease of TOC (55%) and total *E. coli* inactivation (10^5–10^6 CFU mL^{-1}).	Moncayo-Lasso et al. (2009)

HA	Synthetic HA solution (Aldrich HA)	Iron oxides/Fe^0/H_2O_2/HP-UV	TOC, FTIR, DBPFP, fractionation, MWD by UF	The core–shell structure of nanoparticles of iron oxides coated on Fe^0 (nIOCl) produced a highly efficient photo-Fenton catalyst for HA degradation. After 90 min irradiation, about 62% of DOC was removed and DBPFP almost disappeared.	Nie et al. (2010)
NOM	Surface water	Solar photo-Fenton	TOC, THM	Solar photo-Fenton was effective for the transformation and partial mineralization of NOM.	Moncayo-Lasso et al. (2012)

CFU, colony forming units; CHA, hydrophilic charged; DBP, disinfection by-products; DOC, dissolved organic carbon; FTIR, Fourier transform infrared spectroscopy; HA, humic acid; HPSEC, high pressure size exclusion chromatography; HMM, high molecular mass; MWD, molecular weight distribution; NOM, natural organic matter; PEC, photoelectrocatalysis; THM, trihalomethane; THMFP, trihalomethane formation potential; TOC, total organic carbon; UF, ultrafiltration.

environmentally benign properties, though several others have been investigated. TiO_2 in anatase form is stable, efficient, inexpensive and nontoxic. UV irradiation of the TiO_2 induces the excitement of electrons from the valence band to the conduction band, leading to the creation of highly oxidative holes on the valence band and the formation of radicals, in particular $^{\bullet}OH$ radicals. Organic compounds are degraded by holes in the TiO_2 surface and by radicals in the bulk solution (Hoffmann et al., 1995; Carp et al., 2004). The parameters affecting the oxidation are catalyst concentration, light wavelength and intensity, water matrix, solution pH, and the addition of H_2O_2 as an extra oxidant to promote reactions (Hand et al., 1995; Bekbölet and Özkösemen, 1996; Eggins et al., 1997). Uyguner-Demirel and Bekbolet (2011) have recently reviewed the literature on the photocatalytic degradation of NOM. The major advantage of TiO_2 photocatalysis is optical adsorption in the near UV region in comparison to UVC-driven AOPs (Uyguner-Demirel and Bekbolet, 2011).

Photocatalytic oxidation has been proposed as a good candidate for NOM degradation, especially for the removal of DBPs and their precursors (Valencia et al., 2014). This is due to the three-way removal of compounds: oxidation by $^{\bullet}OH$ radicals, reductive dechlorination by superoxide radicals, and physical adsorption by TiO_2 (Hand et al., 1995; Bekbölet and Özkösemen, 1996; Eggins et al., 1997; Wiszniowski et al., 2002; Gerrity et al., 2009). Adsorption has been found to play a prominent role in the photocatalytic degradation of NOM, especially under acidic pH conditions when carboxylate surface groups of NOM are adsorbed onto TiO_2 (Wiszniowski et al., 2002; Liu et al., 2008a). This adsorption of NOM onto the TiO_2 surface seems to occur rapidly (Huang et al., 2008) leading to the conclusion that hydrophobic NOM is preferably adsorbed onto TiO_2 surface, and more hydrophilic intermediate oxidation products are then released. In photocatalytic NOM removal, pH plays a central role, determining the mechanism but not the sequence of photocatalytic degradation (Valencia et al., 2012). Copper greatly inhibits the larger molecular weight fraction of NOM in particular and reduces the formation of formic and oxalic acids (Tercero Espinoza et al., 2011). Several anions lessened photocatalytic NOM removal at elevated concentrations (Uyguner-Demirel et al., 2013). Reductive dechlorination of, e.g., THMs can also be a beneficial use of photocatalytic oxidation in full-scale water treatment at targeted locations (Gerrity et al., 2009). Though TiO_2 based photocatalytic oxidation has significant potential for NOM removal, this process will only become economically viable if its energy requirements are significantly reduced (Philippe et al., 2010). Uyguner-Demirel and Bekbolet (2011) concluded that TiO_2 based photocatalytic oxidation of NOM diminishes DBP formation capacity.

Photocatalytic processes can be conducted by using a slurry of fine particles of the semiconductor. Nevertheless, the use of catalyst in slurry form requires an additional treatment step to remove it from the treated effluent, which can make the process expensive (Murray and Parsons, 2006; Klavarioti et al., 2009). TiO_2 slurry can be separated through a microfiltration (Fu et al., 2006b; Choo et al., 2008; Benotti et al., 2009) or ultrafiltration membrane (Fu et al., 2006a). The powder catalyst may penetrate into the pores of the membrane, however, which in turn reduces the TiO_2 concentration and blocks membrane pores. TiO_2/Al_2O_3 microspheres consisting of nanosized catalyst crystals could present one solution to this problem (Xu et al., 2009).

Another alternative to filtration is to immobilize the catalyst on a suitable support matrix. This, however, can reduce the surface area available for reactions compared to suspended systems. To make the process viable, the catalyst therefore needs to be immobilized on a high surface area material. Different coating matrices have been studied, including nonwoven webs (Gumy et al., 2006), nonwoven porous fabrics such as silica-supported titania (Navalon et al., 2009), and magnetic core matrices with silica layer (Chen et al., 2009). Durable supported films can be obtained in various ways, for example by the sol–gel process (Murray and Parsons, 2006; Chang et al., 2009) and the atomic layer deposition (ALD) technique (Vilhunen et al., 2009). TiO_2 was synthesized by the sol–gel method coupled with a solvothermal technique for photocatalytical NOM degradation (Valencia et al., 2012). The resulting differences in crystallographic and morphological properties did not affect the selectivity of NOM photodegradation, but the synthesized material had higher efficiency than Degussa P-25.

The aromaticity of the organic matter is reduced after limited photocatalysis, as indicated, e.g., by lowered specific UV absorbance (SUVA) values. The applied energy values, however, are insufficient to demonstrate organic mineralization (Gerrity et al., 2009; Liu et al., 2010). Higher TiO_2 concentrations have achieved greater DOC removal rates, but also a possible increase in THMFP (Fu et al., 2006a; Gerrity et al., 2009). Beyond a certain level, additional TiO_2 particles cannot be involved because of increased turbidity and thus decreased UV light penetration. Extended photocatalysis ($>10\,kWh\,m^{-3}$) has yielded high DOC removal and mineralization, and nearly complete reduction of UV_{254} and SUVA values (Gerrity et al., 2009). Even further extended illumination ($160\,kWh\,m^{-3}$) significantly reduced

THMFP, although the energy requirements of this experiment make it infeasible for treatment plant application (Gerrity et al., 2009).

Renewable energy sources such as sunlight can aid photocatalysis. Solar irradiation has been observed to change the characteristics on NOM (Frimmel, 1998; Chow et al., 2004), resulting in several studies of the use of natural sunlight irradiation for water treatment (Malato et al., 2007, 2009). Sunlight can provide a low-cost, sustainable, and simple method of treating drinking waters (Ljubas, 2005), especially in developing countries with consistently sunny climates. Solar photocatalysis has been proposed mainly for water disinfection and detoxification (Malato et al., 2007; Rincón and Pulgarin, 2007; Malato et al., 2009). Tercero Espinoza et al. (2009a, 2009b) have used a solar UV simulator for photocatalytic research. The disadvantage of solar light is that only UVA irradiation can be used for catalyst photoactivation, and the solar irradiation reaching the surface of the earth contains only 3–5% UVA (Ljubas, 2005). Thus, there is a great incentive to extend the absorbance wavelength range of TiO_2 to the visible region without losing photocatalytic activity. This has been done by doping TiO_2 with non-metallic elements such as nitrogen, sulfur, boron, carbon, and selenium (Comninellis et al., 2008; Malato et al., 2009) or by noble metals like Ag, Pt, Ni, Cu and Pd (Chen et al., 2009; Kumar et al., 2009; Malato et al., 2009).

Aqueous H_2O_2 decomposes over heterogeneous catalysts including metals such as Fe, Cu, Pt, Ti, and Ni and metal oxides immobilized on various support materials including sand, silica, zeolites, and alumina (Pirkanniemi and Sillanpää, 2002). As in Fenton reactions, the decomposition of peroxide through interactions with the surface sites of such catalysts results in the formation of strong oxidants including $^{\bullet}OH$ radicals (Kitis and Kaplan, 2007). Contaminated soil and groundwater remediation has been attempted with various oxidants, such as sulfate radical-based AOPs, which use peroxymonosulfate and persulfate as precursor oxidants (Anopsitakis and Dionysiou, 2004). The application of a chelating agent noticeably improved the effectiveness of Fentonlike reactions (Lee and Sedlak, 2009; Rastogi et al., 2009).

Natural pumice has been coated with iron oxides and sand for catalytic oxidation (Kitis and Kaplan, 2007). An investigation into the influence of mineral particles added to the Fenton processes concluded that although bentonite enhanced NOM removal, THMFP also increased (Park and Yoon, 2007). The addition of glass improved both THM removal and the of H_2O_2 decomposition rate.

Catalytic ozonation can double the NOM removal efficiency achieved by ozonation alone (Legube and Vel Leitner, 1999; Kasprzyk-Hordern et al., 2005; Zhang and Jian, 2006), Catalytic ozonation with MnO_2 improved the elimination of NOM measured as TOC and chemical oxygen demand (COD) (Alsheyab and Muñoz, 2007). Furthermore, catalytic wet peroxide oxidation, developed with Al/Fe-pillared clays, has proved to be an efficient Fentonlike method, achieving over 95% DOC removal in 4h (Galeano et al., 2012). Zhou et al. (2014) also synthesized efficient TiO_2–graphene composites with different graphene weight addition ratios by a solvothermal process for the removal of FA under UV light. Catalytic studies are summarized in Table 6.4.

6.6 ULTRASOUND IRRADIATION AND E-BEAM IRRADIATION

Ultrasonic (US) irradiation can also be deployed for NOM removal (Nagata et al., 1996). It is attractive because of its application simplicity. It has been shown that high-power ultrasound produces strong cavitation in aqueous solution, and the violent collapse of the cavitation bubble causes shock waves and reactive free radicals. These effects should contribute to both the physical disruption and inactivation of microbial structures and the decomposition of organic compounds (Naddeo et al., 2007). US also appears to influence the chemical degradation of humic matter in the cavitation bubbles by oxidation and the physical aggregation of HA fragments (Naddeo et al., 2007).

The effect of NOM fouling on membrane filtration may be mitigated by ultrasound (Chen et al., 2006; Tran et al., 2007). A shorter sonication time appears to reduce flux decline and degrade NOM to a less adsorbable form, but this beneficial effect is diminished as the sonication time is extended.

To combat the low rate of sonochemical decomposition of aromatic compounds in aqueous solutions, the combined treatments, such as US/UV and US/O_3, have also been applied. The combination of US with other oxidation methods has resulted in higher rates of NOM oxidation and mineralization (Goncharuk et al., 2008). US and O_3 have proved more effective and energy efficient for HA mineralization in combination than alone (Stepniak et al., 2012). Higher US intensities improved the efficiency. US/O_3 is one of the more highly efficient O_3 based NOM removal methods (Ziylan and Ince, 2013).

HAs were removed by electron beam irradiation with the addition of persulfate (Roshani and Leitner, 2011). The advantages of persulfate include

Table 6.4 Catalytic studies

Target compounds	Matrix	AOP features	Parameters monitored	Summary of results	References
UV/TiO_2					
FA	Synthetic FA solution (Beijing BioChem)	Suspended Degussa TiO_2/LP UV	TOC, UV_{254}	TOC removal improved with low pH, increased additive concentration ($K_2S_2O_8$) and relatively high light intensity.	Fu et al. (2006a)
NOM	Natural raw water and Suwannee river NOM (SRNOM)	LP-UV/suspended Degussa TiO_2	TOC, HPSEC	Operating values are feedwater dependent. With low-NOM water requires a pH 4–5 and low TiO_2 concentration ($0.1\,g\,L^{-1}$).	Le-Clech et al. (2006)
DBP precursors	Synthetic HA solution (Aldrich HA)	UVA at 365 nm/ thin-film-coated TiO_2 slides	DOC, UV_{254}, SUVA, HPSEC	UV_{254} reduced c. 48%. Immobilized catalysts need a larger surface for removal in a slurry reactor.	Murray and Parsons (2006)

NOM	Raw water	Suspended Degussa TiO_2/UV	DOC, UV_{254}, UV_{280}, THMFP, fluorescence	Photocatalytic oxidation after enhanced coagulation increased NOM degradation by <5–32% measured as DOC. UV_{254} and SUVA decreased c.10–33% and 15–47%, respectively. Differences depended on the related properties of NOM.	Uyguner et al. (2007a)
HA	Synthetic HA solution (Aldrich HA)	Suspended Degussa TiO_2/UV	UV_{254}, $Color_{436}$, fluorescence	AOP pretreatment decreased coagulation removal efficiency by 15%, while pretreatment to AC filtration increased removal by c. 45%.	Uyguner et al. (2007b)

(Continued)

Table 6.4 Catalytic studies—cont'd

Target compounds	Matrix	AOP features	Parameters monitored	Summary of results	References
NOM	Natural lake water and synthetic HA water (Aldrich HA)	Degussa TiO_2/8-W UV-lamp/MF membrane/ ferrihydrite particles	TOC, UV_{254}	Addition of ferrihydrite enhanced HA removal in synthetic water, but lake water NOM with low SUVA and high pH values was adversely affected. Photocatalytic degradation controlled membrane fouling well.	Choo et al. (2008)
	Suwannee river NOM (SRNOM)	Suspended Degussa TiO_2/LP UV at 254 nm	TOC, UV_{254}, HPSEC	At 40 min reaction time, DOC and UV_{254} were reduced by 50% and 85%, respectively. At 120 min reaction time, more than 80% of NOM as TOC was removed.	Huang et al. (2008)

HA	Synthetic HA solution (Fluka HA)	Suspended Degussa TiO_2/UVA with maximum centered at 365 nm	DOC, UV_{254}, HPSEC, resin fractionation, THMFP, zeta potential analysis	After 150 min, UV_{254} reduced by over 90%, DOC reduced by only c. 80%. THMFP also lowered, and H_2O_2 addition enhanced the reaction rate. HPSEC showed that AMW shifted toward LMM compounds. Hydrophobic fraction removed efficiently, NEU fraction most difficult to remove.	Liu et al. (2008a)

(Continued)

Table 6.4 Catalytic studies—cont'd

Target compounds	Matrix	AOP features	Parameters monitored	Summary of results	References
NOM	Surface raw water	Suspended Degussa TiO_2/UVA with maximum centered at 365 nm	DOC, UV_{254}, HPSEC, resin fractionation, THMFP, HAAs, aldehydes, and ketones formation	Higher UV_{254} reduction than DOC after 120 min (over 90 and c. 70%, respectively). H_2O_2 addition enhanced initial kinetics of the degradation reactions, but residual concentrations of DOC remained similar. LMM (200–1000 Da) matter remained (hydrophilic charged or neutral molecules). THMFP and HAAFP reduced significantly.	Liu et al. (2008b)
NOM, model compounds, DHBA, salicylic acid and bromide	Lake water and model compound solution	Suspended Degussa TiO_2/LP UV	DOC, IC, AOX, THM, DHBA concentration	No formation of BrO_3^- and THMs in DOC containing water was detected with up to 45 min of irradiation. With higher initial bromide concentrations, adsorbable organic Br increased.	Tercero Espinoza and Frimmel (2008)

HA	Synthetic HA solution (Aldrich HA)	Suspended Degussa TiO_2/UV/MF	TOC	Photocatalytic oxidation and MF process effectively degraded HA, membrane fouling was well controlled.	Bai et al. (2009)
DBP precursors	Surface raw water	Degussa TiO_2 slurry/Photo-Cat Lab® reactor with LP UV bulbs	DOC, UV_{254}, SUVA, TTHM, THMFP	Limited photocatalysis ($<5\,kWh\,m^{-3}$) achieved reductions in DOC and UV_{254} (nearly 50%) but led to increased chlorine demand and THMFP. THMFP also started to decrease when energy consumption exceeded $20\,kWh\,m^{-3}$.	Gerrity et al. (2009)
NOM		Suspended Degussa TiO_2/UVA with maximum centered at 365 nm	DOC, UV_{254}, HPSEC, resin fractionation, THMFP	UV_{254} reduced over 95% after 150 min. DOC reduced only 80–85%. HPSEC showed that AMW shifted toward LMM compounds. The hydrophilic fraction dominated the organic distribution, NEU fraction was the most difficult to remove.	Liu et al. (2010)

(Continued)

Table 6.4 Catalytic studies—cont'd

Target compounds	Matrix	AOP features	Parameters monitored	Summary of results	References
NOM model compound, DHBA and bromide	Model compound and bromide solution	Suspended Degussa TiO_2/solar UV simulator	DOC, THMs, DHBA concentration	Considerable variability in the extent and time scale of brominated by-product formation depending on the raw water and design parameters.	Tercero Espinoza et al. (2009a)
NOM, especially HA	Humic-rich lake water	Suspended Degussa TiO_2/solar UV simulator	SEC-DOC/UV, IC, bacterial incubation	Sequential photocatalytic NOM degradation from the larger to LMM fractions. Bioavailable products formed, e.g., LMM organic acids.	Tercero Espinoza et al. (2009b)
HA	Synthetic HA solution	TiO_2/Al_2O_3 microsphere/LP-UV/ultrafiltration	HA concentration, TMP	TiO_2/Al_2O_3 microspheres showed good photocatalytic activity, mechanical strength, and fouling control capacity.	Xu et al. (2009)

NOM	Natural river water and synthetic HA solution (Aldrich HA)	Degussa TiO_2/8-W UV-lamp/MF membrane/iron oxide particles (IOP)	TOC, UV_{254}, FE-SEM, TMP	Additional IOP precoat layers on the membrane surface did not aggravate membrane fouling, adsorptive removal of photocatalyzed NOM enhanced DOC removal efficiently, especially for high DOC waters.	Yao et al. (2009)
HA	Synthetic HA solution (Aldrich HA)	TiO_2 nanotube/LP-UV/membrane filtration	TOC, UV_{436}	Near complete mineralization (93–98%) of HA. Nanotubes were totally separated and recovered with membrane filtration, and reused efficiently.	Zhang et al. (2009)
NOM	Surface water	Degussa TiO_2	SEC, DOC	Cu and some other cations inhibited photocatalytic degradation of NOM.	Tercero Espinoza et al. (2011)
		Solar UV-light simulator	SEC, AOX, THM, DOC	pH determines the mechanism, but not the sequence of photocatalytic degradation.	Valencia et al. (2012)

(Continued)

Table 6.4 Catalytic studies—cont'd

Target compounds	Matrix	AOP features	Parameters monitored	Summary of results	References
HA	Prepared HA solution	Degussa TiO_2	DOC, UV_{254}, UV_{365}	Several anions at elevated concentrations inhibited photocatalytic NOM degradation.	Uyguner-Demirel et al. (2013)
O_3/UV/TiO_2					
NOM	Natural water, pretreated with coagulation—sand filtration	LP-UV/TiO_2 coated titanium sheet	TOC, UV_{254}, BDOC, SUVA	Combined process was much more effective than ozonation or photocatalysis. HMM compounds decreased faster than LMM compounds. Ozone dosage and retention time were the main parameters affecting removal.	Zhang and Jian (2006)

Photoelectrocatalysis					
FA	Synthetic FA solution (Beijing BioChem)	Ti/TiO_2 electrode/ LP-UV	TOC, SEM	Under optimal process parameters, 57% removal of FA was obtained.	Fu et al. (2007)
HA	Synthetic HA solution (Aldrich HA)	Thin-film TiO_2 photoelectrodes/ Xe–Hg arc lamp	TOC, UV_{254}	PEC improved photo-catalytic degradation of HA 2- to 4-fold, depending on system conditions.	Selcuk and Bekbolet (2008)

AC, activated carbon; AOP, advanced oxidation processes; DOC, dissolved organic carbon; FA, fulvic acid; LP, low pressure; HA, humic acid; HP, high pressure; HPSEC, high pressure size exclusion chromatography; IC, ion chromatography; BDOC, biodegradable dissolved organic carbon; DHBA, 2,4-dihydroxybenzoic acid; NEU, neutral; TMP, transmembrane pressure; FE-SEM, field emission scanning electron microscope; PEC, photoelectrocatalysis; TOC, total organic carbon; NOM, natural organic matter; THM, trihalomethane; THMFP, trihalomethane formation potential.

high solubility, low cost, environmentally friendly end products, and high stability. Persulfate anions can be activated to generate sulfate radicals, one of the strongest oxidizing species (Roshani and Leitner, 2011). Yet aromatic structures limit the positive impact of persulfate on e-beam irradiation. US studies are summarized in Table 6.5.

6.7 CONCLUSIONS

Different raw waters contain variable amounts of NOM. The characteristics of NOM also vary with geological conditions and the surrounding watershed. An increase in the amount of NOM has been observed worldwide in the last few years. Seasonal variability in the same location and natural disasters such as floods or droughts present even more challenges to water purification facilities. New and improved methods are thus urgently needed to reduce NOM, hence DBPs, in drinking water.

Hydrophobic and HMM NOM compounds appear more amenable to oxidation with AOPs such as photocatalysis or UV/H_2O_2. The photo-Fenton process also removes LMM organics (Murray and Parsons, 2004), which are the most difficult to remove in traditional treatment. The efficiencies of AOPs for DOC removal are as follows: UV/H_2O_2 < Fenton < photo-Fenton (Goslan et al., 2006; Jarvis et al., 2008). The addition of chelating agents further improves the efficiency of Fenton-like processes (Lee and Sedlak, 2009), as do various catalyst solutions such as iron oxides coated on Fe^0 core shell nanospheres (Nie et al., 2010).

Photocatalysis has been the subject of a great deal of research (Malato et al., 2009; Choi et al., 2010). Catalyst separation, recovery, and reuse are significant obstacles to the practical applications of TiO_2 powder heterogeneous photocatalysis, mainly due to the small particle size of the TiO_2. More advanced catalyst separation techniques have been developed, such as membrane filtration, although membrane fouling is a problem. Another alternative is to immobilize the catalyst on suitable support matrices. Different matrices and TiO_2 coating techniques such as ALD have emerged recently (Vilhunen et al., 2009; Chen et al., 2009).

Current advances in photocatalytic oxidation include nanotechnology (Choi et al., 2010), which uses, for example, titanium-derived 1D nanomaterials such as titanate nanotubes and nanowires (Doong and Chiang, 2008; Zhang et al., 2009) or high-strength TiO_2 fiber (Giri et al., 2008). The latter technique looks promising for overcoming the reduced efficiency of an immobilized TiO_2 catalyst. TiO_2 doped with noble metals such as platinum

Table 6.5 Ultrasonic studies

Target compounds	Matrix	AOP features	Parameters monitored	Summary of results	References
NOM	Synthetic HA solution (Aldrich HA)	20 kHz ultrasonic probe/membrane filtration	TOC, UV, FTIR	Ultrasound significantly reduced membrane fouling.	Chen et al. (2006)
		Horn sonotrode at 20 kHz	TOC, UV_{254}	20 min sonication at 42 W cm^{-2} provided the best TOC removal.	Naddeo et al. (2007)
	Natural surface water	Horn sonotrode at 20 kHz, sonication at 800, 1400, or 1500 W/MF membrane	Fouling	The shorter sonication produced smaller molecular NOM, and longer sonication may encourage aggregation or readier adsorption.	Tran et al. (2007)
HA	Synthetic HA solution (Aldrich HA)	Ultrasonic generator of 24 kHz	TOC, DOC, SUVA, UV_{254}	US/O_3 enhanced HA mineralization.	Stepniak et al. (2012)

DOC, dissolved organic carbon; FTIR, Fourier transform infrared spectroscopy; HA, humic acid; TOC, total organic carbon.

significantly improves photocatalytic activity (Chen et al., 2009). Alternative semiconductors to TiO_2, such as ZnO (Akyol and Bayramoglu, 2010), have also been considered, although these are not generally considered to be as attractive for application as TiO_2.

Due to the high energy demand of conventional UV lamps, other sources of UV light appear increasingly attractive, and the VUV process has proved its efficiency. Sunlight is environmentally friendly and cost effective but less efficient (Ljubas, 2005; Malato et al., 2007, 2009). To increase efficiency, the absorbance wavelength range of TiO_2 can be extended by doping it with nonmetallic elements such as nitrogen, sulfur, boron, carbon, and selenium (Comninellis et al., 2008; Chang et al., 2009; Chen et al., 2009; Yang et al., 2010). Some new catalyst alternatives more efficient than TiO_2, which are more applicable on the solar spectrum (Shifu et al., 2009; Song et al., 2009), and new photolysis oxidants could advance the process (Anopsitakis and Dionysiou, 2004; Criquet and Vel Leitner, 2009).

Photoelectrocatalysis uses TiO_2 coated surfaces in the electrochemical process to enhance the removal of NOM significantly further than photocatalysis (Selcuk et al., 2006; Selcuk and Bekbolet, 2008). In bromide-containing waters, however, bromate formation was observed during oxidation.

As an alternative to oxidative degradation, enzyme-mediated oxidative coupling processes are a potentially cost effective. These processes reconfigure the NOM molecules to HMM compounds, thus enhancing NOM removal, for example by coagulation, thus reducing DBPFP (Lu and Huang, 2009).

To enable better control of the treatment processes, further research is needed on the oxidation product formation and reaction mechanisms of AOPs in conjunction with other treatments. AOPs in combination with other treatments may enhance the efficiency of NOM removal (Buchanan et al., 2006; Toor and Mohseni, 2007; Uyguner et al., 2007a,b; Huang et al., 2008), and often have better NOM removal efficiency than individual treatments (Lamsal et al., 2011). Coagulation prior to oxidation improves AOP treatment by removing most of the HMM and hydrophilic organic matter present (Uyguner et al., 2007a). AOP treatment may reduce harmful DBPs more effectively when combined with BAC (Toor and Mohseni, 2007). The more biodegradable compounds formed during an AOP can readily be utilized as substrates by microorganisms in the BAC. Moreover, in full-scale application, UV-AOP treatment has proved to be very robust and a reliable barrier against pathogenic microorganisms and organic micropollutants with postfiltration, which was found to efficiently remove residual H_2O_2, biodegradable compounds, and any formed nitrite (Kruithof et al., 2007). AOP combined with membrane filtration has also been found to improve membrane performance efficiently.

AOPs are among the most studied and promising technologies for drinking water purification and disinfection. Although total reduction of NOM has not been achieved with AOPs, several studies have shown efficient NOM reduction and mitigation of DBP formation. Besides AOP process efficiency, the nature of residual NOM after AOP processes also needs consideration to identify the reactive precursors of DBPs (Molnar et al., 2012a,b).

Due to their nonselective nature, AOPs may be suitable processes across the range of NOM (Bond et al., 2011). It should be emphasized that AOPs have site-specific effects, so pilot-scale and full-scale tests must be conducted to obtain the greatest removal efficiency (Rahman et al., 2010). Studies are often conducted under conditions that are not economically feasible in commercial applications (Tuhkanen, 2004; Toor and Mohseni, 2007; Jarvis et al., 2008). The full-scale application of AOPs in drinking water treatment plants is still limited because of high cost, high level of pretreatment required, a lack of experience, and operational difficulties. Nevertheless, the interest attracted by AOPs implies that their application to full-scale treatment will increase (Parsons and Byrne, 2004; Comninellis et al., 2008). The objective of the treatment needs to be defined, whether it is to enhance biodegradability or mineralize organic compounds, and the optimal conditions for each case must be determined according to the type and amount of organic compounds present in the water.

REFERENCES

Agustina, T., Ang, H., Vareek, V., 2005. A review of synergistic effect of photocatalysis and ozonation on wastewater treatment. J. Photochem. Photobiol. Photochem. Rev. 6, 264–273.

Akyol, A., Bayramoglu, M., 2010. Preparation and characterization of supported ZnO photocatalyst by zincate method. J. Hazard. Mater. 175, 484–491.

Alkan, U., Teksoy, A., Ateply, A., Bapkaya, H., 2007. Efficiency of the UV/H_2O_2 process for the disinfection of humic surface waters. J. Environ. Sci. Health A Tox. Hazard. Subst. Environ. Eng. 42, 497–506.

Alsheyab, M., Muñoz, A., 2006. Reducing the formation of trihalomethanes (THMs) by ozone combined with hydrogen peroxide (H_2O_2/O_3). Desalination 194, 121–126.

Alsheyab, M., Muñoz, A., 2007. Comparative study of ozone and MnO_2/O_3 effects on the elimination of TOC and COD of raw water at Valmayor station. Desalination 207, 179–183.

Anopsitakis, G., Dionysiou, D., 2004. Transition metal/UV-based advanced oxidation technologies for water decontamination. Appl. Catal. B Environ. 54, 155–163.
Bai, H., Zhang, X., Pan, J., Sun, D., Shao, J., 2009. Combination of nano TiO_2 photocatalytic oxidation with microfiltration (MF) for natural organic matter removal. Water Sci. Technol. Water Supply 9, 31–37.
Bekbölet, M., Özkösemen, G., 1996. A preliminary investigation on the photocatalytic degradation of a model humic acid. Water Sci. Technol. 33, 189–194.
Benotti, M., Stanford, B., Wert, E., Snyder, S., 2009. Evaluation of a photocatalytic reactor membrane pilot system for the removal of pharmaceuticals and endocrine disrupting compounds from water. Water Res. 43, 1513–1522.
Bond, T., Goslan, E., Jefferson, B., Roddick, F., Fan, L., Parsons, S., 2009. Chemical and biological oxidation of NOM surrogates and effect on HAA formation. Water Res. 43, 2615–2622.
Bond, T., Goslan, E.H., Parsons, S.A., Jefferson, B., 2011. Treatment of disinfection by–product precursors. Environ. Technol. 32, 1–25.
Bose, P., Reckhow, D., 2007. The effect of ozonation on natural organic matter removal by alum coagulation. Water Res. 41, 1516–1524.
Buchanan, W., Roddick, F., Porter, N., 2006. Formation of hazardous by-products resulting from irradiation of natural organic matter: comparison between UV and VUV irradiation. Chemosphere 63, 1130–1141.
Carp, O., Huisman, C., Reller, A., 2004. Photoinduced reactivity of tianium dioxide. Progr. Solid State Chem. 32, 33–177.
Chang, P.-Y., Hsin, C.-H., Doong, R., 2009. Characterization and photocatalytic activity of vanadium-doped titanium dioxide nanocatalyst. Water Sci. Technol. 59, 523–530.
Chen, D., Weavers, L., Walker, H., Lenhart, J., 2006. Ultrasonic control of ceramic membrane fouling caused by natural organic matter and silica particles. J. Membr. Sci. 276, 135–144.
Chen, P., Rosenfeldt, E., Kullman, S., Hinton, D., Linden, K., 2007. Biological assessments of a mixture of endocrine disruptors at environmentally relevant concentrations in water following UV/H_2O_2 oxidation. Sci. Total Environ. 376, 18–26.
Chen, Y.-H., Chen, L.-L., Shang, N.-C., 2009. Photocatalytic degradation of dimethyl phthalate in an aqueous solution with Pt-doped TiO_2-coated magnetic PMMA micropheres. J. Hazard. Mater. 172, 20–29.
Chen, C., Zhang, X., Zhu, L., He, W., Han, H., 2011. Changes in different organic matter fractions during conventional treatment and advanced treatment. J. Environ. Sci. 23, 582–586.
Chin, A., Bérubé, P., 2005. Removal of disinfection by-product precursors with ozone-UV advanced oxidation process. Water Res. 39, 2136–2144.
Choi, H., Al-Abed, S.R., Dionysiou, D.D., Stathatos, E., Lianos, P., 2010. TiO_2-based advanced oxidation nanotechnologies for water purification and reuse. In: Abraham, M.A., Sheldon, R. (Eds.), Sustainability Science and Engineering, Book Series, vol. 2. In: Escobar, I.C., Schäfer, A.I. (Eds), Sustainable Water for the Future: Water Recycling versus Desalination. Elsevier, pp. 229–254.
Choo, K.-W., Tao, R., Kim, M.-J., 2008. Use of photocatalytic membrane reactor for the removal of natural organic matter in water: effect of photoinduced desorption and ferrihydrite adsorption. J. Membr. Sci. 322, 368–374.
Chow, C., Fabris, R., Drikas, M., 2004. A rapid fractionation technique to characterize natural organic matter for the optimization of water treatment processes. J. Water Supply Res. Technol. Aqua 53, 85–92.
Comninellis, C., Kapalka, A., Malato, S., Parsons, S., Poulios, I., Mantzavinos, D., 2008. Advanced oxidation processes for water treatment: advances and trends for R&D. J. Chem. Technol. Biotechnol. 83, 769–776.

Criquet, J., Vel Leitner, K., 2009. Degradation of acetic acid with sulfate radical generated by persulfate ions photolysis. Chemosphere 77, 194–200.
Dobrovic, S., Juretic, H., Ruzinski, N., 2007. Photodegradation of natural organic matter in water with UV irradiation at 185 and 254 nm: importance of hydrodynamic conditions on the decomposition rate. Sep. Sci. Technol. 42, 1421–1432.
Doong, R.-A., Chiang, L.-F., 2008. Coupled removal of organic compounds and heavy metals by titanate/carbon nanotube composites. Water Sci. Technol. 58, 1958–1992.
Duguet, J., Brodard, E., Dussert, B., Mallevialle, J., 1985. Improvement in the effectiveness of ozonation of drinking water through the use of hydrogen peroxide. Ozone Sci. Eng. 7, 241–253.
Eggins, B., Palme, F., Byrne, J., 1997. Photocatalytic treatment of humic substances in drinking water. Water Res. 31, 1223–1226.
Frimmel, F., 1994. Photochemical aspects related to humic substances. Environ. Int. 20, 373–385.
Frimmel, F., 1998. Impact of light on the properties of aquatic natural organic matter. Environ. Int. 24, 559–571.
Fu, J., Ji, M., Zhao, Y., Wang, L., 2006a. Kinetics of aqueous photocatalytic oxidation of fulvic acids in a photocatalysis-ultrafiltration reactor (PUR). Sep. Purif. Technol. 50, 107–113.
Fu, J., Min, J., Wang, Z., Jin, L., An, D., 2006b. A new submerged membrane photocatalysis reactor (SMPR) for fulvic acid removal using nano-structured photocatalyst. J. Hazard. Mater. 131, 238–242.
Fu, J., Zhao, Y., Wu, Q., 2007. Optimising photoelectrocatalytic oxidation of fulvic acid response surface methodology. J. Hazard. Mater. 144, 499–505.
Fujishima, A., 1971. Electrochemical photolysis of water at a semiconductor electrode. Nature 238, 37–38.
Fukushima, M., Tatsumi, K., Nagao, S., 2001. Degradation characteristics of humic acid during photo-Fenton processes. Environ. Sci. Technol. 35, 3683–3690.
Galeano, L., Bravo, P., Luna, C., Vicente, M., Gil, A., 2012. Removal of natural organic matter for drinking water production by Al/Fe-PILC-catalyzed wet peroxide oxidation: effect of the catalyst preparation from concentrated precursors. Appl. Catal. B Environ. 111–112, 527–535.
Gerrity, D., Ryu, H., Crittenden, J., Abbaszadegan, M., 2008. Photocatalytic inactivation of viruses using titanium dioxide and low-pressure UV light. J. Environ. Sci. Health A Tox. Hazard. Subst. Environ. Eng. 43, 1261–1270.
Gerrity, D., Mayer, B., Ryu, H., Crittenden, J., Abbaszadegan, M., 2009. A comparison of pilot-scale photocatalysis and enhanced coagulation for disinfection byproduct mitigation. Water Res. 43, 1597–1610.
Giri, R., Ozaki, H., Takanami, R., Taniguchi, S., 2008. Heterogenous photocatalytic ozonation of 2,4-D in dilute aqeuous solution with TiO_2 fiber. Water Sci. Technol. 58, 207–216.
Gjessing, E., Källqvist, T., 1991. Algicidal and chemical effect of u.v.-radiation of water containing humic substances. Water Res. 25, 491–494.
Glaze, W.H., Beltran, F., Tuhkanen, T., Kang, J.-W., 1992. Chemical models of advanced oxidation processes. Water Pollut. Res. J. Can. 27, 23–42.
Glaze, W., Peyton, G., Lin, S., Huang, R., Burleson, J., 1982. Destruction of pollutants in water with ozone in combination with ultraviolet radiation. 2. Natural trihalomethane precursors. Environ. Sci. Technol. 16, 454–458.
Glaze, W., Kang, J., Chapin, D., 1987. The chemistry of water treatment processes involving ozone, hydrogen peroxide and ultraviolet radiation. Ozone Sci. Eng. 9, 335–352.
Goncharuk, V., Malyarenko, V., Yaremenko, V., 2008. Use of ultrasound in water treatment. J. Water Chem. Technol. 30, 137–150.

Gonzalez, M., Oliveros, E., Wörner, M., Braun, A., 2004. Vacuum-ultraviolet photolysis of aqueous reaction systems. J. Photochem. Photobiol. C. Photochem. Rev. 5, 225–246.

Goslan, E., Gurses, F., Banks, J., Parsons, S., 2006. An investigation into reservoir NOM reduction by UV photolysis and advanced oxidation processes. Chemosphere 65, 1113–1119.

Gumy, A., Rincon, A.G., Hajdu, R., Pulgarin, C., 2006. Solar photocatalysis for detoxification and disinfection of water: different types of suspended and fixed TiO_2 catalysts study. Sol. Energy 80, 1376–1381.

von Gunten, U., 2003. Ozonation of drinking water: part I. Oxidation kinetics and product formation. Water Res. 37, 1443–1467.

Hand, D., Perram, D., Crittenden, J., 1995. Destruction of DBP precursors with catalytic oxidation. J. Am. Water Works Assoc. 87, 84–96.

Hoffmann, M., Martin, S., Choi, W., Bahnemann, D., 1995. Environmental applications of semiconductor photocatalysis. Chem. Rev. 95, 69–96.

Hu, Q., Zhang, C., Wang, Z., Chen, Y., Mao, K., Zhang, X., Xiong, Y., Zhu, M., 2008. Photodegradation of methyl *tert*-butyl ether (MTBE) by UV/H_2O_2 and UV/TiO_2. J. Hazard. Mater. 154, 795–803.

Huang, X., Leal, M., Li, Q., 2008. Degradation of natural organic matter by TiO_2 photocatalytic oxidation and its effect on fouling of low-pressure membranes. Water Res. 42, 1142–1150.

Ikehata, K., El-Din, M., 2005. Aqeuous pesticide degradation by ozonation and ozone-based advanced oxidation processes: a review (part II). Ozone Sci. Eng. 27, 173–202.

Imoberdorf, G., Mohseni, M., 2011. Degradation of natural organic matter in surface water using vacuum-UV irradiation. J. Hazard. Mater. 186, 240–246.

Irabelli, A., Jasim, S., Biswas, N., 2008. Pilot-scale evaluation of ozone vs. peroxone for trihalomethane formation. Ozone Sci. Eng. 30, 356–366.

Jarvis, P., Banks, J., Molinder, R., Stephenson, T., Parsons, S., Jefferson, B., 2008. Processes for enhanced NOM removal: beyond Fe and Al coagulation. Water Sci. Technol. Water Supply 8, 709–716.

Jiang, C., Zhang, J., 2007. Progress and prospect in electro-Fenton process for wastewater treatment. J. Zhejiang Univ. Sci. A 8, 1118–1125.

Kasprzyk-Hordern, B., Raczyk-Stanislawiak, U., Swietlik, J., Nawrocki, J., 2005. Catalytic ozonation of natural organic matter on alumina. Appl. Catal. B Environ. 62, 345–358.

Katsumata, H., Sada, M., Kaneco, S., Suzuki, T., Ohta, K., Yobiko, Y., 2008. Humic acid degradation in aqeuous solution by the photo-Fenton process. Chem. Eng. J. 137, 225–230.

Kitis, M., Kaplan, S., 2007. Advanced oxidation of natural organic matter using hydrogen peroxide and iron-coated pumice particles. Chemosphere 68, 1846–1853.

Klavarioti, M., Mantzavinos, D., Kassinos, D., 2009. Removal of residual pharmaceuticals from aqueous systems by advanced oxidation processes. Environ. Int. 35, 402–417.

Kleiser, G., Frimmel, F., 2000. Removal of precursors for disinfection by-products (DBPs) – differences between ozone- and OH-radical-induced oxidation. Sci. Total Environ. 256, 1–9.

Kruithof, J., Kamp, P., Martijn, B., 2007. UV/H_2O_2 treatment: a practical solution for organic contaminant control and primary disinfection. Ozone Sci. Eng. 29, 273–280.

Kumar, P., Manivel, A., Anandan, S., 2009. Syntesis of Ag-ZnO nanoparticles for enhanced photocatalytic degradation of acid red 88 in aqeuous environment. Water Sci. Technol. 59, 1423–1430.

Kutschera, K., Börnick, H., Worch, E., 2009. Photoionitiated oxidation of geosmin and 2-methylisoborneol by irradiation with 254 nm and 185 nm UV light. Water Res. 43, 2224–2232.

Lamsal, R., Walsh, M., Gagnon, G., 2011. Comparison of advanced oxidation processes for the removal of natural organic matter. Water Res. 45, 3263–3269.

Lanao, M., Ormad, M., Ibarz, C., Miguel, N., Ovelleiro, J., 2008. Bactericidal effectiveness of O_3, O_3/H_2O and O_3/TiO_2 on *Clostridium perfringens*. Ozone Sci. Eng. 30, 431–438.
Langlais, B., Reckhow, D.A., Brink, D.R., 1991. Ozone in Water Treatment: Application and Engineering. Lewis Publishers, Chelsea, Michigan, USA.
Le-Clech, P., Lee, E.-K., Chen, V., 2006. Hybrid photocatalysis/membrane treatment for surface waters containing low concentrations of natural organic matters. Water Res. 40, 323–330.
Lee, C., Sedlak, D., 2009. A novel homogenous Fenton-like system with Fe(III)-phosphotungstate for oxidation of organic compounds at neutral pH values. J. Mol. Catal. A Chem. 311, 1–6.
Legrini, O., Oliveros, E., Braun, A., 1993. Photochemical processes for water treatment. Chem. Rev. 83, 671–698.
Legube, B., Vel Leitner, N.K., 1999. Catalytic ozonation: a promising advanced oxidation technology for water treatment. Catal. Today 53, 61–72.
Li, K., Hokanson, D., Crittenden, J., Trussell, R., Minakata, D., 2008. Evaluating UV/H_2O_2 processes for methyl *tert*-butyl ether and tertiary butyl alcohol removal: effect of pre-treatment options and light sources. Water Res. 42, 5045–5053.
Lin, H.-C., Wang, G.-S., 2011. Effects of UV/H_2O_2 on NOM fractionation and corresponding DBPs formation. Desalination 270, 221–226.
Liu, S., Lim, M., Fabris, R., Chow, C., Chiang, K., Drikas, M., Amal, R., 2008a. Removal of humic acid using TiO_2 photocatalytic process – fractionation and molecular weight characterization studies. Chemosphere 72, 263–271.
Liu, S., Lim, M., Fabris, R., Chow, C., Drikas, M., Amal, R., 2008b. TiO_2 photocatalysis of natural organic matter in surface water: Impact on trihalomethane and haloacetic acid formation potential. Environ. Sci. Technol. 42, 6218–6223.
Liu, S., Lim, M., Fabris, R., Chow, C., Drikas, M., Amal, R., 2010. Comparison of photocatalytic degradation of natural organic matter in two Australian surface waters using multiple analytical techniques. Org. Geochem. 41, 124–129.
Ljubas, D., 2005. Solar photocatalysis-a possible step in drinking water treatment. Energy 30, 1699–1710.
Lu, J., Huang, Q., 2009. Removal of acetaminophen using enzyme-mediated oxidative coupling processes: II. Cross-coupling with natural organic matter. Environ. Sci. Technol. 43, 7068–7073.
Malato, S., Blanco, J., Alarcón, D., Maldonado, M., Fernández-Ibáñez, P., Gernjak, W., 2007. Photocatalytic decontamination and disinfection of water with solar collectors. Catal. Today 122, 137–149.
Malato, S., Fernández-Ibáñez, P., Maldonado, M., Blanco, J., Gernjak, W., 2009. Decontamination and disinfection of water by solar photocatalysis: recent overview and trends. Catal. Today 147, 1–59.
Matilainen, A., Sillanpää, M., 2010. Removal of natural organic matter from drinking water by advanced oxidation processes: a review. Chemosphere 80, 351–365.
Metz, D.M., Meyer, M., Dotson, A., Beerendonk, E., Dionysiou, D., 2011. The effect of UV/H_2O_2 treatment on disinfection by-product formation potential under simulated distribution system conditions. Water Res. 45, 3969–3980.
Molnar, J., Agbaba, J., Dalmacija, B., Klašnja, M., Dalmacija, M., Kragulj, M., 2012a. A comparative study of the effects of ozonation and TiO_2-catalyzed ozonation on the selected chlorine disinfection by-product precursor content and structure. Sci. Total Environ. 425, 169–175.
Molnar, J., Agbaba, J., Dalmacija, B., Rončević, S., Prica, M., Tubić, A., 2012b. Influence of pH and ozone dose on the content and structure of haloacetic acid precursors in groundwater. Environ. Sci. Pollut. Res. 19, 3079–3086.

Molnar, J., Agbaba, J., Dalmacija, B., Tubić, A., Krčmar, D., Maletić, S., Tomašević, D., 2013. The effects of matrices and ozone dose on changes in the characteristics of natural organic matter. Chem. Eng. J. 222, 435–443.

Moncayo-Lasso, A., Pulgarin, C., Benítez, N., 2008. Degradation of DBPs' precursors in river water before and after slow sand filtration by photo-Fenton process at pH 5 in solar CPC reactor. Water Res. 42, 4125–4132.

Moncayo-Lasso, A., Sanabria, J., Pulgarin, C., Benítez, N., 2009. Simultaneous *E. coli* inactivation and NOM degradation in river water via photo-Fenton process at natural pH in solar CPC reactor: a new way for enhancing solar disinfection of natural water. Chemosphere 77, 296–300.

Moncayo-Lasso, A., Rincon, A.-G., Pulgarin, C., Benitez, N., 2012. Significant decrease of THMs generated during chlorination of river water by previous photo-Fenton treatment at near neutral pH. J. Photochem. Photobiol. A Chem. 229, 46–52.

Mosteo, R., Miguel, N., Martin-Muniesa, S., Ormad, M., Ovelleiro, J., 2009. Evaluation of trihalomethane formation potential in function of oxidation processes used during the drinking water production process. J. Hazard. Mater. 172, 661–666.

Murray, C., Parsons, S., 2004. Removal of NOM from drinking water: Fenton's and photo-Fenton's processes. Chemosphere 54, 1017–1023.

Murray, C., Parsons, S., 2006. Preliminary laboratory investigation of disinfection by-product precursor removal using an advanced oxidation process. Water Environ. J. 20, 123–129.

Naddeo, V., Belgiorno, V., Napoli, R., 2007. Behaviour of natural organic matter during ultrasonic irradiation. Desalination 210, 175–182.

Nagata, Y., Hirai, K., Bandow, H., Maeda, Y., 1996. Decomposition of hydrobencoic and humic acids in water by ultrasonic irradiation. Environ. Sci. Technol. 30, 1133–1138.

Navalon, S., Alvaro, M., Garcia, H., Escrig, D., Costa, V., 2009. Photocatalytic water disinfection of *Cryptosporidium parvum* and *Giardia lamblia* using a fibrous ceramic TiO_2 photocatalysis. Water Sci. Technol. 59, 639–645.

Nie, Y., Hu, C., Zhou, L., Qu, J., Wei, Q., Wang, D., 2010. Degradation characteristics of humic acid over iron oxides/Fe^0 core-shell nanoparticles with UVA/H_2O_2. J. Hazard. Mater. 173, 474–479.

Park, S., Yoon, T., 2007. The effects of iron species and mineral particles on advanced oxidation processes for the removal of humic acids. Desalination 208, 181–191.

Parsons, S., Byrne, A., 2004. Water treatment applications. In: Parsons, S. (Ed.), Advanced Oxidation Processes for Water and Wastewater Treatment. IWA Publishing, London, pp. 329–346.

Pera-Titus, M., Garcia-Molina, V., Baños, M., Giménes, J., Esplugas, S., 2004. Degradation of chlorophenols by means of advanced oxidation processes: a general review. Appl. Catal. B Environ. 47, 219–256.

Pereira, V., Weinberg, H., Linden, K., Singer, P., 2007. UV degradation kinetics and modeling of pharmaceutical compounds in laboratory grade and surface water via direct and indirect photolysis at 254 nm. Environ. Sci. Technol. 41, 1682–1688.

Philippe, K., Hans, C., MacAdam, J., Jefferson, B., Hart, J., Parsons, S., 2010. Photocatalytic oxidation of natural organic matter surrogates and the impact on trihalomethane formation potential. Chemosphere 81, 1509–1516.

Pirkanniemi, K., Sillanpää, M., 2002. Heterogeneous water phase catalysis as an environmental application: a review. Chemosphere 48, 1047–1060.

Rahman, M., Jasim, S., Yanful, E., Ndiongue, S., Borikar, D., 2010. Advanced oxidation treatment of drinking water: part II. Turbidity, particles and organics removal from Lake Huron water. Ozone Sci. Eng. 32, 295–304.

Rastogi, A., Al-Abed, S., Dionysiou, D., 2009. Effect of inorganic, synthetic and naturally occurring chelating agents on Fe(II) mediated advanced oxidation of chlorophenols. Water Res. 43, 684–694.

Rincón, A.-G., Pulgarin, C., 2007. Fe^{3+} and TiO_2 solar-light-assisted inactivation of *E.coli* at field scale. Implications in solar disinfection at low temperature of large quantities of water. Catal. Today 122, 128–136.

Rosenfeldt, E., Linden, K., Canonica, S., von Gunten, U., 2006. Comparison of the efficiency of OH radical formation during ozonation and advanced oxidation processes O_3/H_2O_2 and UV/H_2O_2. Water Res. 40, 3695–3704.

Rosenfeldt, E.J., Linden, K.G., 2007. The ROH, UV concept to characterize and the model UV/H_2O_2 process in natural waters. Environ. Sci. Technol. 41, 2548–2551.

Roshani, B., Leitner, N., 2011. Effect of persulfate on the oxidation of benzotriazole and humic acid by e-beam irradiation. J. Hazard. Mater. 190, 403–408.

Sanly, Lim, M., Chiang, K., Amal, R., Fabris, R., Chow, C., Drikas, M., 2007. A study on the removal of humic acid using advanced oxidation process. Sep. Sci. Technol. 42, 1391–1404.

Sarathy, S., Mohseni, M., 2007. The impact of UV/H_2O_2 advanced oxidation on molecular size distribution of chromophoric natural organic matter. Environ. Sci. Technol. 41, 8315–8320.

Sarathy, S.R., Mohseni, M., 2009. The fate of natural organic matter during UV/H_2O_2 advanced oxidation of drinking water. Can. J. Civ. Eng. 36, 160–169.

Sarathy, S., Mohseni, M., 2010. Effects of UV/H_2O advanced oxidation on chemical characteristics and chlorine reactivity of surface water natural organic matter. Water Res. 44, 4087–4096.

Sarathy, S., Stefan, M., Royce, A., Mohseni, M., 2011. Pilot-scale UV/H_2O_2 advanced oxidation process for surface water treatment and downstream biological treatment: effects on natural organic matter characteristics and DBP formation potential. Environ. Technol. 32, 1709–1718.

Selcuk, H., Bekbolet, M., 2008. Photocatalytic and photoelectrocatalytic humic acid removal and selectivity of TiO_2 coated photoanode. Chemosphere 73, 854–858.

Selcuk, H., Sarikaya, H., Bekbolet, M., Anderson, M., 2006. Bromate formation on the nonporous TiO_2 photoanode in the photoelectrocatalytic system. Chemosphere 62, 715–721.

Shifu, C., Wei, Z., Wei, L. Huaye, Z., Xiaoling, Y., Yinghao, C., 2009. Preparation, characterization and activity evaluation of p-n junction photocatalyst p-$CaFe_2O_4$/n-Ag_3VO_4 under visible light irradiation. J. Hazard. Mater. 172, 1415–1423.

Sindelar, H., Brown, M., Boyer, T., 2014. Evaluating UV/H_2O_2, UV/percarbonate, and UV/perborate for natural organic matter reduction from alternative water sources. Chemosphere 105, 112–118.

Song, W., Ravindran, V., Pirbazari, M., 2008. Process optimization using a kinetic model for the ultraviolet radiation-hydrogen peroxide decomposition of natural and synthetic organic compounds in groundwater. Chem. Eng. Sci. 63, 3249–3270.

Song, H., Orr, O., Hong, Y., Karanfil, T., 2009. Isolation and fractionation of natural organic matter: evaluation of reverse osmosis performance and impact of fractionation parameters. Environ. Monit. Assess. 153, 307–321.

Stepniak, L., Stańczyk-Mazanek, E., Kusiak, M., 2012. Ultrasounds and ozone in removal of humic substances from water. Inżynieria Ochrona Środowiska 15, 143–154.

Suty, H., De Traversay, C., Cost, M., 2004. Applications of advanced oxidation processes: present and future. Water Sci. Technol. 49, 227–233.

Swaim, P., Royce, A., Smith, T., Maloney, T., Ehlen, D., Carter, B., 2008. Effectiveness of UV advanced oxidation for destruction of mirco-pollutants. Ozone Sci. Eng. 30, 34–42.

Tercero Espinoza, L., Frimmel, F., 2008. Formation of brominated products in irradiated titanium dioxide suspensions containing bromide and dissolved organic carbon. Water Res. 42, 1778–1784.

Tercero Espinoza, L., Frimmel, F., 2009. A simple simulation of the degradation of natural organic matter in homogenous and heterogenous advanced oxidation processes. Water Res. 43, 3902–3909.
Tercero Espinoza, L., Rembor, M., Matesanz, C., Heidt, A., Frimmel, F., 2009a. Formation of bromoform in irradiated titanium dioxide suspensions with varying photocatalyst, dissolved organic carbon and bromide concentrations. Water Res. 43, 4143–4148.
Tercero Espinoza, L., ter Hazeborg, E., Weber, M., Frimmel, F., 2009b. Investigation of the photocatalytic degradation of brown water natural organic matter by size exclusion chromatography. Appl. Catal. B Environ. 87, 56–62.
Tercero Espinoza, L., ter Haseborg, E., Weber, M., Karle, E., Peschke, R., Frimmel, F., 2011. Effect of selected metal ions on the photocatalytic degradation of bog lake water natural organic matter. Water Res. 45, 1039–1048.
Thomson, J., Roddick, F., Drikas; M., 2002. Natural organic matter removal by enhanced photo-oxidation using low pressure mercury vapour lamps. Water Sci. Technol. Water Supply 2, 435–443.
Thomson, J., Roddick, F.A., Drikas, M., 2004. Vacuum ultraviolet irradiation for natural organic matter removal. J. Water Supply Res. Technol. Aqua 53, 193–206.
Toor, R., Mohseni, M., 2007. UV-H_2O_2 based AOP and its integration with biological activated carbon treatment for DBP reduction in drinking water. Chemosphere 66, 2087–2095.
Tran, T., Gray, S., Bolto, B., Farmer, T., Collings, T., 2007. Ultrasound enhancement of microfiltration performance for natural organic matter removal. Org. Geochem. 38, 1091–1096.
Treguer, R., Tatin, R., Couvert, A., Wolbert, D., Tazi-Pain, A., 2010. Ozonation effect on natural organic matter adsorption and biodegradation – application to a membrane bioreactor containing activated carbon for drinking water production. Water Res. 44, 781–788.
Tubić, A., Agbaba, J., Dalmacija, B., Perović, S., Klašnja, M., Rončević, S., Ivančev-Tumbas, I., 2011. Removal of natural organic matter from groundwater using advanced oxidation processes at a pilot scale drinking water treatment plant in the Central Banat Region (Serbia). Ozone Sci. Eng. 33, 267–278.
Tuhkanen, T.A., 2004. UV/H_2O_2 processes. In: Parsons, S. (Ed.), Advanced Oxidation Processes for Water and Wastewater Treatment. IWA Publishing, London, pp. 86–110.
Uyguner, C., Bekbolet, M., Selcuk, H., 2007a. A comparative approach to the application of a physico-chemical and advanced oxidation combined system to natural water samples. Sep. Sci. Technol. 42, 1405–1419.
Uyguner, C., Suphandag, S., Kerc, A., Bekbolet, M., 2007b. Evaluation of adsorption and coagulation characteristics of humic acids preceded by alternative advanced oxidation techniques. Desalination 210, 183–193.
Uyguner-Demirel, C., Bekbolet, M., 2011. Significance of analytical parameters for the understanding of natural organic matter in relation to photocatalytic oxidation. Chemosphere 84, 1009–1031.
Uyguner-Demirel, C., Birben, C., Bekbolet, M., 2013. Key role of common anions on the photocatalytic degradation profiles of the molecular size fractions of humic acid. Catal. Today 209, 122–126.
Valencia, S.H., Marín, J.A., Restrepo, G.M., 2012a. Evolution of natural organic matter by size exclusion chromatography during photocatalytic degradation by solvothermal-synthesized titanium dioxide. J. Hazard. Mater. 213–214, 318–324.
Valencia, S., Marína, J., Velásqueza, J., Restrepoa, G., Frimmel, F., 2012b. Study of pH effects on the evolution of properties of brown-water natural organic matter as revealed by size-exclusion chromatography during photocatalytic degradation. Water Res. 46, 1198–1206.

Valencia, S., Marín, J.M., Restrepo, G., Frimmel, F.H., 2014. Evaluation of natural organic matter changes from Lake Hohloh by three-dimensional excitation-emission matrix fluorescence spectroscopy during TiO_2/UV process. Water Res. 51, 124–133.
Vilhunen, S., Bosund, M., Kääriäinen, M.-L., Cameron, D., Sillanpää, M., 2009. Atomic layer deposited TiO_2 films in photodegradation of aqueous salicylic acid. Sep. Purif. Technol. 66, 130–134.
Voelker, B., Sulzberger, B., 1996. Effects of fulvic acid on Fe(II) oxidation by hydrogen peroxide. Environ. Sci. Technol. 30, 1106–1114.
Wang, G.-S., Liao, C.-H., Chen, H.-W., Yang, H.-C., 2006. Characteristics of natural organic matter degradation in water by UV/H_2O_2 treatment. Environ. Technol. 27, 277–287.
Wang, K., Guo, J., Yang, M., Junji, H., Deng, R., 2009. Decomposition of two haloacetic acids in water using UV radiation, ozone and advanced oxidation processes. J. Hazard. Mater. 162, 1243–1248.
Wiszniowski, J., Robert, D., Surmacz-Gorska, J., Miksch, K., Weber, J.-V., 2002. Photocatalytic decomposition of humic acids on TiO_2 part I: discussion of adsorption and mechanism. J. Photochem. Photobiol. A Chem. 152, 267–273.
Xu, S., Zhang, X., Ng, J., Sun, D., 2009. Preparation and application of TiO_2/Al_2O_3 microsperical photocatalyst for water treatment. Water Sci. Technol. Water Supply 9, 39–44.
Yan, M., Wang, D., Shi, B., Wang, M., Yan, Y., 2007. Effect of pre-ozonation on optimized coagulation of a typical North-China source water. Chemosphere 69, 1695–1702.
Yang, S., Sun, C., Li, X., Gong, Z., Quan, X., 2010. Enhanced photocatalytic activity for titanium dioxide by co-modifying with silica and fluorine. J. Hazard. Mater. 175, 258–266.
Yao, P., Choo, K., Kim, M., 2009. A hydridized photocatalysis-microfiltration system with iron oxide-coated membranes for the removal of natural organic matter in water treatment: effects of iron oxide layers and colloids. Water Res. 43, 4238–4248.
Zepp, R., Faust, B., Hoigné, J., 1992. Hydroxyl radical formation in aqueous reactions (pH 3–8) of iron(II) with hydrogen peroxide: the photo-Fenton reaction. Environ. Sci. Technol. 26, 313–319.
Zhang, P., Jian, L., 2006. Ozone-enhanced photocatalytic degradation of natural organic matter in water. Water Sci. Technol. Water Supply 6, 53–61.
Zhang, X., Pan, J.H., Fu, W., Du, A.J., Sun, D.D., 2009. TiO_2 nanotube photocatalytic oxidation for water treatment. Water Sci. Technol. Water Supply 9, 45–49.
Zhao, Y.-Y., Boyd, J., Woodbeck, M., Andrews, R., Qin, F., Hrudey, S., Li, X.-F., 2008. Formation of *N*-nitrosamines from eleven disinfection treatments of seven different surface waters. Environ. Sci. Technol. 42, 4857–4862.
Zhao, G., Lu, X., Zhou, Y., Gu, Q., 2013. Simultaneous humic acid removal and bromate control by O_3 and UV/O_3 processes. Chem. Eng. J. 232, 74–80.
Zhou, Q., Zhong, Y.-H., Chen, X., Liu, J.-H., Huang, X.-J., Wu, Y.-C., 2014. Adsorption and photocatalysis removal of fulvic acid by TiO_2–graphene composites. J. Mater. Sci. 49, 1066–1075.
Ziylan, A., Ince, N., 2013. Ozonation-based advanced oxidation for pre-treatment of water with residuals of anti-inflammatory medication. Chem. Eng. J. 220, 151–160.

CHAPTER 7

NOM Removal by Adsorption

Mika Sillanpää*, **Amit Bhatnagar****

*Lappeenranta University of Technology, LUT Faculty of Technology, LUT Chemtech, Laboratory of Green Chemistry, Sammonkatu 12, 50130 Mikkeli, Finland; **University of Eastern Finland, Department of Environmental Science, Faculty of Science and Forestry, P.O. Box 1627, 70211 Kuopio, Finland

Abstract

The removal of natural organic matter (NOM) and its constituents from water is an emerging issue, and a robust treatment technology is needed to achieve this goal. Interest in research to overcome this problem has been growing steadily, and many technologies have been proposed for the removal of NOM from water. This chapter focuses on the removal of NOM and its constituents from water using adsorption technology, which has proved to be one of the best technologies for the removal of diverse types of aquatic pollutants. The principal NOM removal results obtained with different adsorbents have been summarized.

Keywords: Adsorption; Carbon adsorbent; Hydroxide adsorbent; Metal oxide adsorbent; Nanoadsorbent; Natural organic matter (NOM); Water treatment.

ABBREVIATIONS

AOPs Advanced oxidation processes
CNT Carbon nanotube
CSZ Chitosan/zeolite composite
CUR Carbon usage rate
DBP Disinfection by-product
DBPFP Disinfection by-product formation potential
DOC Dissolved organic carbon
DOM Dissolved natural organic matter
EBCT Empty bed contact time
FTIR Fourier transform infrared
GAC Granular activated carbon
HA Humic acids
HAAs Haloacetic acids
HPIA-DEAE Hydrophilic acid-weak ionic exchange resin
HPLC High performance liquid chromatography
HPSEC High performance size exclusion chromatography
IOCS Iron oxide-coated sand
IOP Iron oxide particle
LDH Layered double hydroxide
Mag-PCMAs Magnetic permanently confined micelle arrays

MWCNT Multiwalled carbon nanotube
NMR Nuclear magnetic resonance
NOM Natural organic matter
PAC Powdered activated carbon
ROM Residual organic matter
SAM Self-assembled monolayer
SEC Size exclusion chromatography
SPTHMFP Simulated plant trihalomethane formation potential
SUVA Specific UV absorbance
THMFP Trihalomethane formation potential
THM Trihalomethane
TOC Total organic carbon

7.1 INTRODUCTION

The choice of water treatment process depends on the characteristics of the water and the treatment method costs (Richardson et al., 2007). An improved understanding of the chemical properties of aquatic natural organic matter (NOM) sheds light on choosing the most appropriate treatment strategies for removing NOM from raw drinking waters (Owen et al., 1985). A great variety of NOM removal processes have been investigated and employed, such as coagulation (Bell-Ajy et al., 2000; Crozes et al., 1995; Matilainen et al., 2010), membrane filtration (Siddiqui et al., 2000; Metsämuuronen et al., 2014), advanced oxidation processes (Matilainen and Sillanpää, 2010), and adsorption (Karanfil et al., 1999; Genz et al., 2008). However, all these methods have their own shortcomings or limitations; for example, inorganic coagulants such as aluminum- or iron-based salts remove only a portion of NOM (Chow et al., 2009). Furthermore, the removal of NOM by conventional and advanced treatment processes (flocculation, oxidation, and membrane processes) is not always adequate with regard to removal efficiency, chemical and energy consumption, and management of residuals (Genz et al., 2008). The adsorption process, the focus of this chapter, is generally considered one of the best water treatment technologies because of its convenience, ease of operation, and simplicity of design.

7.2 ADSORPTION

Adsorption can be described as the adhesion of atoms, ions, or molecules from a gas, liquid, or dissolved solid to a surface. In other words, adsorption is a mass transfer process that includes the accumulation of substances at the interface of two phases, such as the liquid–liquid, gas–liquid, gas–solid, or liquid–solid interface (Dąbrowski, 2001). The substance being adsorbed is the adsorbate, and the adsorbing material is the adsorbent. In a real adsorption system, equilibrium is achieved between the adsorbent, which is in contact with the bulk phase, and the so-called interfacial layer. In contrast, desorption denotes the converse process. Adsorption is different from absorption, where the adsorbate molecules penetrate the bulk solid phase. The term *sorption*—together with the terms *sorbent*, *sorbate*, and *sorptive*—is also used to represent both adsorption and absorption, when both occur simultaneously or cannot be distinguished (Dąbrowski, 2001). An adsorption isotherm—the equilibrium relation between the quantity of the adsorbed material and the pressure (for gas) or concentration (for liquids) in the bulk fluid phase at constant temperature—is widely used to explain the adsorption process. In the literature, readers can find several models describing the adsorption process, namely the Freundlich isotherm (Freundlich, 1906), Langmuir isotherm (Langmuir, 1916), and Brunauer–Emmett–Teller (BET) isotherm (Brunauer et al., 1938). Adsorption can be distinguished as either physical adsorption (or physisorption, from the universal van der Waals interactions) or chemical adsorption (or chemisorption, which can have the character of a chemical process) (Dąbrowski, 2001). It may also occur due to electrostatic attraction. Physical adsorption is a reversible process that takes place at a temperature lower or close to the critical temperature of an adsorbed substance. In contrast, chemical adsorption usually occurs at temperatures much higher than the critical temperature, and is a specific process that can only take place on some solid surfaces for a given gas/liquid (Dąbrowski, 2001). Under favorable conditions, both processes can occur simultaneously or alternately. Physical adsorption is accompanied by a decrease in free energy and entropy of the adsorption system, so this process is exothermic (Dąbrowski, 2001). It has been argued that in view of the higher bonding strength in chemisorption, it is difficult to remove chemisorbed species from the solid surface. A brief discussion of historical developments in adsorption science and engineering is presented below.

The credit for the first quantitative observations on adsorption goes to Scheele (Mantell, 1951) in 1773, who reported some experiments on the uptake of gases by charcoal and clays. This was followed by observations made by Lowitz in 1785 on the reversible removal of color and odor-producing compounds from water by wood charcoal. Larvitz in 1792 and Kehl in 1793 observed similar phenomena with vegetable and animal charcoals, respectively. The term *adsorption* was proposed by du Bois-Reymond but introduced into the literature by Kayser (Dąbrowski, 2001). Kayser also developed some theoretical concepts that became basic to monomolecular adsorption theory (Dąbrowski, 2001). Systematic studies of adsorption were conducted by de Saussure and, based on the results of these studies, he concluded that all types of gases are taken up by porous substances (sea foam, cork, charcoal, asbestos), and this process is accompanied by the evolution of heat (Dąbrowski, 2001). Thus, he discovered the exothermic character of adsorption processes, and he was the first to note the commonness of adsorption. The term *absorption* was introduced by McBain (McBain, 1909) to determine an uptake of hydrogen by carbon much slower than adsorption. He proposed the term *sorption* for adsorption and absorption. Practical application of adsorption processes is based mainly on the selective uptake of individual components from their mixtures by other substances. Selective adsorption, which has many industrial applications, was discovered by Tswett in 1903 (Dąbrowski, 2001). Later, he took advantage of this phenomenon to separate chlorophyll and other plant pigments by means of silica materials. The technique proposed by Tswett has been called column solid/liquid adsorption chromatography. This discovery was not only the beginning of a new analytical technique, but also the origin of a new field of surface science. Readers interested in a detailed discussion of the theory and applications of adsorption should refer to an excellent comprehensive review by Dąbrowski (Dąbrowski, 2001).

7.3 ADSORBENTS AND THEIR CHARACTERISTICS

To achieve the best performance in adsorption systems, knowledge of various types of adsorbents is essential. The most important properties of a good adsorbent are a well-developed porous structure and high surface area. Another feature of a good adsorbent is a short equilibration time so that it can be used to remove contaminants more quickly. Thus, the best adsorbents for pollutant removal have a high surface area and porosity, and demonstrate fast adsorption kinetics. Commercial sorbents used in cyclic adsorption processes should ideally meet several important requirements (Deng, 2006), including high selectivity derived from equilibrium, kinetic, or steric effects; large adsorption capacity; fast adsorption kinetics; easy regenerability; good mechanical strength; and low cost. The above adsorbent performance requirements can be expressed simply as adsorbent characteristic requirements as follows (Deng, 2006): large internal pore volume; large internal surface area; controlled surface properties through selected functional groups; controlled pore size distribution, preferably in the micropore range; weak interactions between adsorbate and adsorbent (mostly on physical sorbents); inorganic or ceramic materials to enhance chemical and mechanical stability; and low-cost raw materials. In recent decades, a number of materials from different sources have been extensively investigated as adsorbents in water pollution control. Some of the important ones include silica gel, activated alumina, zeolites, and activated carbon.

7.3.1 Silica Gel

Silica gel is most commonly encountered in everyday life as beads in a small paper packet. In this form, it is used as a desiccant to control local humidity to avoid spoilage or degradation of some goods. Silica gels are classified into three types: regular, intermediate, and low density gels. Regular density silica gel is prepared in an acid medium and has a high surface area (e.g., $750\,m^2\,g^{-1}$). Intermediate and low density silica gels have low surface areas (300–350 and $100–200\,m^2\,g^{-1}$, respectively). Modified forms of silica have also been widely explored for the removal of different pollutants (Najafi et al., 2011; Wang et al., 2009; Zaitseva et al., 2013).

7.3.2 Activated Alumina

Activated alumina comprises a series of nonequilibrium forms of partially hydroxylated alumina oxide, Al_2O_3. In general, as a hydrous alumina precursor is heated, hydroxyl groups are driven off, leaving a porous solid structure of activated alumina. This process is also used to remove water from organic liquids, including gasoline, kerosene, oils, aromatic hydrocarbons, and many chlorinated hydrocarbons, with a surface area ranging from 200 to $300\,m^2\,g^{-1}$. Activated alumina has received wide attention as an adsorbent, and a number of reports are available on its adsorption characteristics. Readers interested in a detailed discussion of the application of alumina in water treatment should refer to comprehensive reviews by Kasprzyk-Hordern (2004) and Kumar et al. (2014).

7.3.3 Zeolites

Zeolites are microporous, aluminosilicate minerals commonly used as commercial adsorbents. About 40 natural and over 200 synthetic zeolites have been identified so far. Zeolite based materials are extremely versatile, and their main applications include ion-exchange resins, catalytic applications in the petroleum industry, separation process, and as an adsorbent for water, carbon dioxide, and hydrogen sulfide. Zeolites have been extensively used for the removal of diverse pollutants (Grieco and Ramarao, 2013; Xie et al., 2013; Martucci et al., 2012; Lin et al., 2013). Wang and Peng (2010) presented a review on the role of natural zeolites as effective adsorbents in water and wastewater treatment.

7.3.4 Activated Carbon

Activated carbon has undoubtedly been the most popular and widely used adsorbent in wastewater treatment worldwide. It is produced by a process consisting of raw material dehydration and carbonization followed by activation. The product obtained generally has a very porous structure with a large surface area ranging from 600 to $2000\,m^2\,g^{-1}$. Activated carbon has proved a versatile adsorbent that can remove diverse types of pollutants from water and wastewater (Marsh and Rodríguez-Reinoso, 2006; Bansal and Goyal, 2005; Bhatnagar et al., 2013).

The performance of the adsorption process depends on many factors, including (1) surface area, (2) nature and initial concentration of the adsorbate, (3) solution pH, (4) temperature, (5) interfering substances, and (6) nature and dose of the adsorbent. The physicochemical nature of the adsorbent significantly affects both adsorption rate and capacity. The solubility of the solute also considerably influences the adsorption equilibrium. In general, an inverse relationship can be expected between the extent of adsorption of a solute and its solubility in the solvent where the adsorption takes place. The presence of (in)organic compounds can also limit the adsorption process.

7.4 NOM REMOVAL FROM WATER BY ADSORPTION

7.4.1 Carbon Adsorbents

Granular activated carbon (GAC) adsorption is one of the best technologies employed for the removal of NOM and other diverse aquatic pollutants. The most common adsorbent used for water treatment is activated carbon, which can be applied as powdered activated carbon (PAC) or GAC. While PAC can be applied at various stages of water treatment, GAC is typically utilized after coagulation–filtration/sedimentation but before postdisinfection (Bond et al., 2011). Residual humic substances after the coagulation process are most likely fragments of lower size and charge. Owing to the hydrophobicity of remaining NOM, in situations where they retain significant disinfection by-product formation potential (DBPFP), activated carbon adsorption is recommended for their removal (Bond et al., 2011). Vidic and Suidan (1991) demonstrated the role of dissolved oxygen in the adsorptive capacity of activated carbon for both synthetic matter and NOM. The adsorption isotherm tests were conducted using an initial NOM concentration of 7.6 mg of DOC L^{-1}. The results indicate that the adsorptive capacity of GAC for this particular NOM increased as much as twofold as a result of the presence of molecular oxygen. The effectiveness of GAC adsorption for the removal of NOM and trihalomethanes (THM) from the drinking water treated at Ivedik Water Treatment Plant in Ankara City was investigated (Capar and Yetis, 2001). The Freundlich Isotherm constants, K and n, were determined as 17.61 $(mg\,g^{-1})\,(L\,mg^{-1})1 = n$ and 1.66, respectively. Bench-scale GAC columns were run with empty bed contact times (EBCT) varying from 0.40 to 2.67 min to evaluate adsorption performance and to investigate the effect of EBCT on service life. The treated volumes of water increased with EBCT, showing a linear increase in GAC service life. A 5-fold increase in EBCT resulted in an almost 16-fold increase in service life. Correspondingly, the carbon usage rate decreased, and the optimal bed depth was observed to be 10 cm. The capacities calculated by the isotherm equation and achieved in columns were also compared. The column capacities were within 43–65% of the isotherm capacities at complete breakthrough. However, they were only within 8–17% of the isotherm capacities at 50% breakthrough. GAC was also studied by Iriarte-Velasco et al. (2008) for the adsorption of NOM. Three different carbons of different origin were initially used. Their physical properties were studied by means of N_2 adsorption. Their chemical properties were studied by means of thermogravimetric analysis, acid–base titrations, and Fourier transform infrared spectroscopy (FTIR). Only one of the carbons showed a broad absorption band in the 1300–1000 cm^{-1} region in FTIR spectra, which was assigned to the C–O stretching and O–H bending modes of alcoholic, phenolic, and carboxylic groups. Adsorption of NOM was studied by batch adsorption experiments, and uptake of NOM by the different carbons was evaluated by UV absorbance, DBPFP tests, and high-performance liquid chromatography–size exclusion chromatography (HPLC–SEC). The Freundlich equation was used to fit equilibrium data. Point of zero charge (pHPZc) and overall surface basicity were shown to improve the removal of THM precursors. Differences were reported in the molecular weight distribution of the adsorbed material by different carbons. A clear correlation was found between a reduction in the THM formation capacity of the sample and the reduction in intensity of a specific peak in SEC chromatograms. Furthermore, trihalomethane formation potential (THMFP) tests showed the existence of some fractions of NOM that were not adsorbable with activated carbons and undetected by measurement of DOC.

The removal of NOM from lake water was studied in two pilot-scale adsorbers containing GAC with different physical properties (Velten et al.,

2011). In both GAC adsorbers, the adsorbability of the remaining NOM fractions, compared on the basis of partition coefficients, increased with decreasing molecular size, suggesting that increasingly large portions of the internal GAC surface area could be accessed as the size of NOM decreased. Overall DOC uptake at pseudosteady-state differed between the two tested GACs (18.9 and 28.6 g-C kg^{-1} GAC); the percentage difference in DOC uptake closely matched the percentage difference in the volume of pores with widths in the 1–50 nm range that was measured for the two fresh GACs. Despite the differences in NOM uptake capacity, the two GACs removed individual NOM fractions in similar proportions.

Besides GACs, PACs were also applied in combination with aluminum sulfate to improve NOM removal from high-DOC source water (Fabris et al., 2004). The treated water quality was analyzed using absorbance at 254 nm, DOC, THMFP, rapid fractionation, C13 NMR, and molecular weight distribution by high performance size exclusion chromatography. Carbon C (steam activated, coal based carbon), and B (chemically activated, wood based carbon, which is primarily mesoporous (2–50 nm pore width)) offered equivalent DOC removal, but carbon C was considered superior because it reduced treated water THMFP. It is postulated that an organic fraction of around 1000 Da is responsible for the differences in THMFP shown in the treated waters.

Modified carbons have also been used for NOM removal. Several pathways have been employed to systematically modify two GACs, F400 (coal based) and Macro (wood based), for examining the adsorption of dissolved natural organic matter (DOM) from natural waters (Dastgheib et al., 2004). A total of 24 activated carbons with different physical and chemical characteristics were produced. The impact of carbon treatment on DOM adsorption was examined by conducting isotherm experiments at a neutral pH using the modified carbons and a DOM isolated from the influent to a drinking water treatment plant in Myrtle Beach, South Carolina, USA. Adsorption of the DOM by two activated carbon fibers (ACFs) with relatively uniform pore size distributions showed that only pores with widths larger than 1 nm were accessible to the DOM macromolecules. Increases in the carbon supermicropore and mesopore volume (i.e., >1 nm) increased the DOM uptake, if the surface chemistry was favorable. Normalized on a surface area basis, the isotherms showed the significance of carbon surface chemistry for DOM uptake. At neutral pH, the adsorption of negatively charged DOM molecules was favored by basic and positively charged surfaces, while DOM uptake was minimized when the surface had acidic characteristics. High-temperature ammonia treatment of oxidized carbons considerably enhanced DOM uptake, mainly due to the increase in accessible surface area and surface basicity. Iron-impregnated carbons indicated an enhanced affinity of the iron-laden carbon surface toward the DOM species, if the surface was not negatively charged. DOM adsorption by virgin and modified GACs was studied by Cheng et al. (2005). DOM samples were obtained from two water treatment plants before (i.e., raw water) and after coagulation/flocculation/sedimentation processes (i.e., treated water). GACs were modified by high-temperature helium or ammonia treatment, or iron impregnation followed by high-temperature ammonia treatment. Two ACFs were also used, with no modification, to examine the effect of carbon porosity on DOM adsorption. Size exclusion chromatography (SEC) and specific ultraviolet absorbance ($SUVA_{254}$) were employed to characterize the DOMs before and after adsorption. Iron-impregnated (HDFe) and ammonia-treated (HDN) activated carbons showed significantly higher DOM uptakes than the virgin GAC. The enhanced DOM uptake by HDFe was due to the presence of iron species on the carbon surface. The higher uptake of HDN was attributed to the enlarged carbon pores and basic surface created during ammonia treatment. The SEC and $SUVA_{254}$ results showed no specific selectivity in the removal of different DOM components as a result of carbon modification. The removal of DOM from both raw and treated waters was negligible by ACF10, with 96% of its surface area in pores smaller than 1 nm. Low molecular weight (MW) DOM components were preferentially removed by ACF20H, with 33% of its surface area in 1–3 nm pores. ACF20H pores excluded DOM components with MWs larger than 1600, 2000, and 2700 Da of Charleston raw, Charleston treated, and Spartanburg treated waters, respectively. In contrast to carbon fibers, DOM components from the entire MW range were removed from waters by virgin and modified GACs.

7.4.2 Nanoadsorbents

Nanotechnology has emerged as one of the promising technologies in water treatment. In light of this, nanoscale carbon black was employed to remove NOM in water in the presence and absence of coagulation (Wang et al., 2010). In the absence of coagulation, more than 60% of NOM removal was achieved by carbon black adsorption. A lower pH (3–5) was favorable for NOM removal. More than 35% of NOM was removed by carbon black adsorption in the first 20 min, and the adsorption of NOM onto carbon black occurred within about 2 h. Mixing intensity was found to be essential for the mixture of NOM and carbon black, while insufficient stirring or over-stirring decreased NOM removal efficiency. When low dosages of coagulants were used in combination with carbon black at pH 6–7, the removal efficiency of NOM increased significantly. Depending on the coagulant, the sequencing of adsorption and coagulation can be important. Almost 90% of NOM was removed in 15 min by carbon black adsorption and alum coagulation, which is a higher removal rate than for conventional treatment. This study indicated that carbon black might be an important adsorbent for NOM removal in water treatment in combination with low doses of alum.

Multiwalled carbon nanotubes (MWCNTs) were thermally treated and employed as adsorbents to study their characterizations and adsorption performance for NOM in aqueous solutions (Lu and Su, 2007). The physiochemical properties of CNTs, such as the structure and nature of the carbon surface, were changed after thermal treatment, which made CNTs adsorb more NOM. The amount of NOM adsorbed onto CNTs increased with a rise in initial NOM concentration and solution ionic strength, but decreased with a rise in solution pH. A comparative analysis of the NOM adsorption of CNTs and GAC was also conducted. Under the same conditions, treated CNTs have the best NOM adsorption performance, followed by raw CNTs and then GAC, suggesting that the CNTs are efficient NOM adsorbents and possess good potential for applications to maintain high-quality water. Further research on the toxicity of CNTs and CNT-related nanomaterials is needed to promote safe and optimized applications of CNTs in water treatment. MWCNTs were also used by Naghizadeh et al. (2013) for the adsorption of hydrophobic NOM from aqueous solution under different operational conditions, including contact time, pH, initial concentration of NOM, and temperature. MWCNTs with an average diameter of 10–50 nm were synthesized via chemical vapor deposition. The results illustrated that both as-prepared and functionalized MWCNTs had high adsorption capacity for the NOM studied. Functionalization of MWCNT affected the surface area and introduced oxygen-containing functional groups onto its surface, which depressed the adsorption of NOM onto MWCNTs–COOH. The obtained data had the best conformity to the Freundlich isotherm. Kinetic studies were performed and the adsorption kinetics successfully followed the pseudosecond-order kinetic model. Adsorption of three NOM analogues, including salicylic acid, phthalic acid, and catechol on MWCNTs as well as commercial PAC were investigated (Liu et al., 2013). All adsorption isotherms fit the Freundlich isotherm model well. PAC possessed 2–10 times higher adsorption capacities toward the three compounds than MWCNTs, indicating a pore-filling effect. Hydrophobic interaction, electrostatic interaction, hydrogen bond and π–π interaction acted simultaneously, but made different contributions to the adsorption of the three NOM analogues on PAC and MWCNTs. Also, magnetic chitosan nanoparticles have proven effective in humic acid (HA) removal (Dong et al., 2014).

7.4.3 Metal Oxide and Hydroxide Based Adsorbents

Besides carbon adsorbents, metal oxides and hydroxides have also been investigated for NOM removal. Fe and Al oxyhydroxides have been found effective in adsorbing NOM from solutions. Three types of iron oxide particles (IOPs) viz. ferrihydrite, goethite, and hematite, were prepared to remove residual organic matter (ROM) from secondary effluents (Choo and Kang, 2003). Of the three types of IOPs tested, the most amorphous, ferrihydrite, was found to be the most effective adsorbent in the removal of ROM from the secondary effluent. Regarding the ROM removal efficiency of ferrihydrite, the chemical oxygen demand and DOC removal were relatively low and were stabilized at a relatively low dose of IOPs; UV removal, however, gradually increased, even at higher IOP dosages, which might be associated with the IOP removal of phenolic compounds from the ROM pool. The optimal pH level (approximately pH 6.0) for maximum ROM removal by iron oxide adsorption could be attributed to the association and dissociation of ferrihydrite and ROM depending on solution pH.

Mechanisms contributing to the adsorption of NOM on surfactant-modified iron oxide-coated sand (IOCS) were explored by microscopic surface characterization techniques and adsorption tests (Ding and Shang, 2010). Electrostatic interactions that were thought to originate from the positively charged, surfactant-coated surface, hexadecyl trimethyl ammonium (HDTMA) seemed to be unimportant, probably because the outward-pointing tail groups of the surface-coated HDTMA monolayers hindered the interactions. Improved hydrophobic interactions followed by ligand exchange are believed to be the dominant mechanisms. Atomic force microscopy analysis with chemically modified tips was used to explore the adsorption mechanisms between NOM and IOCS, where an iron oxide-coated mica surface was utilized as a substitute for the IOCS surface. This demonstrated the changes in pull-on forces and the increases in hydrophobic interactions from the modification of IOCS with HDTMA. IOCS was modified with HDTMA and tested as an adsorbent for the removal of NOM from water (Ding et al., 2010). The modification did not change the physical properties of the IOCS but coated HDTMA onto its surface. The HDTMA-modified IOCS displayed a faster initial NOM adsorption and substantially higher capacity than the unmodified IOCS over a wide pH range in both batch and column adsorption. The enhancement was more pronounced at higher pH. Compared to unmodified IOCS, the HDTMA-modified IOCS removed more hydrophobic and larger NOM molecules and its NOM adsorption was less sensitive to changes in ionic strength. The adsorption capacity of the modified IOCS was regenerated in situ with NaOH solution and ex situ with HDTMA solution. HDTMA-modified IOCS adsorption may be a promising alternative technology for NOM removal.

Bolto et al. (2001) investigated the effect of adding suspended matter in the form of clay or metal oxide when a cationic polymer was employed as the primary coagulant. The solids provide both an adsorbent for NOM and a nucleating species for precipitating the NOM-polymer complex. Metal oxides in conjunction with a cationic polymer were more promising than clay, with effectiveness in the order $Fe_2O_3 > Fe_3O_4 > Al_2O_3 > MnO_2$. Magnesium oxide at a much lower dose was nearly as effective as ferric oxide, but raised the pH level significantly. Using simple self-assembled monolayer (SAM) techniques, a silica substrate was modified as an adsorbent and tested for its potential for reduction of water quality parameters such as UV absorbance, color, and DOC (Chow et al., 2009). Silica particles that were coated with an aminosiloxane SAM (NH_2-SAM) were evaluated in both a high surface area powder form and a more realistic granular sand form. Initial results using direct stirred contact with powdered NH_2-SAM showed promising results, with 60% reduction of UV_{254} after 1 h and up to 70% removal of DOC with increased doses and contact times. NH_2-SAM powder removed NOM in a broader and less selective MW range than coagulation treatment, and this removal was enhanced by pH control at 6, especially for medium MW components. When NH_2-SAM sand was applied, the significantly reduced effective surface area resulted in lower DOC removal, but color removal was still considerable for realistic treatment plant contact times. Attempted regeneration with acidic solutions showed greater effectiveness at lower applied pH, but recovery of adsorption capacity reduced with successive adsorption/regeneration cycles, highlighting the need for further refinement of operating conditions for more effective application of this relatively simple water treatment technology.

The effect of mixed oxidants and PAC on the removal of NOM was studied (Álvarez-Uriarte et al., 2010). Results obtained in this work indicated that mixed oxidants promote conversion of humic matter to hydrophilic matter. Hydrophilic acid-weak ionic exchange resin (HPIA-DEAE) fraction accounted for 8% in raw water whereas it increased to 20% in finished water, based on THMFP results. The results also indicated that hydrophilic fraction was an important source of THM formation in water treatment plant effluent. Extraction and quantification of humic substances from surface waters was more effective with DEAE ionic resin. The authors pointed out that the addition of PAC during the coagulation of natural surface waters preoxidized with electrochemically generated oxidants could significantly improve process efficiency for the removal of low MW and hydrophilic NOM fractions. Based on THMFP and simulated plant trihalomethane formation potential (SPTHMFP) results, this would allow an important reduction in THM formation during the final disinfection step. The addition of a small amount of PAC ($50\,mg\,L^{-1}$) allowed for a 50% reduction of the coagulant dose, whereas the removal of high MW fractions of NOM was maintained and that of small fractions was significantly increased. Chlorination tests confirmed the trend that showed improvement in the removal of THMFPF and SPTHMFP. PAC was especially effective for the removal of fast-reacting precursors, as deduced from the SPTHMFP results. The presence of PAC in suspension aids in the formation of the flocs. This way, the minimum dose needed for coagulation can be reduced without affecting process performance.

7.4.4 Composite, Modified and/or Miscellaneous Adsorbents

A composite adsorbent was prepared with chitosan and PAC (Zhang and Liu, 2010). Jar tests were carried out to investigate effects of pH, adsorption time, temperature, and initial concentration of NOM on the adsorbent's NOM removal efficiency. UV absorbance at 254 nm wavelength (UV_{254}) was used as a surrogate parameter for NOM concentration. It was demonstrated that NOM removal by chitosan-PAC composite adsorbent could be as high as 69% under optimal conditions. Pseudofirst-order rate expression and pseudosecond-order rate expression were fitted to the experimental results, and the latter was found to fit the experimental results quite well. The adsorption isotherm of NOM onto the adsorbent under various initial NOM concentrations was also experimentally determined. The Freundlich isotherm was found to fit the adsorption data well. NOM and its interaction with anion exchanger and adsorber resins were investigated to optimize the uptake of organics (Pürschel and Ender, 2008). Four different starches (one of them 14C-labelled) with different molecular size distributions and L-phenylalanine (L-Phe) were selected as model substances for the high-MW biopolymer and the low-MW neutral/amphiphilic fractions of NOM. The uptake of starches by various ion exchangers and adsorbers was measured in column experiments. Results were discussed in terms of size exclusion, anion exchange, adsorption, and hydrophilic/hydrophobic repulsion. In summary, at neutral pH, starch has been most effectively removed by size exclusion followed by adsorption, whereas ion exchange resins show higher uptake capacities than "pure" adsorber due to the stronger attraction between starch and the polar functional groups of the ion exchangers. At acidic pH, the uptake of sulfate, as a competitive adsorptive, leads to an earlier starch breakthrough at ion exchangers. Therefore, adsorbers are more effective. For L-Phe, ion exchange is the main uptake mechanism. It was found for both organics that the higher the water content of the resins, the more effective the uptake.

Natural pumice particles were used as granular support media and coated with iron oxides to investigate their adsorptive NOM removal from waters (Kitis et al., 2007). The study examined the impacts of natural pumice source, particle size fraction, pumice dose, pumice surface chemistry and specific surface area, and NOM source on the ultimate extent and rate of NOM removal. All adsorption isotherm experiments were conducted using the variable dose completely mixed batch reactor bottle point method. Iron oxide coating overwhelmed the surface electrical properties of the underlying pumice particles. Surface areas as high as $20.6\,m^2\,g^{-1}$ were achieved after iron coating of pumice samples, which are greater than those of IOC samples reported in the literature. Iron coating of natural pumices significantly increased the NOM uptakes both on an adsorbent mass- and surface area basis. The smallest size fractions (<63 μm) of coated pumice generally exhibited the highest NOM uptakes. A strong linear correlation between the iron content of coated pumices and their Freundlich affinity parameters (KF) indicated that the enhanced NOM uptake is due to iron oxides bound on the pumice surfaces. Iron oxide-coated pumice surfaces preferentially removed high UV-absorbing fractions of NOM, with UV absorbance reductions of up to 90%. Control experiments indicated that iron oxide species bound on pumice surfaces are stable, and potential iron release to the solution is not a concern at the pH values of typical natural waters. Based on high NOM adsorption capacities, iron oxide-coated pumice may be a promising novel adsorbent for NOM removal from waters. Furthermore, due to preferential removal of high UV-absorbing NOM fractions, iron oxide-coated pumice may also be effective in controlling the formation of disinfection by-products in drinking water treatment.

Natural organic polyelectrolytes, such as HAs and fulvic acids (FAs), were removed by adsorption onto silicate rocks (Kaneco et al., 2003). Tobermorite, zeolite, and molecular sieves 5A were used as the adsorbents. Tobermorite was more efficient by 40–50% in the removal of FAs, and by 30–50% for HA than zeolite or molecular sieves, respectively. HA removal from the solution by adsorption onto silicate rocks occurred more readily than FA removal. From the determined heat of adsorption, the adsorption process in the present study may be chemisorption (ligand exchange). Metal/HA complexes were effectively removed by adsorption onto tobermorite. Because tobermorite (a silicate rock) can be easily synthesized and obtained commercially, the adsorption method of removal of FAs and HAs is superior to their precipitation. Magnetic permanently confined micelle arrays (Mag-PCMAs) were synthesized by coating the surface of Fe_3O_4 particles with a silica/surfactant mesostructured hybrid layer for NOM removal (Wang et al., 2011a). It was determined that NOM removal efficiency by Mag-PCMAs could be as high as 80% at a wide range of initial pH values (~6.0–10.0). The adsorption isotherm fitted a Langmuir model well. Although Fe_3O_4 had a high positive charge and Mag-PCMAs a small negative charge, Mag-PCMAs had a higher NOM removal efficiency than uncoated Fe_3O_4 particles (which are also magnetic), indicating that the adsorption of NOM onto Mag-PCMAs was not dominated by electrostatic interactions. Possible mechanisms for the adsorption of NOM onto Mag-PCMAs were hydrophobic interactions and hydrogen bonding. It was feasible to reuse Mag-PCMAs after regeneration.

The main NOM component in several waters is attributed to humic substances (HS), which have several characteristics that influence how NOM may be removed from water (Fabris et al., 2008; Eikebrokk et al., 2007). Many researchers specifically focused on the removal of HS by adsorption processes. Granules with positive surface charges were prepared by coating glass beads with polypyrrole (PPy) (Bai and Zhang, 2001). Studies were conducted with batch and fixed-bed HA adsorption experiments using coated glass beads as adsorbents. X-ray photoelectron spectroscopy showed that 28% of the nitrogen atoms in the PPy coating were protonated, leading to a highly positively charged surface at $pH < 10.5$. The results also showed that the amount of protonated nitrogen atoms decreased by up to 25% due to HA adsorption, suggesting that HA uptake by the PPy-coated glass beads was affected at least partly by charge neutralization. HA adsorption also resulted in a reverse of the positive zeta potential of the PPy coating, indicating the importance of macromolecular adsorption in the process. Both pH and ionic concentration were found to affect the extent of HA adsorption by the PPy-coated granules. Chitosan was coated on the surface of polyethylene terephthalate granules through a dip and phase inversion process, and was examined for HA removal by batch adsorption experiments (Zhang and Bai, 2003). A zeta potential study indicated that the chitosan-coated granules had positive zeta potentials at $pH < 6.6$ and negative zeta potentials at $pH > 6.6$. Adsorption of HA onto the chitosan-coated granules was found to be strongly pH dependent. Significant amounts of HA were adsorbed under acidic and neutral pH conditions, but the adsorption capacity was reduced remarkably with increasing solution pH values. X-ray photoelectron spectroscopy revealed that the amino groups of the chitosan layer were protonated due to HA adsorption, suggesting the formation of an organic complex between the protonated amino groups and HA. A kinetic study indicated that the adsorption process was transport limited at low-solution pH values, but became both transport and attachment limited at high-solution pH values.

Aminated polyacrylonitrile fibers (APANFs) were prepared by surface modification of polyacrylonitrile fibers (PANFs) with diethylenetriamine in a solution, and the APANFs were studied as an adsorbent for HA removal (Deng and Bai, 2003). The surface modification reaction introduced the amine groups on the surface of the fibers, and the APANFs had a zero point of ζ potentials at pH 8.1, in contrast to pH 3.5 for the PANFs. Adsorption experiments indicated that the APANFs were very effective in removing HA from aqueous solutions in the pH range of 2–10, whereas the PANFs did not adsorb HA at all under the same conditions. It was found that both electrostatic interaction and surface complexation mechanisms played important roles in HA adsorption on the APANFs, although the relative importance of each of the adsorption mechanisms varied with solution pH values.

A chitosan/zeolite composite (CSZ) and surfactant-modified CSZ (SMCSZ) were prepared and tested for HA adsorption (Lin and Zhan, 2012). SMCSZ exhibited a higher HA adsorption capacity than CSZ. The HA adsorption capacities for CSZ and SMCSZ decreased with increasing solution pH from 4 to 12. The HA molecules adsorbed on CSZ and SMCSZ could only be partially desorbed in $1\,mol\,L^{-1}$ NaOH solution. The mechanisms for the adsorption of HA on CSZ at pH 7 might include electrostatic interaction and hydrophobic interaction. The mechanisms for the adsorption of HA on SMCSZ at pH 7 might include electrostatic interaction, organic partitioning, and hydrogen bonding. Polyaniline/attapulgite composite (ATP–PANI) was prepared by in situ chemical oxidation and studied for HA adsorption (Wang et al., 2011b). The maximum adsorption amounts were found to be 43.01, 52.91, and 61.35 $mg\,g^{-1}$ at 15, 25, and 35 °C, respectively. The HA molecules adsorbed on ATP–PANI can be effectively desorbed in 2 M NaOH solution, and regenerated adsorbent can be used repeatedly in the subsequent four adsorption–regeneration cycles with little loss of HA adsorption amount.

The use of rice husk ash (RHA) as an adsorbent of HA from water was studied (Imyim and Prapalimrungsi, 2010). RHA was also functionalized with 3-aminopropyltriethoxysilane. The adsorption capacity of $RHA\text{-}NH_2$ was higher than that of RHA. Experimental adsorption data fitted well with the Langmuir equation and the maximum adsorption capacity was 8.2 $mg\,g^{-1}$ at pH 6. Pillared bentonite was prepared for the removal of HA from water (Peng et al., 2005). It was found to be effective for the removal of HA with a high adsorption capacity of 537 $mg\,g^{-1}$, and adsorption was favored under acid conditions. Pillared bentonite was regenerated with NaOH, and the regeneration efficiency reached 83% and 85% when the concentration of NaOH reached 0.025 and 0.05 $mol\,L^{-1}$. Surfactant-modified bentonite (SMB) in removing HA from wastewaters was evaluated (Anirudhan and Ramachandran, 2007). HDTMA chloride was used to modify the surface of the clay mineral. The SMB exhibits very high adsorption potential for HA, and at pH 3.0 more than 99% removal was achieved from an initial concentration of 25 $\mu mol\,L^{-1}$. The maximum adsorption capacity was 73.52 $\mu mol\,g^{-1}$ with the binding constant, $b = 0.155\,L\,\mu mol^{-1}$ at 30 °C and pH 3.0. The adsorbent was suitable for repeated use (more than three cycles) without any noticeable loss of capacity. The ability of organo-layered double hydroxides (LDHs) was studied to remove HA from an aqueous medium (Zhang et al., 2012). Organo-LDHs with different Mg/Al molar ratios (2:1, 3:1 and 4:1) containing dodecyl sulfonate (DSO) as the interlamellar anions were prepared through coprecipitation. The maximum HA removal capacities for 2:1, 3:1, and 4:1 type LDH-DSO were 594.3, 646.7, and 428.4 $mg\,g^{-1}$, respectively, while a carbonate-based LDH only removed 23.77 $mg\,g^{-1}$. The adsorption of HA on cross-linked chitosan-epichlorohydrin (chitosan–ECH) beads was investigated by Wan Ngah et al. (2008). The maximum adsorption capacity determined from the Langmuir model was 44.84 mg/g. More than 60% of HA could be desorbed from the adsorbent using 1.0 M HCl for 180 min. Bhattacharya et al. (2012) examined the suitability of poly(amidoamine) (PAMAM) dendrimers for HA

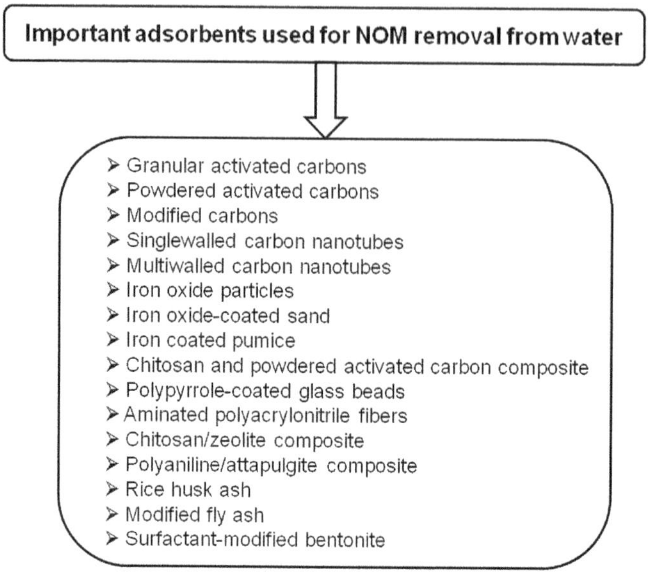

Figure 7.1 List of some of the important adsorbents used for natural organic matter (NOM) removal from water.

removal. Researchers have demonstrated efficient removal of dissolved HA using biocompatible PAMAM dendrimers. A study was conducted using ATR-FTIR to identify the specific chemical groups involved in the dendrimer–HA complexation. Upon raising the pH of the dendrimer-bound HA to 10, wherein dendrimers become neutral (pKa of primary amine) and HA remains highly negatively charged, regeneration of the dendrimers is possible as a result of the much weakened electrostatic interaction between the two species. Besides the above mentioned studies, other researchers have also used various adsorbents for HA (Shaker et al., 2012; Moussavi et al., 2013; Omri et al., 2013; Wang et al., 2012; Tang et al., 2012; Liu et al., 2011a,b; Zhan et al., 2010, 2011; Doulia et al., 2009; Zhao et al., 2008) or for NOM removal from water (Humbert et al., 2008; Matilainen et al., 2006; Korshin et al., 1997; Gu et al., 1994; Smith et al., 2012; Hyung and Kim, 2008; Wu et al., 2013; Al-Naseri and Abbas, 2009; Lambert and Graham, 1995; Smith, 1994; Jung et al., 2007; Shi et al., 2009; Matsui et al., 1998, 2004; Bjelopavlic et al., 1999; Qi and Schideman, 2008; Fettig, 2005; Day et al., 1994; McMeen and Benjamin, 1997; Heijman et al., 1999; Davis, 1982; Joseph et al., 2012; Kim et al., 2013; Ng et al., 2014). Figure 7.1 shows some of the important adsorbents used for NOM removal from water. Table 7.1

Table 7.1 Summary of the main findings on natural organic matter (NOM) removal from water by adsorption

Adsorbent	Main findings	References
Iron-impregnated activated carbons	Higher DOM uptake due to the presence of iron species on the carbon surface	Cheng et al. (2005)
Ammonia-treated activated carbons	Higher DOM uptake due to the enlargement of carbon pores and basic surface creation during ammonia treatment	Cheng et al. (2005)
Nanoscale carbon black	A lower pH (3–5) was found to be favorable for NOM removal; more than 35% of NOM was removed in the first 20 min	Wang et al. (2010)
Iron oxide particles	Optimal pH 6.0 helped to remove a maximum amount of residual organic matter by iron oxide adsorption, which was attributed to the association and dissociation of ferrihydrite and residual organic matter depending on solution pH	Choo and Kang (2003)
Iron oxide-coated sand (IOCS) was modified with HDTMA	A faster initial NOM adsorption and substantially higher capacity was observed than the unmodified IOCS over a wide pH range in both batch and column adsorption	Ding et al. (2010)
Silica particles coated with an aminosiloxane SAM (NH_2-SAM)	60% reduction of UV_{254} after 1 h and up to 70% removal of dissolved organic carbon with higher doses and contact times	Chow et al. (2009)
Chitosan-PAC composite adsorbent	Removal of NOM was as high as 69% under optimal conditions	Zhang and Liu (2010)
Tobermorite, zeolite, and molecular sieves 5A	Tobermorite was more efficient by 40–50% in the removal of fulvic acids, and by 30–50% for humic acid (HA)	Kaneco et al. (2003)
Magnetic permanently confined micelle arrays	NOM removal efficiency could be as high as 80% at a wide range of initial pH values (~6.0–10.0)	Wang et al. (2011a)

Continued

Table 7.1 Summary of the main findings on natural organic matter (NOM) removal from water by adsorption—cont'd

Adsorbent	Main findings	References
Aminated polyacrylonitrile fibers (APANFs)	Found to be very effective in removing HA from aqueous solutions in the pH range of 2–10; both electrostatic interaction and surface complexation mechanisms played important roles in HA adsorption on the APANFs	Deng and Bai (2003)
Polyaniline/attapulgite composite	HA adsorption was found to be in the range of 43–61 $mg\,g^{-1}$ at different temperatures	Wang et al. (2011b)
Rice husk ash functionalized with 3-aminopropyltriethoxysilane	Adsorption capacity was reported as 8.2 $mg\,g^{-1}$ for HA at pH 6	Imyim and Prapalimrungsi (2010)
Pillared bentonite	Adsorption capacity was reported as 537 $mg\,g^{-1}$ for HA	Peng et al. (2005)
Surfactant-modified bentonite	Showed very high adsorption potential for HA; more than 99% removal was achieved at pH 3.0 from an initial concentration of 25 $\mu mol\,L^{-1}$. The maximum adsorption capacity was 73.52 $\mu mol\,g^{-1}$ with binding constant, $b = 0.155\,L\,\mu mol^{-1}$ at 30 °C and pH 3.0	Anirudhan and Ramachandran (2007)
LDH-DSO	HA removal capacities were reported as 428–646 $mg\,g^{-1}$	Zhang et al. (2012)
Cross-linked chitosan-epichlorohydrin (chitosan–ECH) beads	HA adsorption was 44 $mg\,g^{-1}$	Wan Ngah et al. (2008)
Almond shell activated carbon	HA adsorption was reported as 169 $mg\,g^{-1}$	Omri et al. (2013)
Amine functionalized magnetic mesoporous composite microspheres	HA adsorption was found to be 128 $mg\,g^{-1}$	Tang et al. (2012)

DOM, dissolved natural organic matter; PAC, powdered activated carbon; LDH, layered double hydroxide; DSO, dodecyl sulfonate.

lists the summary of the main findings of NOM removal from water by adsorption.

7.5 CONCLUSIONS

This chapter has attempted to cover a wide range of existing literature on NOM removal from water by adsorption technology. It is evident that a great variety of adsorbents (GACs, PACs, modified carbons, single walled and multiwalled carbon nanotubes, IOPs (ferrihydrite, goethite, and hematite), IOCS, chitosan and PAC composite, polypyrrole-coated glass beads, APANFs, CSZ, RHA, SMB, etc.) have been tested thus far, with varying degrees of success, for the removal of NOM and its constituents from water. The literature also reveals that in some cases the modification of the adsorbent increased NOM removal efficiency. However, very little work has been carried out in this direction, especially on the mechanism of NOM adsorption by the modified adsorbents. A variety of adsorbents are used for the removal of NOM and its constituents from water; therefore, the type of adsorbent that is best for a specific purpose is a critical issue requiring detailed and systematic research. No definite conclusion can yet be reached, since each of the adsorbents has its own advantages and disadvantages. There is great potential for improvement in this field in the hope that new and selective adsorbents can be applied for NOM removal not only on a laboratory scale, but also commercially.

REFERENCES

Al-Naseri, S.K., Abbas, T.R., 2009. Predicting NOM removal by fixed-bed GAC adsorbers. Jordan J. Civ. Eng. 3, 172–183.

Álvarez-Uriarte, J.I., Iriarte-Velasco, U., Chimeno-Alanís, N., González-Velasco, J.R., 2010. The effect of mixed oxidants and powdered activated carbon on the removal of natural organic matter. J. Hazard. Mater. 181, 426–431.

Anirudhan, T.S., Ramachandran, M., 2007. Surfactant-modified bentonite as adsorbent for the removal of humic acid from wastewaters. Appl. Clay Sci. 35, 276–281.

Bai, R., Zhang, X., 2001. Polypyrrole-coated granules for humic acid removal. J. Colloid Interface Sci. 243, 52–60.

Bansal, R.C., Goyal, M., 2005. Activated Carbon Adsorption. CRC Press, Boca Raton.

Bell-Ajy, K., Abbaszadegan, M., Ibrahim, E., Verges LeChevallier, D., 2000. Conventional and optimized coagulation for NOM removal. J. Am. Water Works Assoc. 92, 44–58.

Bhatnagar, A., Hogland, W., Marques, M., Sillanpää, M., 2013. An overview of the modification methods of activated carbon for its water treatment applications. Chem. Eng. J. 219, 499–511.

Bhattacharya, P., Conroy, N., Rao, A.M., Powell, B.A., Ladner, D.A., Ke, P.C., 2012. PAMAM dendrimer for mitigating humic foulant. RSC Adv. 2, 7997–8001.

Bjelopavlic, M., Newcombe, G., Hayes, R., 1999. Adsorption of NOM onto activated carbon: effect of surface charge, ionic strength, and pore volume distribution. J. Colloid Interface Sci. 210, 271–280.
Bolto, B., Dixon, D., Eldridge, R., King, S., 2001. Cationic polymer and clay or metal oxide combinations for natural organic matter removal. Water Res. 35, 2669–2676.
Bond, T., Goslan, E.H., Parsons, S.A., Jefferson, B., 2011. Treatment of disinfection by–product precursors. Environ. Technol. 32, 1–25.
Brunauer, S., Emmett, P.H., Teller, E., 1938. Adsorption of gases in multimolecular layers. J. Am. Chem. Soc. 60, 309–319.
Capar, G., Yetis, U., 2001. Removal of natural organic matter and trihalomethanes from the drinking water of Ankara City. Turk. J. Eng. Environ. Sci. 25, 527–535.
Cheng, W., Dastgheib, S.A., Karanfil, T., 2005. Adsorption of dissolved natural organic matter by modified activated carbons. Water Res. 39, 2281–2290.
Choo, K.-H., Kang, S.-K., 2003. Removal of residual organic matter from secondary effluent by iron oxides adsorption. Desalination 154, 139–146.
Chow, C.W.K., Majewski, P., Bauer, S., Fabris, R., Drikas, M., 2009. Removal of natural organic matter using self-assembled monolayer technology. Desalin. Water Treat. 12, 344–351.
Crozes, G., White, P., Marshall, M., 1995. Enhanced coagulation: its effect on NOM removal and chemical costs. J. Am. Water Works Assoc. 87, 78–89.
Dąbrowski, A., 2001. Adsorption — from theory to practice. Adv. Colloid Interface Sci. 93, 135–224.
Dastgheib, S.A., Karanfil, T., Cheng, W., 2004. Tailoring activated carbons for enhanced removal of natural organic matter from natural waters. Carbon 42, 547–557.
Davis, J.A., 1982. Adsorption of natural dissolved organic matter at the oxide/water interface. Geochim. Cosmochim. Acta 46, 2381–2393.
Day, G.M., Hart, B.T., McKelvie, I.D., Beckett, R., 1994. Adsorption of natural organic matter onto goethite. Colloids Surf. A Physicochem. Eng. Aspects 89, 1–13.
Deng, S., 2006. Sorbent technology. In: Encyclopedia of Chemical Processing. Taylor & Francis, New York.
Deng, S., Bai, R.B., 2003. Aminated polyacrylonitrile fibers for humic acid adsorption: behaviors and mechanisms. Environ. Sci. Technol. 37, 5799–5805.
Ding, C., Shang, C., 2010. Mechanisms controlling adsorption of natural organic matter on surfactant-modified iron oxide-coated sand. Water Res. 44, 3651–3658.
Ding, C., Yang, X., Liu, W., Chang, Y., Shang, C., 2010. Removal of natural organic matter using surfactant-modified iron oxide-coated sand. J. Hazard. Mater. 174, 567–572.
Dong, C., Chen, W., Liu, C., 2014. Preparation of novel magnetic chitosan nanoparticle and its application for removal of humic acid from aqueous solution. Appl. Surf. Sci. 292, 1067–1076.
Doulia, D., Leodopoulos, C., Gimouhopoulos, K., Rigas, F., 2009. Adsorption of humic acid on acid-activated Greek bentonite. J. Colloid Interface Sci. 340, 131–141.
Eikebrokk, B., Fabris, R., Drikas, M., Chow, C., 2007. NOM characteristics and treatability by coagulation. In: Hahn, H.H., Hoffman, E., Ødegaard, H. (Eds.), Chemical Water and Wastewater Treatment IX. IWA Publishing, London.
Fabris, R., Chow, C.W.K., Drikas, M., 2004. Practical application of a combined treatment process for removal of recalcitrant NOM – alum and PAC. Water Sci. Technol. Water Supply 4, 89–94.
Fabris, R., Chow, C.W.K., Drikas, M., Eikebrokk, B., 2008. Comparison of NOM character in selected Australian and Norwegian drinking waters. Water Res. 42, 4188–4196.
Fettig, J., 2005. Modelling the uptake of natural organic matter (NOM) by different granular sorbent media. J. Water Supply Res. Technol. Aqua 54, 83–93.
Freundlich, H.M.F., 1906. Over the adsorption in solution. J. Phys. Chem. 57, 385–471.

Genz, A., Baumgarten, B., Goernitz, M., Jekel, M., 2008. NOM removal by adsorption onto granular ferric hydroxide: equilibrium, kinetics, filter and regeneration studies. Water Res. 42, 238–248.
Grieco, S.A., Ramarao, B.V., 2013. Removal of TCEP from aqueous solutions by adsorption with zeolites. Colloids Surf. A Physicochem. Eng. Aspects 434, 329–338.
Gu, B., Schmitt, J., Chen, Z., Liang, L., McCarthy, J.F., 1994. Adsorption and desorption of natural organic matter on iron oxide: mechanisms and models. Environ. Sci. Technol. 28, 38–46.
Heijman, S.G.J., van Paassen, A.M., van der Meer, W.G.J., Hopman, R., 1999. Adsorptive removal of natural organic matter during drinking water treatment. Water Sci. Technol. 40, 183–190.
Humbert, H., Gallard, H., Suty, H., Croué, J.-P., 2008. Natural organic matter (NOM) and pesticides removal using a combination of ion exchange resin and powdered activated carbon (PAC). Water Res. 42, 1635–1643.
Hyung, H., Kim, J.-H., 2008. Natural organic matter (NOM) adsorption to multi-walled carbon nanotubes: effect of NOM characteristics and water quality parameters. Environ. Sci. Technol. 42, 4416–4421.
Imyim, A., Prapalimrungsi, E., 2010. Humic acids removal from water by aminopropyl functionalized rice husk ash. J. Hazard. Mater. 184, 775–781.
Iriarte-Velasco, U., Álvarez-Uriarte, J.I., Chimeno-Alanís, N., González-Velasco, J.R., 2008. Natural organic matter adsorption onto granular activated carbons: implications in the molecular weight and disinfection byproducts formation. Ind. Eng. Chem. Res. 47, 7868–7876.
Joseph, L., Flora, J.R.V., Park, Y.-G., Badawy, M., Saleh, H., Yoon, Y., 2012. Removal of natural organic matter from potential drinking water sources by combined coagulation and adsorption using carbon nanomaterials. Sep. Purif. Technol. 95, 64–72.
Jung, H.R., Seo, S.D., Choi, H.C., 2007. Preparation of mesostructured carbon (MSC) for removal of natural organic matter (NOM) in aqueous solution. Mater. Sci. Forum 544–545, 123–126.
Kaneco, S., Itoh, K., Katsumata, H., Suzuki, T., Masuyama, K., Funasaka, K., Hatano, K., Ohta, K., 2003. Removal of natural organic polyelectrolytes by adsorption onto tobermorite. Environ. Sci. Technol. 37, 1448–1451.
Karanfil, T., Kitis, M., Kilduff, J.E., Wigton, A., 1999. Role of granular activated carbon surface chemistry on the adsorption of organic compounds. 2. Natural organic matter. Environ. Sci. Technol. 33, 3225–3233.
Kasprzyk-Hordern, B., 2004. Chemistry of alumina, reactions in aqueous solution and its application in water treatment. Adv. Colloid Interface Sci. 110, 19–48.
Kim, S., Kim, J., Seo, G., 2013. Iron oxide nanoparticle-impregnated powder-activated carbon (IPAC) for NOM removal in MF membrane water treatment system. Desalin. Water Treat. 51, 1–9.
Kitis, M., Kaplan, S.S., Karakaya, E., Yigit, N.O., Civelekoglu, G., 2007. Adsorption of natural organic matter from waters by iron coated pumice. Chemosphere 66, 130–138.
Korshin, G.V., Benjamin, M.M., Sletten, R.S., 1997. Adsorption of natural organic matter (NOM) on iron oxide: effects on NOM composition and formation of organo-halide compounds during chlorination. Water Res. 31, 1643–1650.
Kumar, E., Bhatnagar, A., Hogland, W., Marques, M., Sillanpää, M., 2014. Interaction of anionic pollutants with Al-based adsorbents in aqueous media – a review. Chem. Eng. J. 241, 443–456.
Lambert, S.D., Graham, N.J.D., 1995. Removal of non-specific dissolved organic matter from upland potable water supplies—I. Adsorption. Water Res. 29, 2421–2426.
Langmuir, I., 1916. The adsorption of gases on plane surface of glass, mica and platinum. J. Am. Chem. Soc. 40, 1361–1368.

Lin, J., Zhan, Y., 2012. Adsorption of humic acid from aqueous solution onto unmodified and surfactant-modified chitosan/zeolite composites. Chem. Eng. J. 200–202, 202–213.
Lin, L., Lei, Z., Wang, L., Liu, X., Zhang, Y., Wan, C., Lee, D.-J., Tay, J.H., 2013. Adsorption mechanisms of high-levels of ammonium onto natural and NaCl-modified zeolites. Sep. Purif. Technol. 103, 15–20.
Liu, F., Xu, Z., Wan, H., Wan, Y., Zheng, S., Zhu, D., 2011a. Enhanced adsorption of humic acids on ordered mesoporous carbon compared with microporous activated carbon. Environ. Toxicol. Chem. 30, 793–800.
Liu, F.-F., Fan, J.-L., Wang, S.-G., Ma, G.-H., 2013. Adsorption of natural organic matter analogues by multi-walled carbon nanotubes: comparison with powdered activated carbon. Chem. Eng. J. 219, 450–458.
Liu, Z., Zu, Y., Meng, R., Xing, Z., Tan, S., Zhao, L., Sun, T., Zhou, Z., 2011b. Adsorption of humic acid onto carbonaceous surfaces: atomic force microscopy study. Microsc. Microanal. 17, 1015–1021.
Lu, C., Su, F., 2007. Adsorption of natural organic matter by carbon nanotubes. Sep. Purif. Technol. 58, 113–121.
Mantell, C.L., 1951. Adsorption. McGraw-Hill Book Company, New York.
Marsh, H., Rodríguez-Reinoso, F., 2006. Activated Carbon. Elsevier, Oxford.
Martucci, A., Pasti, L., Marchetti, N., Cavazzini, A., Dondi, F., Alberti, A., 2012. Adsorption of pharmaceuticals from aqueous solutions on synthetic zeolites. Microporous Mesoporous Mater. 148, 174–183.
Matilainen, A., Sillanpää, M., 2010. Removal of natural organic matter from drinking water by advanced oxidation processes. Chemosphere 80, 351–365.
Matilainen, A., Vepsäläinen, M., Sillanpää, M., 2010. Natural organic matter removal by coagulation during drinking water treatment: a review. Adv. Colloid Interface Sci. 159, 189–197.
Matilainen, A., Vieno, N., Tuhkanen, T., 2006. Efficiency of the activated carbon filtration in the natural organic matter removal. Environ. Int. 32, 324–331.
Matsui, Y., Murase, R., Sanogawa, T., Aoki, N., Mima, S., Inoue, T., Matsushita, T., 2004. Micro-ground powdered activated carbon for effective removal of natural organic matter during water treatment. Water Sci. Technol. Water Supply 4, 155–163.
Matsui, Y., Yuasa, A., Li, F., 1998. Overall adsorption isotherm of natural organic matter. J. Environ. Eng. 124, 1099–1107.
McBain, J.W., 1909. The mechanism of the adsorption ("sorption") of hydrogen by carbon. Philos. Mag. Ser. 6, 916–935.
McMeen, C.R., Benjamin, M.M., 1997. NOM removal by slow sand filtration through iron oxide-coated olivine. J. Am. Water Works Assoc. 89, 57–71.
Metsämuuronen, S., Sillanpää, M., Bhatnagar, A., Mänttäri, M., 2014. Natural organic matter removal from drinking water by membrane technology. Sep. Purif. Rev. 43, 1–61.
Moussavi, S.P., Ehrampoush, M.H., Mahvi, A.H., Ahmadian, M., Rahimi, S., 2013. Adsorption of humic acid from aqueous solution on single-walled carbon nanotubes. Asian J. Chem. 25, 5319–5324.
Naghizadeh, A., Nasseri, S., Rashidi, A.M., Kalantary, R.R., Nabizadeh, R., Mahvi, A.H., 2013. Adsorption kinetics and thermodynamics of hydrophobic natural organic matter (NOM) removal from aqueous solution by multi-wall carbon nanotubes. Water Sci. Technol. Water Supply 13, 273–285.
Najafi, M., Rostamian, R., Rafati, A.A., 2011. Chemically modified silica gel with thiol group as an adsorbent for retention of some toxic soft metal ions from water and industrial effluent. Chem. Eng. J. 168, 426–432.
Ng, M., Kho, E., Liu, S., Lim, M., Amal, R., 2014. Highly adsorptive and regenerative magnetic TiO_2 for natural organic matter (NOM) removal in water. Chem. Eng. J. 246, 196–203.

Omri, A., Benzina, M., Trabelsi, W., Ammar, N., 2013. Adsorptive removal of humic acid on activated carbon prepared from almond shell: approach for the treatment of industrial phosphoric acid solution. Desalin. Water Treat., 1–12.
Owen, D.M., Amy, G.L., Chowdhury, Z.K., Paode, R., McCoy, G., Viscosil, K., 1985. NOM characterization and treatability. J. Am. Water Works Assoc. 87, 46–63.
Peng, X., Luan, Z., Chen, F., Tian, B., Jia, Z., 2005. Adsorption of humic acid onto pillared bentonite. Desalination 174, 135–143.
Pürschel, M., Ender, V., 2008. Sorption of natural organic matter by adsorber and ion exchange resins – investigations with starch and phenylalanine as model substances. In: 15th International Conference on the Properties of Water and Steam (15th ICPWS), Radisson SAS Hotel Berlin.
Qi, S., Schideman, L.C., 2008. An overall isotherm for activated carbon adsorption of dissolved natural organic matter in water. Water Res. 42, 3353–3360.
Richardson, S.D., Plewa, M.J., Wagner, E.D., Schoeny, R., DeMarini, D.M., 2007. Occurrence, genotoxicity, and carcinogenicity of regulated and emerging disinfection by-products in drinking water: a review and roadmap for research. Mutat. Res. 636, 178–242.
Shaker, A.M., Komy, Z.R., Heggy, S.E.M., El-Sayed, M.E.A., 2012. Kinetic study for adsorption humic acid on soil minerals. J. Phys. Chem. A 116, 10889–10896.
Shi, D., Xie, S., Wen, D., Xi, D., 2009. Removal of bromate and natural organic matter by using biologically activated carbon. Int. J. Environ. Pollut. 38, 180–192.
Siddiqui, M., Amy, G., Ryan, J., Odem, W., 2000. Membranes for the control of natural organic matter from surface waters. Water Res. 34, 3355–3370.
Smith, B., Yang, J., Bitter, J.L., Ball, W.P., Fairbrother, D.H., 2012. Influence of surface oxygen on the interactions of carbon nanotubes with natural organic matter. Environ. Sci. Technol. 46, 12839–12847.
Smith, E.H., 1994. Bench-scale tests and modeling of adsorption of natural organic matter by activated carbon. Water Res. 28, 1693–1702.
Tang, Y., Liang, S., Yu, S., Gao, N., Zhang, J., Guo, H., Wang, Y., 2012. Enhanced adsorption of humic acid on amine functionalized magnetic mesoporous composite microspheres. Colloids Surf. A Physicochem. Eng. Aspects 406, 61–67.
Velten, S., Knappe, D.R.U., Traber, J., Kaiser, H.-P., von Gunten, U., Boller, M., Meylan, S., 2011. Characterization of natural organic matter adsorption in granular activated carbon adsorbers. Water Res. 45, 3951–3959.
Vidic, R.D., Suidan, M.T., 1991. Role of dissolved oxygen on the adsorptive capacity of activated carbon for synthetic and natural organic matter. Environ. Sci. Technol. 25, 1612–1618.
Wan Ngah, W.S., Hanafiah, M.A.K.M., Yong, S.S., 2008. Adsorption of humic acid from aqueous solutions on crosslinked chitosan–epichlorohydrin beads: kinetics and isotherm studies. Colloids Surf. B Biointerfaces 65, 18–24.
Wang, H., Kang, J., Liu, H., Qu, J., 2009. Preparation of organically functionalized silica gel as adsorbent for copper ion adsorption. J. Environ. Sci. 21, 1473–1479.
Wang, H., Keller, A., Li, F., 2010. Natural organic matter removal by adsorption onto carbonaceous nanoparticles and coagulation. J. Environ. Eng. 136, 1075–1081.
Wang, H., Keller, A.A., Clark, K.K., 2011a. Natural organic matter removal by adsorption onto magnetic permanently confined micelle arrays. J. Hazard. Mater. 194, 156–161.
Wang, J., Han, X., Ma, H., Ji, Y., Bi, L., 2011b. Adsorptive removal of humic acid from aqueous solution on polyaniline/attapulgite composite. Chem. Eng. J. 173, 171–177.
Wang, M., Liao, L., Zhang, Z., Li, Z., 2012. Adsorption of low concentration humic acid from water by palygorskite. Appl. Clay Sci. 67–68, 164–168.
Wang, S., Peng, Y., 2010. Natural zeolites as effective adsorbents in water and wastewater treatment. Chem. Eng. J. 156, 11–24.

Wu, C.N., Wang, L., Ling, Q., Tang, Y.C., Huang, J., Pan, F.K., 2013. Removal of natural organic matter of artificial lake by modified fly ash. Adv. Mater. Res. 690–693, 1020–1023.
Xie, Q., Xie, J., Wang, Z., Wu, D., Zhang, Z., Kong, H., 2013. Adsorption of organic pollutants by surfactant modified zeolite as controlled by surfactant chain length. Microporous Mesoporous Mater. 179, 144–150.
Zaitseva, N., Zaitsev, V., Walcarius, A., 2013. Chromium(VI) removal via reduction–sorption on bi-functional silica adsorbents. J. Hazard. Mater. 250–251, 454–461.
Zhan, Y., Lin, J., Qiu, Y., Gao, N., Zhu, Z., 2011. Adsorption of humic acid from aqueous solution on bilayer hexadecyltrimethyl ammonium bromide-modified zeolite. Front. Environ. Sci. Eng. China 5, 65–75.
Zhan, Y., Zhu, Z., Lin, J., Qiu, Y., Zhao, J., 2010. Removal of humic acid from aqueous solution by cetylpyridinium bromide modified zeolite. J. Environ. Sci. 22, 1327–1334.
Zhang, G., Wu, T., Li, Y., Huang, X., Wang, Y., Wang, G., 2012. Sorption of humic acid to organo layered double hydroxides in aqueous solution. Chem. Eng. J. 191, 306–313.
Zhang, J., Liu, F., 2010. Adsorption of natural organic matter onto a composite adsorbent prepared with chitosan and powdered activated carbon. Desalin. Water Treat. 20, 291–296.
Zhang, X., Bai, R., 2003. Mechanisms and kinetics of humic acid adsorption onto chitosan-coated granules. J. Colloid Interface Sci. 264, 30–38.
Zhao, L., Luo, F., Wasikiewicz, J.M., Mitomo, H., Nagasawa, N., Yagi, T., Tamada, M., Yoshii, F., 2008. Adsorption of humic acid from aqueous solution onto irradiation-crosslinked carboxymethylchitosan. Bioresour. Technol. 99, 1911–1917.

CHAPTER 8

Ion Exchange

Mika Sillanpää, Irina Levchuk
Lappeenranta University of Technology, LUT Faculty of Technology, LUT Chemtech, Laboratory of Green Chemistry, Sammonkatu 12, 50130 Mikkeli, Finland

Abstract

The presence of natural organic matter (NOM) in drinking water causes difficulties during conventional water treatment processes such as coagulation. Also, problems arise when alternative treatment techniques for NOM removal, such as nanofiltration (NF), are applied. The most common problem with NF is membrane fouling. Ion exchange (IE) processes proved to be an efficient NOM removal technology, and are recommended for use at the beginning of the treatment process. This approach not only significantly decreases the concentration of NOM but also prevents the formation of disinfection by-products, such as trihalomethanes. This chapter reviews recent studies conducted on NOM removal from water by IE.

Keywords: Ion exchange; Natural organic matter (NOM); Water treatment.

ABBREVIATIONS

BDOC Biodegradable dissolved organic carbon
DBP Disinfection by-product
DOC Dissolved organic carbon
DOM Dissolved organic matter
ERD Effective resin dose
FIX Fluidized ion exchange
FP Formation potential
HAAs Haloacetic acids
HRT Hydraulic residence time
IE Ion exchange
MIEX® Magnetic ion exchange resin
NF Nanofiltration
NOM Natural organic matter
PAC Powdered activated carbon
SHA Slightly hydrophobic acid
SUVA Specific UV absorbance
THMFP Trihalomethane formation potential
THM Trihalomethane
TOC Total organic carbon
VHA Very hydrophobic acid

8.1 INTRODUCTION

This chapter introduces the basics of ion exchange (IE), and briefly describes types of ion exchangers and recent findings on NOM removal from water by the IE process. An *ion exchange process* is commonly defined as the reversible exchange of ions between the solid and liquid phases (Harland, 1994; Wachinski and Etzel, 1997). The solid phase is also known as *ion exchanger* or *ion exchange resin* and, being insoluble in the liquid phase, it carries exchangeable ions and does not undergo substantive structural changes during the reaction (Helfferich, 1995). Depending on exchangeable ions, ion exchangers can be divided into cation and anion exchangers. Also, simultaneous cation and anion exchange can be attributed to certain types of materials, known as *amphoteric ion exchangers*. Typical cation and anion exchange reactions are shown in Eqn (8.1) and (8.2), respectively, where X is the structural unit of the ion exchanger.

$$2NaX + CaCl_2\ (aq) \leftrightarrow CaX_2 + 2NaCl\ (aq) \tag{8.1}$$

$$2XCl + Na_2SO_4\ (aq) \leftrightarrow X_2SO_4 + 2NaCl\ (aq) \tag{8.2}$$

Reaction (8.1) is attributed to the softening of water. During the typical water softening process, hard water is pumped through an IE column (NaX) and cations such as calcium are removed from the water and replaced by an equivalent amount of sodium. When all sodium ions in the cation exchanger are replaced, it is considered to be "exhausted"; however, "regeneration" of IE resin is easily achievable by pumping in a sodium salt solution (Helfferich, 1995). Schematic presentation of this process is shown in Figure 8.1. The left part of Figure 8.1 demonstrates the initial step of the process, when contact between ion exchanger and water containing calcium cations starts. On the right side of Figure 8.1, achieved equilibrium is shown.

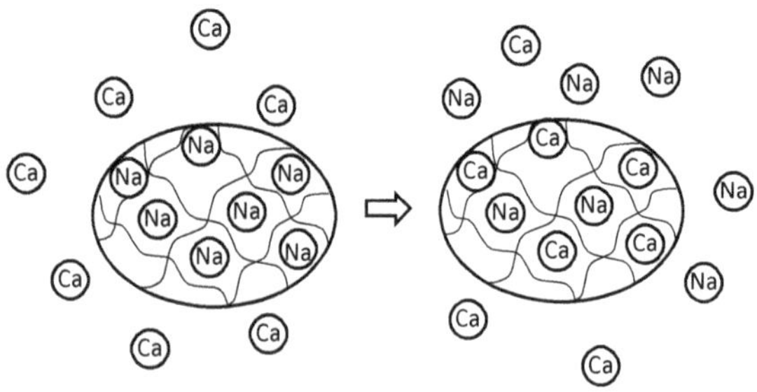

Figure 8.1 Schematic representation of Na^+–Ca^{2+} ion exchange (IE) process. *Adapted from Helfferich (1995).*

The unique properties of ion exchangers can be explained by their structure, which, according to the common definition, is a framework carrying a positive or negative surplus charge, compensated by opposite ions also known as counter-ions. Since the counter-ions move freely, they can easily be replaced by other ions that have the same sign. The content of counter-ions in an ion exchanger is known as *ion-exchange capacity*. Another important characteristic of ion exchangers is selectivity, which is commonly defined as the ability of an ion exchanger to distinguish between different species of counter-ions (Helfferich, 1995).

Thus, the ideal ion exchanger should meet following requirements (Harland, 1994):

- Hydrophilicity
- Chemical and physical stability
- Relatively high speed of IE
- Sufficient IE capacity
- Particle size and effective surface area that are adequate to the application.

An IE process can be conducted in a batch or column method. In the batch process, IE resin is mixed with solution in a tank, after which it is separated. The main drawback of this method is the inefficiency of batch-wise resin regeneration. Hence, the number of applications for the batch IE process is limited, and the column method is widely used on an industrial scale (Wachinski and Etzel, 1997).

IE properties can be attributed to a wide range of natural and synthetic materials. Among the natural materials, aluminosilicates are well known mainly as cation exchangers; however, some of them exhibit anion exchanging properties (apatite and hydroxylapatite; Wachinski and Etzel, 1997). For instance, recently it was demonstrated that natural zeolite, consisting mainly of clinoptilolite obtained from the Semnan province in Iran, is a suitable ion exchanger for $NH4^+$-ion removal from water. Thus, it has potential for application in wastewater treatment and the agricultural industry, since it is an environmentally friendly and relatively inexpensive ion exchanger (Malekian et al., 2011). Other examples of natural ion exchangers are different types of coal, alumina, alginic acid, collodine, and keratin. Some natural materials gain IE properties after certain chemical treatment, for instance, sulfonation and phosphorylation is successfully applied to wood, tar, nut shells, and cotton (Wachinski and Etzel, 1997).

Most of the ion exchangers used in wastewater treatment and other industrial applications are synthetic resins. Different types of IE resins can be distinguished

1. Strong-acid cation resins
2. Weak-acid cation resins
3. Strong-base anion resins
4. Weak-base anion resins
5. Heavy metal selective chelating resins.

The chemical behavior of strong-acid cation resins resembles the behavior of strong acids, which explains the origin of their name. The chemical matrix of this type of resin consists of styrene and divinylbenzene and the functional groups are sulfonic acid radicals. The behavior of weak-acid cation resin is similar to that of weak organic acids that are weakly dissociated. The functional group is carboxylic acid.

Strong-base anion resins can be divided into two main types. Resins of the first type contain a functional group of three methyl groups, while in resins of the second type an ethanol group replaces one of the methyl groups. Strong-base exchangers of the first type are more stable, whereas exchangers of the second type have a higher capacity and regeneration efficiency (Wachinski and Etzel, 1997).

The weak-base anion exchangers are normally based on phenol–formaldehyde or epoxy matrices with a functional group of secondary or tertiary amines (Wachinski and Etzel, 1997). Heavy metal selective chelating resins, such as ethylenediaminetetraacetic acid, behave in a similar way to weak-acid cation resins and have a high selectivity for metals Eckenfelder (2000).

8.2 REMOVAL OF NOM FROM WATER BY ION EXCHANGE

What follows is a brief overview of recent developments in NOM removal from water by IE. The presence of NOM in treated water implies difficulties during the purification process and problems with water quality after treatment. For instance, NOM causes membrane fouling and decreases activated carbon efficiency (Kennedy et al., 2008; Kim and Dempsey, 2013; Lee et al., 2004; Teixeira and Sousa, 2013; Zularisam et al., 2007; Gibert et al., 2013a,b). Also, NOM can react with chemicals used during the water treatment process, leading to the formation of disinfection by-products (DBPs), for instance, potentially carcinogenic chlorinated hydrocarbons can be formed after water chlorination (Yang et al., 2013; Tian et al., 2013; Bolto et al., 2004).

Anion exchange water treatment is well known as an alternative to conventional water treatment in cases of high NOM concentration in raw water (Hongve et al., 1999). It was reported that NOM concentration in drinking water is usually in the range of 2–10 ppm, of which only 10–30% are defined species (Bolto et al., 2002). According to some estimations, from 10% to 40% of NOM is not removed after IE, which probably correlates with the amount of uncharged species in NOM (Bolto et al., 2004).

Three hydrophobic and three hydrophilic fractions of NOM can be distinguished (Nkambule et al., 2009). The main constituents of hydrophobic fractions are humic acids (HAs) and fulvic acids (FAs), while low molecular weight (MW) carbohydrates, proteins, and amino acids are often found in the hydrophilic fraction of NOM. The distribution of organic compounds for average river water can be presented in following way: 40% FAs, 10% HAs, 30% hydrophilic acids, 10% carbohydrates, 6% carboxylic acids, and 4% amino acids (Bolto et al., 2004). Depending on the fraction of NOM, deviations in DBP formation can be observed. One example is trihalomethane (THM) formation, which is caused by all six fractions of NOM. The hydrophilic basic fraction of NOM is the most reactive of all six fractions, and it was reported to result in higher THM formation (Nkambule et al., 2009).

Humic substances are known as "problematic" compounds because they can interact with nonpolar contaminants, such as pesticides or polychlorinated biphenyls (Bolto et al., 2004). Hence, to choose the appropriate water treatment method for NOM removal, it is critical to define the species contained in NOM in each specific case. For instance, it was demonstrated that for the removal of charged organic compounds from water, IE is more efficient than coagulation (Bolto et al., 2002). A comparative study of humic substance removal using adsorption onto activated carbon, anion exchange resins, carbonaceous resins, and metal oxides claimed that the IE process was the most efficient (Fettig, 1999).

Strongly basic anion exchange resins have been investigated as a means of NOM removal from drinking water (Bolto et al., 2004; Brattebø et al., 1987). Widely studied resins include quaternary ammonium resins in chlorine form. NOM removal from water can be presented with the following reaction, where R^- is charged dissolved organic carbon (DOC) (Bolto et al., 2004):

$$\text{Resin-NMe}_3^+\text{Cl}^- + \text{R}^- \leftrightarrow \text{Resin-NMe}_3^+\text{R}^- + \text{Cl}^- \tag{8.3}$$

Regeneration of this type of resin is conducted with an excess of brine or caustic brine. The effect of physical and chemical structure on NOM elimination was studied as well, and it was concluded that the highest NOM removal is attributed to resins with a high water content (Bolto et al., 2002). Also, it was reported that polystyrene resins have a higher selectivity than acrylic. Studies conducted with strongly basic anion exchangers prior to 2004 are comprehensively reviewed by Bolto et al. (2004). They suggested that NOM removal is not as efficient with weakly basic resins as with strongly basic resins or their acrylic analogues.

Commercially available anionic exchange resins were tested for NOM acid removal from water, and the best performance was observed for the smallest resins and/or those with the highest water content. The resins were not named in this work, however, for reasons of confidentiality (Cornelissen et al., 2008).

One of the main problems of IE implementation for NOM removal from water on an industrial scale is the need to use a packed bed of IE resin, which is possible in the later stages of the water treatment process. As an example of DOC removal from water on a large scale, a 2.5 ML day^{-1} plant in Germany can be considered (Bolto et al., 2002). Macroporous (MP) styrene resin successfully removed DOC from water (initial concentration 6.5 mgL^{-1}). Within a contact time of 1.2 min, 50% of DOC was removed. Nevertheless, the plant stopped this process due to difficulties in disposing of the regeneration solution.

A magnetic ion exchange resin (MIEX®) and MIEX®-DOC process developed by the Australian Water Quality Centre, in collaboration with Orica Australia Pty Ltd., used MIEX® in a slurry reactor prior to coagulation (Cornelissen et al., 2008; Drikas et al., 2002). MIEX® is a strong base anion exchange resin with a MP polyacrylic matrix in chloride form, containing magnetic iron oxide particles (IOP) within its core (Boyer and Singer, 2006; Drikas et al., 2011). Incorporated into resin, IOPs simplify separation and recycling of the resin in a continuous process. The MIEX® particle size is known to be 2–5 times smaller in diameter than that of traditional resins (Hsu and Singer, 2010), which leads to faster IE kinetics. A flow chart of the MIEX® process is shown in Figure 8.2.

The MIEX® process consists of two main steps, the first being contact of raw water with MIEX® slurry within 10–30 min, after which the resin is separated in a clarifier and the supernatant is sent to subsequent treatment, usually coagulation. Collected MIEX® is regenerated and reused (Singer and Bilyk, 2002). MIEX® can be used without pretreatment and is very stable, making it a highly beneficial option for water treatment (Kitis et al., 2007).

A comparison of NOM removal efficiency by MIEX® and by conventional methods was conducted on both laboratory and pilot scales. Laboratory-scale studies demonstrated that alum in combination with MIEX® has a

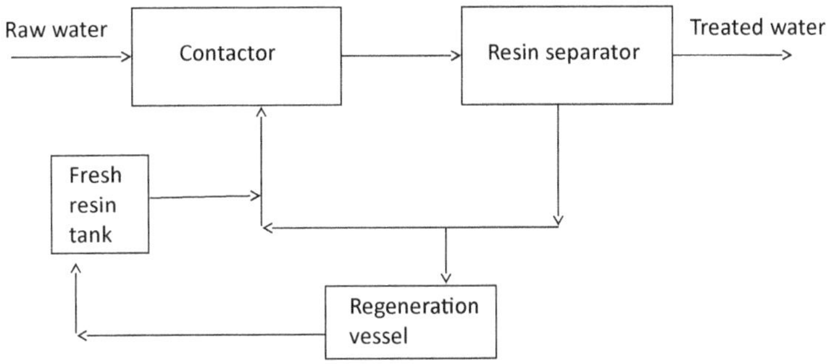

Figure 8.2 A flow chart of the magnetic ion exchange resin (MIEX®) process. *Adapted from Boyer and Singer (2006).*

higher efficiency of DOC removal from water than alum alone. Also it was observed that a broader range of compounds was removed and the formation of THMs was lower when MIEX® was included in the treatment scheme (Drikas et al., 2003). Singer and Bilyk (2002) compared alum with combined MIEX® and alum treatment of nine surface waters, and observed that values of total organic carbon (TOC) concentration, UV absorbance, THM, and haloacetic acid concentrations in treated water were lower when MIEX® resin was also used. The ability of MIEX® to remove hydrophobic and hydrophilic acids was observed as well. The effect of water hydrophobicity on DOC removal by MIEX® was studied, and it was found that with increased water hydrophobicity, DOC removal decreases. Thus, DOC elimination of 56, 33, and 25% was observed for waters containing 21, 50, and 75% of hydrophobic NOM, respectively (Mergen et al., 2008).

Monitoring of different parameters at a water treatment plant in South Australia suggested that changes in water quality and the character of the NOM affect the performance of MIEX® significantly (Morran et al., 2004) and confirm the efficiency of a combined MIEX® and coagulation process. Improved NOM removal efficiency using MIEX® in comparison with coagulation was also reported by Fearing et al. (2004) and Humbert et al. (2005). Apart from improving NOM removal from water, the incorporation of the MIEX® process to a conventional water treatment scheme (coagulation–flocculation) leads to a decrease in the required dose of coagulants and other chemicals. For instance, five different raw drinking water samples from the City of Istanbul were treated with a MIEX® process prior to coagulation on laboratory scale, to investigate the effect of different doses of MIEX® and contact time, as well as coagulant dose on the quality of treated water (Kitis et al., 2007). After treatment with a resin dose of 5–10 mL settled resin L^{-1} and a contact time 10–20 min, the values of DOC and specific UV absorbance ($SUVA_{254}$) were <1.5 mg L^{-1} and <21 mg^{-1} DOC-m, respectively. Moreover, removal of nitrates and sulfates from all raw water samples was detected at 17–42% and 9–24%, respectively; the dose of MIEX® was 10 mL settled resin L^{-1} at 10 min of contact time. Depending on the quality of raw water, the decrease in the required coagulant (alum) dose by treating water with MIEX® prior to coagulation was estimated to be 0–30 mg L^{-1} (Kitis et al., 2007). Another study was conducted with four types of drinking water from California characterized by low turbidity, moderate organic carbon concentration, moderate to high concentration of bromide, and a wide range of alkalinities (Boyer and Singer, 2005), to determine if MIEX® treatment is more efficient than coagulation with alum. The findings suggested that MIEX® treatment and MIEX® treatment followed by coagulation performed in a similar way, while DOC and SUVA removal efficiency by coagulation was significantly lower, especially for raw waters with high SUVA and low anionic strength (Boyer and Singer, 2005). Interestingly, bromide was removed from all tested water samples with MIEX®, and it was established that bromide removal efficiency was dependent on the alkalinity of the raw water and initial bromide concentration. No changes in bromide concentration after coagulation were determined (Boyer and Singer, 2005).

The effect of bicarbonate, chloride, and bromide presence in water on DOC removal with MIEX® and polystyrene resins A-641 and IRA910 was studied by Hsu and Singer (2010). Bromide removal was reported to be more efficient with use of A-641 and IRA910 resins, but the level of DOC elimination using MIEX® was higher: DOC removal efficiencies by MIEX®, A-641, and IRA910 resins after 30 min were 46, 13, and 7%, respectively (Hsu and Singer, 2010). Increasing the contact time to 5 h resulted in higher DOC elimination for all tested resins but MIEX® still performed better. The simultaneous removal of NOM, nitrates, sulfates, bromides, and pesticides from high DOC-content surface water was performed using four commercial strong anion exchange resins: MIEX®, DOWEX-11, DOWEX-MSA, and IRA938 (Humbert et al., 2005). All the tested resins demonstrated significant DOC removal, but the kinetics of DOC removal from water with MIEX® and IRA938 were faster than that of DOWEX-11 and DOWEX-MSA. After only 5 min of contact time with 8 mL L^{-1} of MIEX® and IRA938, the residual DOC concentration in water varied from 1 to 2 mg L^{-1}. Both high and partially low MW UV-absorbing NOM were removed with all tested resins. Pesticide removal by MIEX®, DOWEX-11, and DOWEX-MSA was estimated to be 10%, while 35% was eliminated by

IRA938. The optimal contact time for DOC removal by MIEX® with doses 2–8 mL L^{-1} was estimated to be 10–20 min (Humbert et al., 2005).

Recently, the efficiency of NOM removal was tested by MIEX® in bicarbonate form and compared with MIEX® chloride form, which has been thoroughly studied (Walker and Boyer, 2011). It was observed that the performance of both resins was similar for removal of DOC, UV_{254}, and sulfate, but bicarbonate-form MIEX® demonstrated higher performance for bromide removal. The sequence of affinity for dissolved organic matter (DOM) and anion removal was similar for fresh MIEX®-HCO3 and virgin MIEX®-Cl, and can be presented in following way: sulfate ~UV_{254} > DOC > bromide. A decrease in removal efficiency for both resins was noted with an increased number of regeneration cycles. The DOC and bromide elimination with the use of MIEX®-HCO_3 resin was reported to be 69% and 34%, correspondingly, from average initial concentrations of 18 mg C L^{-1} and 780 μg L^{-1}, respectively.

The results of NOM removal studies with MIEX® for batch and continuous operations conducted before 2008 are excellently summarized by Mergen et al. (2008). The effect of MIEX® on membrane fouling reduction was demonstrated in a number of studies (Fabris et al., 2007; Humbert et al., 2007; Huang et al., 2011; Kim and Dempsey, 2010).

The first water plant utilizing a MIEX®-DOC process was launched at Mount Pleasant, South Australia, in 2001, with a capacity of 2.5 ML day^{-1} (Drikas et al., 2011) for the treatment of potable water. This plant was divided into two streams with a capacity of 1.25 ML day^{-1} each. The first stream consists of a MIEX®-DOC process followed by coagulation, flocculation, sedimentation, and filtration, while the second stream contains a submerged microfiltration step that is implemented after the MIEX®-DOC process. Monitoring of different parameters and comparison with similar water treatment schemes without the MIEX®-DOC were conducted over a two-year period at this plant (Drikas et al., 2011). The results showed a significant improvement in DOC removal from treated water when MIEX® was used. Also, water quality after treatment was less affected by DOC changes in raw water when MIEX® was used.

Experiments in NOM removal from water conducted with MIEX® and a continuous-flow pilot plant operated at 7.6 L min^{-1} demonstrated substantial removal of UV-absorbing substances and DOC, and the efficient dose of MIEX® was estimated to be 0.16 mL L^{-1} (Boyer and Singer, 2006). Another MIEX® pilot with a flow of 1 ML day^{-1} was successfully tested, and a plant scale-up to 225 ML day^{-1} was then conducted (Bolto et al., 2002).

Fluidized ion exchange (FIX) is another type of process using IE resin for NOM removal (Cornelissen et al., 2009). The main principle of this method is that the rate of sedimentation of suspended solids in feed water is lower than that of IE resin. Thus, when water is pumped into an upflow reactor configuration at a certain speed, IE resin is fluidized, while suspended solids present in raw water are removed. A test was conducted with real surface water and FIX followed by ultrafiltration, demonstrating the feasibility of this process (Cornelissen et al., 2009). It was also shown that use of the FIX process prior to ultrafiltration significantly decreases membrane fouling.

A process consisting of a fluidized bed reactor with MIEX® followed by hydrophilic (NE70) and hydrophobic (NE90) nanofiltration (NF) membranes was tested for NOM removal from raw water from the Nakdong River, Korea (Kaewsuk and Seo, 2011). The raw water was characterized by low-MW NOM content with hydrophilic compounds, such as carboxylic, amino sugar, and aliphatic hydrocarbon compounds. With MIEX®, the removal efficiency of phenolic compounds and compounds with low SUVA values (aliphatic hydrocarbon and amino sugar) was reported to be significant, while the removal of carboxylic compounds was not that efficient (Kaewsuk and Seo, 2011). The best performance for NOM elimination from raw Nakdong River water was achieved with NE90 membrane filtration followed by MIEX®.

Several investigations have been conducted into MIEX® processes combined with other water treatment techniques, such as coagulation (Singer and Bilyk, 2002; Fearing et al., 2004; Allpike et al., 2005), lime softening (Hsu and Singer, 2009), activated carbon (Drikas et al., 2009; Humbert et al., 2008), and filtration (Fabris et al., 2007; Huang et al., 2012). For instance, the effect of MIEX® pretreatment on ozone demand, ozone exposure for disinfection, and bromate formation was studied for three water samples collected from raw water supplies impacted by the San Francisco Bay Delta (Kingsbury and Singer, 2013). After treatment with MIEX®, 41–68% of TOC was removed from raw water, while alum treatment eliminated only 12–44%. The reduction of bromide concentration in water by MIEX® was estimated to be 20–50%. Ozone demand was significantly decreased after MIEX® and alum pretreatment. Chlorination of raw water and of samples after MIEX®, alum, and ozonation treatment was conducted to estimate trihalomethane formation potential (THMFP). The results demonstrate that MIEX® performed better than alum and ozonation, attaining THMFP removal efficiencies of 39–85, 16–56, and 35–45%, respectively (Kingsbury and Singer, 2013). Based on these results, it was concluded that pretreatment with MIEX® prior to ozonation is more efficient than coagulation for bromide-rich waters (Kingsbury and Singer, 2013).

An attempt to simultaneously remove DOC and hardness from water was made by Apell and Boyer (2010). For this purpose, MIEX® was used together with cation IE resin in a single completely mixed reactor, and removal of DOC and hardness was estimated to be 70% and >55%, respectively. Thus, the feasibility of combined IE water treatment was demonstrated, and it was recommended as a possible pretreatment technique to decrease membrane fouling during the subsequent filtration step. In a recent study of a combined IE process (MIEX® and cation exchange resin in a single completely mixed vessel), DOC and hardness removal from groundwater was reported to be 76% and 97%, respectively (Comstock and Boyer, 2014). The successful regeneration of exhausted anion and cation exchange resins was obtained by using NaCl in a single vessel (Table 8.1) Summary of studies conducted in last 10–15 years for NOM removal from water by IE is presented in Table 8.1.

8.3 CONCLUSION

When designing a drinking water treatment plant, one of the most crucial issues to be addressed is the presence of NOM. It reduces water quality in terms of color, odor, and taste, and increases the dose of coagulant required for treatment, which in turn leads to high volumes of sludge formation. Another problem resulting from the presence of NOM in water is bacterial proliferation in drinking water distribution systems.

The complex matrix of water, often containing humic and/or fulvic acids as well as different organic compounds coming from urban or industrial activities, is a major reason for difficulties that arise during NOM characterization. Despite this, it is important to know the fractions of NOM present in treated water to increase the efficiency of the treatment process. Yet, usually only 10–30% of the species of NOM present in water are identified.

The most common technique for NOM removal from water is coagulation and flocculation followed by sedimentation and sand filtration. Yet, the coagulant dose strongly depends on the nature of the NOM present. Filtration, which is sometimes used as an alternative treatment method for NOM removal, often suffers from membrane fouling.

IE is another example of an alternative treatment for water that has a relatively high content of NOM. Depending on the nature of the water,

Table 8.1 Summary of studies on NOM removal by ion exchange (IE) conducted in the last 10–15 years

Treatment	Water quality	NOM removal (%)	Reaction conditions	References
MIEX-C	Wanneroo raw water appeared to be similar to northern European swamp and lake waters, such as water from some lakes in Finland.	After MIEX® treatment: DOC = 2.46 mg L^{-1}; color (at 400 nm) = 29; UV_{254} = 0.175 cm^{-1}; SUVA = 7.1 mg C L^{-1}; pH = 7.46.	Data were obtained from a full-scale (110 ML day^{-1}) potable water treatment plant operating in Perth, Western Australia.	Allpike et al. (2005)
MIEX® combined with coagulation	The following parameters of raw water are available: DOC = 5.70 mg L^{-1}; color (at 400 nm) = 45; UV_{254} = 0.326 cm^{-1}; SUVA = 5.7 mg C L^{-1}; pH = 7.14; M_n (method A DOC) = 2405; M_n (method A UV_{254}) = 4273; M_n (method B UV_{254}) = 3407; M_w (method A DOC) = 5319; M_w (method A UV_{254}) = 6371; M_w (method B UV_{254}) = 5294.	After MIEX® treatment combined with coagulation: alum dose = 55 mg L^{-1}; polyelectrolyte dose = 0.75 mg L^{-1}; DOC = 1.59 mg L^{-1}; color (at 400 nm) = 2; UV_{254} = 0.025 cm^{-1}; SUVA = 1.6 mg C L^{-1}; pH = 6.61.		

Combined IE treatment (MIEX®-Cl and MIEX®-Na)	Cedar key groundwater characteristics: DOC = 5.6 $mg\,L^{-1}$; UV_{254} = 0.171 cm^{-1}; pH = 7.6; alkalinity = 244 $mg\,L^{-1}$ $CaCO_3$; hardness = 275 $mg\,L^{-1}$ $CaCO_3$; Cl^- = 11.8 $mg\,L^{-1}$; SO_4^{2-} = 20.9 $mg\,L^{-1}$; Ca^{2+} = 103 $mg\,L^{-1}$; >90% of the hardness was as calcium.	Water quality after combined IE treatment: DOC = 1.7 $mg\,L^{-1}$; pH = 7.7; Cl^- = 48.8 $mg\,L^{-1}$; SO_4^{2-} = 3.1 $mg\,L^{-1}$. The following trends were observed for DOC and UV_{254} removal when fresh resins were used: sequence 1 ~ sequence 2 > simultaneous IE. After the third regeneration cycle, no difference between simultaneous and sequential IE treatment was observed.	A Phipps & Bird PB 700 jar tester with 2 L square jars was used for experiments. Simultaneous performance of anion and cation exchange resins, sequential anion exchange followed by cation exchange (sequence 1) and sequential cation exchange followed by anion exchange (sequence 2) were conducted.	Apell and Boyer (2010)
20 different resins were tested	Two types of waters were prepared using reverse osmosis to concentrate NOM in surface water samples from the east Moorabool river near Anakie, Victoria, and from the Mount Zero reservoir at Horsham, Victoria. Also, regeneration effluent obtained from MIEX® plant at Hope Valley, South Australia, was used.	NOM removal from Hope Valley water was 98–100%, whether gel or macroporous (MP) styrene or MP acrylic resins was used.	Normally, dose of a resin equal to 3 $g\,L^{-1}$ was used in all experiments and contact time was 16 h.	Bolto et al. (2002)

Continued

Table 8.1 Summary of studies on NOM removal by ion exchange (IE) conducted in the last 10–15 years—cont'd

Treatment	Water quality	NOM removal (%)	Reaction conditions	References
	Fractionation and percentage of UV-absorbing contribution for various NOM fractions were performed for tested water samples. Four fractions were distinguished: very hydrophobic acids (VHA), slightly hydrophobic acids (SHA), charged hydrophilic acids, and neutral hydrophilic compounds. UV-absorbance contribution for each fraction is presented below.	NOM removal efficiencies from Moorabool water (based on UV absorber removal): ResinTech SIR 22P—85%; Amberlite IRA 458–90%; Amberlite IRA 958–90%.		
	Water from Hope Valley: VHA—57%; SHA—17%; charged—26%; neutral—0%.	NOM removal efficiencies from Hope Valley water (based on UV absorber removal): ResinTech SIR 22P—98%; Amberlite IRA 958—99%.		
	Water from Moorabool: VHA—40%; SHA—17%; charged—38%; neutral—5%. Water from Horsham: VHA—70%; SHA—13%; charged—11%; neutral—5%.	NOM removal efficiencies from Horsham water (based on UV absorber removal): ResinTech SIR 22P—84%; Amberlite IRA 958–73%.		

Lewatit MP 500; Lewatit S6328A; Lewatit SP112;	Water samples from Seine river (France) with following characteristics were used: DOC = 1.9–2.7 mg L^{-1}; BDOC = 0.5–0.8 mg L^{-1}; UV_{254} = 6.8–8.6 m^{-1}; pH = 7.8–8.3; conductivity = 460 μS cm^{-1}; ammonia ≤0.1 mg L^{-1}.	Percentage of NOM removal, from DOC: MP 500–59%; S6328—73%; SP112—not significant. Percentage of NOM removal, from UV_{254} (m^{-1}): MP 500–90%; S6328—93%; SP112—not significant.		Afcharian et al. (1997)
MIEX®	Raw water from Little River Reservoir was used. Characteristics of water (mean value calculated from may to december): turbidity = 4.3 NTU; pH = 6.8; alkalinity (as $CaCO_3$) = 24.1 mg L^{-1}; hardness (as $CaCO_3$) = 28.5 mg L^{-1}; conductivity = 95 μS cm^{-1}; Cl^- = 9.5 mg L^{-1}; UV_{254} = 0.205 cm^{-1}; DOC = 5.9 mg L^{-1}; SUVA = 3.4 mg C L^{-1}.	Variations in removal efficiency of DOC and UV_{254} were observed depending on operational conditions. Operational conditions: 20 min HRT; 10% regeneration ratio; steady-state MIEX® concentration of 20 mL L^{-1} (ERD = 2 mL L^{-1}); removal efficiency: DOC = 74%; UV_{254} = 79%.	Experiments were conducted on continuous-flow pilot scale.	Boyer and Singer (2006)

Continued

Table 8.1 Summary of studies on NOM removal by ion exchange (IE) conducted in the last 10–15 years—cont'd

Treatment	Water quality	NOM removal (%)	Reaction conditions	References
		Operational conditions: 20 min HRT; 1% regeneration ratio; ERD = 0.16 mL L^{-1}; MIEX® concentration in contactors = 16 mL L^{-1}; removal efficiency: DOC = 64%; UV_{254} = 71%. Operational conditions: 20 min HRT; 2% regeneration ratio; ERD = 0.2 mL L^{-1}; MIEX® concentration in contactors = 10 mL L^{-1}; removal efficiency: DOC = 63%; UV_{254} = 65%. Operational conditions: 20 min HRT; 2% regeneration ratio; ERD = 0.3 mL L^{-1}; MIEX® concentration in contactors = 15 mL L^{-1}; removal efficiency: DOC = 67%; UV_{254} = 65%.		

		Operational conditions: 20 min HRT; 2% regeneration ratio; ERD = 0.4 mL L^{-1}; MIEX® concentration in contactors = 20 mL L^{-1}; removal efficiency: DOC = 69%; UV_{254} = 72%. Operational conditions: 20 min HRT; 5% regeneration ratio; ERD = 0.75 mL L^{-1}; MIEX® concentration in contactors = 15 mL L^{-1}; removal efficiency: DOC = 73%; UV_{254} = 72%.		

Continued

Table 8.1 Summary of studies on NOM removal by ion exchange (IE) conducted in the last 10–15 years—cont'd

Treatment	Water quality	NOM removal (%)	Reaction conditions	References
MIEX®	Raw water samples were obtained from New Port, Richey, FL; Charleston, SC; Palmdale, CA; Northern KY; Southern NV.	Removal of NOM was evaluated as DOC removal in percent. Data are shown below.	Experiments were conducted on continuous-flow pilot scale.	Budd et al. (2005) and Fonseca et al. (2005)
	Raw water characteristics are presented below. New Port, Richey, FL: DOC = 1.9 $mg\,L^{-1}$; SUVA = 5.6 $mg\,C\,L^{-1}$; sulfate = 18 $mg\,L^{-1}$; alkalinity (as $CaCO_3$) = 200 $mg\,L^{-1}$.	New Port, Richey, FL: DOC = 65–77%.		
	Charleston, SC: DOC = 5.0 $mg\,L^{-1}$; SUVA = 4.3 $mg\,C\,L^{-1}$; sulfate = 15 $mg\,L^{-1}$; alkalinity (as $CaCO_3$) = 27 $mg\,L^{-1}$.	Charleston, SC: DOC = 58–70%.		
	Palmdale, CA: DOC = 4.2 $mg\,L^{-1}$; SUVA = 2.1 $mg\,C\,L^{-1}$; sulfate = 29–41 $mg\,L^{-1}$; alkalinity (as $CaCO_3$) = 126 $mg\,L^{-1}$.	Palmdale, CA: DOC = 52–60%.		
	Northern KY: DOC = 1.8 $mg\,L^{-1}$; SUVA = 4.6 $mg\,C\,L^{-1}$; sulfate = 75 $mg\,L^{-1}$; alkalinity (as $CaCO_3$) = 113 $mg\,L^{-1}$.	Northern KY: DOC = 47–61%.		
	Southern NV: DOC = 2.3 $mg\,L^{-1}$; SUVA = 1.2 $mg\,C\,L^{-1}$; sulfate = 246 $mg\,L^{-1}$; alkalinity (as $CaCO_3$) = 288 $mg\,L^{-1}$.	Southern NV: DOC = 23–44%.		

MIEX®	Four types of drinking water in California were used: North Bay Aqueduct (NBA), South Bay Aqueduct (SBA), Castaic Lake (CL), and Sweetwater Lake (SL). Characteristics of water are shown below.	Removal of DOC and UV_{254} from NBA water: DOC = 70%; UV_{254} ~80%.		Boyer and Singer (2005)
	NBA water: pH = 7.5; turbidity = 20 NTU; alkalinity (as $CaCO_3$) = 149 mg L^{-1}; Br^- = 76 μg L^{-1}; UV_{254} = 0.193 cm^{-1}; TOC = 5.5 mg L^{-1}; DOC = 5.1 mg L^{-1}; SUVA = 3.8 mg C L^{-1}; THM4FP = 294 μg L^{-1}; HAA9FP = 224 μg L^{-1}.	Removal of DOC and UV_{254} from SBA water: DOC = 50%; UV_{254} ~60%.		
	SBA water: pH = 7.8; turbidity = 6 NTU; alkalinity (as $CaCO_3$) = 57 mg L^{-1}; Br^- = 83 μg L^{-1}; UV_{254} = 0.064 cm^{-1}; TOC = 1.9 mg L^{-1}; DOC = 1.9 mg L^{-1}; SUVA = 3.4 mg C L^{-1}; THM4FP = 111 μg L^{-1}; HAA9FP = 90.5 μg L^{-1}.	Removal of DOC and UV_{254} from CL water: DOC = 30%; UV_{254} ~50%.		

Continued

Table 8.1 Summary of studies on NOM removal by ion exchange (IE) conducted in the last 10–15 years—cont'd

Treatment	Water quality	NOM removal (%)	Reaction conditions	References
	CL water: pH = 8.6; turbidity = 7 NTU; alkalinity (as $CaCO_3$) = 92 mg L^{-1}; Br^- = 240 μg L^{-1}; UV_{254} = 0.074 cm^{-1}; TOC = 2.3 mg L^{-1}; DOC = 2.5 mg L^{-1}; SUVA = 3.0 mg C L^{-1}; THM4FP = 150 μg L^{-1}; HAA9FP = 65.3 μg L^{-1}. SL water: pH = 8.1; turbidity = 6 NTU; alkalinity (as $CaCO_3$) = 188 mg L^{-1}; Br^- = 540 μg L^{-1}; UV_{254} = 0.102 cm^{-1}; TOC = 5.2 mg L^{-1}; DOC = 5.1 mg L^{-1}; SUVA = 2.0 mg C L^{-1}; THM4FP = 283 μg L^{-1}; HAA9FP = 127 μg L^{-1}.	Removal of DOC and UV_{254} from SL water: DOC ~40%; UV_{254} ~50%.		

Combined IE ($MIEX^{®}$-Cl and amberlite 200C-Na)	Groundwater was used. Characteristics of water are shown below. Cedar key water (average): pH = 7.4; UV_{254} = 0.222 cm^{-1}; DOC = 6.3 mg L^{-1}; SUVA = 3.6 mg C L^{-1}; total hardness (as $CaCO_3$) = 315 mg L^{-1}; calcium hardness (as $CaCO_3$) = 250 mg L^{-1}; alkalinity (as $CaCO_3$) = 245 mg L^{-1}; sulfate = 17 mg L^{-1}.	The best performance was achieved at 200 BV, therefore DOC removal at 200 BV only is shown below. DOC removal efficiency was obtained for fresh and regenerated resin, but the results were very close. Cedar key water: DOC ~84%.	A Phipps & Bird PB 700 jar tester was used for experiments. Multiple loading was conducted. Resin: 5 mL $MIEX^{®}$-Cl/20 mL A200C-Na; regeneration with 2% NaCl and 20% NaCl. Five 1 L water in series; 30 min mixing at 200 rpm; 5 min settling.	Comstock and Boyer (2014)

Continued

Table 8.1 Summary of studies on NOM removal by ion exchange (IE) conducted in the last 10–15 years—cont'd

Treatment	Water quality	NOM removal (%)	Reaction conditions	References
	Yankeetown water: pH = 7.2; UV_{254} = 0.081 cm^{-1}; DOC = 2.9 mg L^{-1}; SUVA = 2.8 mg C L^{-1}; total hardness (as $CaCO_3$) = 440 mg L^{-1}; calcium hardness (as $CaCO_3$) = 420 mg L^{-1}; alkalinity (as $CaCO_3$) = 320 mg L^{-1}; sulfate = 130 mg L^{-1}. Palm Springs water: pH = 7.6; UV_{254} = 0.370 cm^{-1}; DOC = 9.7 mg L^{-1}; SUVA = 3.8 mg C L^{-1}; total hardness (as $CaCO_3$) = 300 mg L^{-1}; calcium hardness (as $CaCO_3$) = 295 mg L^{-1}; alkalinity (as $CaCO_3$) = 280 mg L^{-1}; sulfate = 10 mg L^{-1}.	Yankeetown water: DOC ~83%. Palm springs water: DOC ~95%.		

Combined treatment by IE resin (MIEX®; IRA938; IRA958; DOWEX-11; DOWEX-MSA; AMBER-SORB) and powdered activated carbon (PAC)	Raw water from the Villejean/Rennes drinking water treatment plant was used. The main quality parameters are: pH = 7.0–7.9; conductivity = 220–250 μS cm^{-1}; alkalinity (mg L^{-1} as $CaCO_3$) = 20–30; NO_3^- = 7–29 mg L^{-1}; SO_4^{2-} = 13–29 mg L^{-1}; DOC = 5.6–6.7 mg L^{-1}; UV_{254} = 0.14–0.16 cm^{-1}; SUVA = 2.2–2.9 mg C L^{-1}.	All tested resins removed more than 75% of DOC (resin dose 8 mL L^{-1}) after pseudoequilibrium time (time varied from 15 min to more than 1 h depending on resin). Less than 28% of DOC was removed by PAC (20–40 mg L^{-1} doses, pseudoequilibrium time of about 2 h). Implementation of PAC treatment after resins led to a slight improvement in DOC removal and higher uptake of pesticides.	Batch tests	Humbert et al. (2008)

Continued

Table 8.1 Summary of studies on NOM removal by ion exchange (IE) conducted in the last 10–15 years—cont'd

Treatment	Water quality	NOM removal (%)	Reaction conditions	References
MIEX®; DOWEX-11; DOWEX-MSA; IRA-938	Raw, clarified and postozonated water from the Villejean/Rennes drinking water treatment plant was used. Characteristics of clarified water: pH = 7.7; alkalinity ($mg\,L^{-1}$ as $CaCO_3$) = 29; NO_3^- = 8 $mg\,L^{-1}$; SO_4^{2-} = 22 $mg\,L^{-1}$; DOC = 4.8 $mg\,L^{-1}$; UV_{254} = 0.1 cm^{-1}; SUVA = 2.0 $mg\,C\,L^{-1}$. Characteristics of postozonated water: pH = 7.1; alkalinity ($mg\,L^{-1}$ as $CaCO_3$) = 13; DOC = 2.15 $mg\,L^{-1}$; UV_{254} = 0.021 cm^{-1}; SUVA = 1.0 $mg\,C\,L^{-1}$.	DOC was successfully removed by all studied resins. MIEX® and IRA938 demonstrated faster kinetics after 5 min of contact time with dose of these resins of 8 $mL\,L^{-1}$; level of DOC was 1–2 $mg\,L^{-1}$. The studied resins were able to remove high- and partly low-MW UV-absorbing NOM.	Bench-scale experiments	Humbert et al. (2005)
MIEX®–NF combined process	Influent of a water treatment plant (Changwon city, Korea) was used. Characteristics the of water: turbidity = 10 NTU; pH = 7.0; DOC = 2.5 $mg\,L^{-1}$.	Water treated with MIEX® was reported to have the following parameters: DOC = 1.28 $mg\,L^{-1}$, UVA = 0.010 cm^{-1}, SUVA = 0.39 $mg\,C\,L^{-1}$. Better performance was attributed to NE90 after MIEX® treatment.	The fluidized bed MIEX® column (150 mL of resin) was operated at flow rate 50 $mL\,min^{-1}$.	Kaewsuk and Seo (2011)

DAX-8; IRA958; DEAE (preswollen); MIEX®	Tertiary-treated wastewater effluent collected from an aeration tank discharge with TOC ranging from 7 to 18 mg L^{-1} was used.	Tests in fixed bed column demonstrated 23–67% of the initial TOC removal, with removal increasing in the order DAX-8 < IRA958 < DEAE < MIEX®. It should be noted that all tested experimental set-ups were equally efficient for TOC removal.	Packed column, in-line addition and addition to complete-mix tank tests were conducted.	Kim and Dempsey (2010) and Kennedy et al. (2008)

Continued

Table 8.1 Summary of studies on NOM removal by ion exchange (IE) conducted in the last 10–15 years—cont'd

Treatment	Water quality	NOM removal (%)	Reaction conditions	References
MIEX®	Raw waters from the North Bay Aqueduct, South Bay Aqueduct, spiked and nonspiked water from Lake Campbell were tested. Water characteristics are shown below. North Bay Aqueduct (NBA): TOC = 3.7 mg L^{-1}; DOC = 3.5 mg L^{-1}; UV_{254} = 0.113 cm^{-1}; SUVA = 3.2 mg C L^{-1}; fluorescence index = 1.39; pH = 7.9; bromide = 40 μg L^{-1}.	In NBA water after treatment with 2.0 mL L^{-1} of MIEX®/Cl, concentration of TOC decreased by 46% and UV absorbance by 58%. In SBA water treated with 2.0 mL L^{-1} of MIEX®/Cl and MIEX®/HCO_3, TOC was decreased by 36% and 25% and UV absorbance by 44 and 42%, respectively. For lake water, MIEX®/Cl and MIEX®/HCO_3 at the doses of 6.0 mL L^{-1} reduced TOC by 63 and 54%, respectively, and UV absorbance by 80%. It was demonstrated that for treatment of bromide-rich waters, MIEX® pretreatment prior to ozonation is more efficient than alum.	Bulk tests were performed in a 16-L glass carboy fitted with a variable-speed motor and an approximately 20-cm by 3-cm rectangular paddle.	Kingsbury and Singer (2013)

	South Bay Aqueduct (SBA): TOC = 2.4 mg L^{-1}; DOC = 2.4 mg L^{-1}; UV_{254} = 0.071 cm^{-1}; SUVA = 3.0 mg C L^{-1}; fluorescence index = 1.43; pH = 8.2; bromide = 360 μg L^{-1}. Lake Campbell: TOC = 8.7 mg L^{-1}; DOC = 8.5 mg L^{-1}; UV_{254} = 0.256 cm^{-1}; SUVA = 3.0 mg C L^{-1}; fluorescence index = 1.43; pH = 7.8; bromide = 33 μg L^{-1}. Lake Campbell, spiked: TOC = 8.7 mg L^{-1}; DOC = 8.5 mg L^{-1}; UV_{254} = 0.256 cm^{-1}; SUVA = 3.0 mg C L^{-1}; fluorescence index = 1.43; pH = 7.8; bromide = 380 μg L^{-1}.			

Continued

Table 8.1 Summary of studies on NOM removal by ion exchange (IE) conducted in the last 10–15 years—cont'd

Treatment	Water quality	NOM removal (%)	Reaction conditions	References
MIEX®	Raw waters from five drinking water plants in city of Istanbul were tested. The physicochemical characteristics are shown below. Elmali: DOC = 4.3 mg L^{-1}; UV_{254} = 0.220 cm^{-1}; SUVA = 5.11 mg C L^{-1}; alkalinity = 30 mg $CaCO_3$ L^{-1}; total hardness = 110 mg $CaCO_3$ L^{-1}; TDS = 140 mg L^{-1}; conductivity = 279 μS cm^{-1}; pH = 7.62; turbidity = 21.2 NTU. B.Cekmece: DOC = 3.1 mg L^{-1}; UV_{254} = 0.084 cm^{-1}; SUVA = 2.71 mg C L^{-1}; alkalinity = 35 mg $CaCO_3$ L^{-1}; total hardness = 150 mg $CaCO_3$ L^{-1}; TDS = 230 mg L^{-1}; conductivity = 460 μS cm^{-1}; pH = 7.92; turbidity = 1.4 NTU.	Optimal dose and contact time for NOM removal from all waters was 10 min and 10 mL settled resin per liter.	Multiple-loading jar tests were conducted.	Kitis et al. (2007)

	Ömerli: DOC = 2.6 mg L^{-1}; UV_{254} = 0.102 cm^{-1}; SUVA = 3.92 mg C L^{-1}; alkalinity = 35 mg $CaCO_3$ L^{-1}; total hardness = 90 mg $CaCO_3$ L^{-1}; TDS = 113 mg L^{-1}; conductivity = 225 μS cm^{-1}; pH = 7.52; turbidity = 4.2 NTU. Ikitelli: DOC = 3.1 mg L^{-1}; UV_{254} = 0.107 cm^{-1}; SUVA = 3.45 mg C L^{-1}; alkalinity = 38 mg $CaCO_3$ L^{-1}; total hardness = 137 mg $CaCO_3$ L^{-1}; TDS = 170 mg L^{-1}; conductivity = 339 μS cm^{-1}; pH = 8.05; turbidity = 1.0 NTU. Kagithane: DOC = 2.8 mg L^{-1}; UV_{254} = 0.108 cm^{-1}; SUVA = 3.86 mg C L^{-1}; alkalinity = 37 mg $CaCO_3$ L^{-1}; total hardness = 123 mg $CaCO_3$ L^{-1}; TDS = 221 mg L^{-1}; conductivity = 441 μS cm^{-1}; pH = 7.63; turbidity = 21.5 NTU.			

Continued

Table 8.1 Summary of studies on NOM removal by ion exchange (IE) conducted in the last 10–15 years—cont'd

Treatment	Water quality	NOM removal (%)	Reaction conditions	References
MIEX®	Three types of water were used: Barcombe, Draycote, and Albert. Barcombe: DOC = 9.6 mg L^{-1}; UV_{254} = 16.5 m^{-1}; SUVA = 1.73 mg C L^{-1}; pH = 7.7; turbidity = 13.1 NTU; conductivity = NA; Alkalinity = 112 mg $CaCO_3$ L^{-1}; zeta potential = −12.7; charge density = 1.9 meq g^{-1} DOC. Draycote: DOC = 10.7 mg L^{-1}; UV_{254} = 13.9 m^{-1}; SUVA = 1.3 mg C L^{-1}; pH = 8.1; turbidity = 1.4 NTU; conductivity = 692 μS m^{-1}; alkalinity = 157 mg $CaCO_3$ L^{-1}; zeta potential = −10.8; charge density = 0.3 meq g^{-1} DOC. Albert: DOC = 9.4 mg L^{-1}; UV_{254} = 60.1 m^{-1}; SUVA = 6.4 mg C L^{-1}; pH = 5.9; turbidity = 1.8 NTU; conductivity = 70 μS m^{-1}; alkalinity <10 mg $CaCO_3$ L^{-1}; zeta potential = −12.4; charge density = 6.4 meq g^{-1} DOC.	Removal of DOC in Draycote, Barcome, and Albert waters was 56%, 33%, and 25% respectively. It was observed that both hydrophilic and hydrophobic NOM can be removed by resin. However, high-MW NOM quickly saturated or blocked the resin.	Bench-scale tests	Mergen et al. (2008)

MIEX® for enhanced coagulation	Nine surface waters were tested. Manchester water works (TOC >2–4 mg L^{-1} and alkalinity 0–60 mg L^{-1} as $CaCO_3$); Metropolitan Water District of Southern California (TOC >2–4 mg L^{-1} and alkalinity >60–120 mg L^{-1} as $CaCO_3$); Davis water treatment plant (TOC >2–4 mg L^{-1} and alkalinity >120 mg L^{-1} as $CaCO_3$); Brown water treatment plant (TOC >4–8 mg L^{-1} and alkalinity 0–60 mg L^{-1} as $CaCO_3$); Haworth water treatment plant (TOC >4–8 mg L^{-1} and alkalinity >60–120 mg L^{-1} as $CaCO_3$); White river filtration plant (TOC >4–8 mg L^{-1} and alkalinity >120 mg L^{-1} as $CaCO_3$); Manatee water treatment plant (TOC >8 mg L^{-1} and alkalinity 0–60 mg L^{-1} as $CaCO_3$); Tampa water department (TOC >8 mg L^{-1} and alkalinity >60–120 mg L^{-1} as $CaCO_3$); Sioux falls water purification plant (TOC >8 mg L^{-1} and alkalinity >120 mg L^{-1} as $CaCO_3$).	Enhanced coagulation was reported to be more efficient than coagulation only. Thus, THM and HAA formation potential was reduced by more than 60% in all studied waters. TOC and UV absorbance of MIEX® pretreated waters were subsequently lower than that of water treated by coagulation.	Batch MIEX® treatment, followed by alum coagulation.	Singer and Bilyk (2002)

MIEX®, magnetic ion exchange resin; TOC, total organic carbon; DOC, dissolved organic carbon; SUVA, specific UV absorbance; NOM, natural organic matter; PAC, powdered activated carbon; MW, molecular weight; UVA, UV absorbance; THM, trihalomethane; HAA, haloacetic acid; BDOC, biodegradable dissolved organic carbon; ERD, effective resin dose; HRT, hydraulic residence time.

10–40% of NOM was reported to remain unremoved after IE treatment; this was speculated to be composed of uncharged species. As various studies demonstrate, IE is more efficient than coagulation with alum for the elimination of charged species contained in NOM. IE is proposed to be an effective choice for their treatment, given its efficiency in treating the transphilic fraction of NOM (Bond et al., 2011).

Taking into account the fact that polar components of NOM can lead to high formations of DBPs, IE can be considered more beneficial than coagulation. Yet, industrialization of the IE process normally requires the use of IE resin in a packed bed, which is only possible at the polishing stages of treatment.

Since the development of magnetic ion exchange resin, known as MIEX®, it has become possible to utilize IE resin in slurry mode prior to coagulation. Due to the presence of magnetic IOPs within the core of MIEX® resin, the separation step is significantly simplified, so this resin can be utilized with different reactor designs. The high stability of this resin and the fact that it can be used without pretreatment makes MIEX® an extremely promising option. It is not surprising that, currently, MIEX® is the most studied ion exchanger for NOM removal from water. Already in 2001, the first potable water treatment plant using a MIEX®-DOC process was launched in Australia. In this plant, the MIEX®-DOC step was introduced prior to conventional treatment, and a significant improvement in water quality was observed. It should be noted that MIEX® is efficient for NOM removal and the prevention of chlorinated and brominated DBP formation (Hsu and Singer, 2010). Yet, NOM removal efficiency is affected by the presence of other anions in water. Therefore, further studies should be conducted to optimize the performance of IE processes and decrease the costs for industrial-scale operations.

REFERENCES

Afcharian, A., Levi, Y., Kiene, L., Scribe, P., 1997. Fractionation of dissolved organic matter from surface waters using macroporous resins. Water Res. 31, 2989–2996.

Allpike, B.P., Heitz, A., Joll, C.A., Kagi, R.I., Abbt-Braun, G., Frimmel, F.H., Brinkmann, T., Her, N., Amy, G., 2005. Size exclusion chromatography to characterize DOC removal in drinking water treatment. Environ. Sci. Technol. 39, 2334–2342.

Apell, J.N., Boyer, T.H., 2010. Combined ion exchange treatment for removal of dissolved organic matter and hardness. Water Res. 44, 2419–2430.

Bolto, B., Dixon, D., Eldridge, R., King, S., Linge, K., 2002. Removal of natural organic matter by ion exchange. Water Res. 36, 5057–5065.

Bolto, B., Dixon, D., Eldridge, R., 2004. Ion exchange for the removal of natural organic matter. React. Funct. Polym. 60, 171–182.
Bond, T., Goslan, E.H., Parsons, S.A., Jefferson, B., 2011. Treatment of disinfection by-product precursors. Environ. Technol. 32, 1–25.
Boyer, T.H., Singer, P.C., 2006. A pilot-scale evaluation of magnetic ion exchange treatment for removal of natural organic material and inorganic anions. Water Res. 40, 2865–2876.
Boyer, T.H., Singer, P.C., 2005. Bench-scale testing of a magnetic ion exchange resin for removal of disinfection by-product precursors. Water Res. 39, 1265–1276.
Brattebø, H., Ødegaard, H., Halle, O., 1987. Ion exchange for the removal of humic acids in water treatment. Water Res. 21, 1045–1052.
Budd, G.C., Long, B. Edwards, J.C., Singer, P.C., Meisch, M., 2005. Evaluation of MIEX process impacts on different source waters. American Water Works Association Research Foundation, 91067F.
Comstock, S.E.H., Boyer, T.H., 2014. Combined magnetic ion exchange and cation exchange for removal of DOC and hardness. Chem. Eng. J. 241, 366–375.
Cornelissen, E.R., Beerendonk, E.F., Nederlof, M.N., van der Hoek, J.P., Wessels, L.P., 2009. Fluidized ion exchange (FIX) to control NOM fouling in ultrafiltration. Desalination 236, 334–341.
Cornelissen, E.R., Moreau, N., Siegers, W.G., Abrahamse, A.J., Rietveld, L.C., Grefte, A., Dignum, M., Amy, G., Wessels, L.P., 2008. Selection of anionic exchange resins for removal of natural organic matter (NOM) fractions. Water Res. 42, 413–423.
Drikas, M., Chow, C.W.K., Cook, D., 2003. The impact of recalcitrant organic character on disinfection stability, trihalomethane formation and bacterial regrowth: an evaluation of magnetic ion exchange resin (MIEX®) and alum coagulation. J. Water Supply Res. Technol. Aqua 52, 475–487.
Drikas, M., Morran, J.Y., Pelekani, C., Hepplewhite, C., Bursill, D.B., 2002. Removal of natural organic matter – a fresh approach. Water Sci. Technol. Water Supply 2, 71–79.
Drikas, M., Dixon, M., Morran, J., 2011. Long term case study of MIEX® pre-treatment in drinking water; understanding NOM removal. Water Res. 45, 1539–1548.
Drikas, M., Dixon, M., Morran, J., 2009. Removal of MIB and geosmin using granular activated carbon with and without MIEX® pre-treatment. Water Res. 43, 5151–5159.
Eckenfelder, W., 2000. Industrial Water Pollution Control, third ed. McGraw-Hill Series in Water Resources and Environmental Engineering.
Fabris, R., Lee, E.K., Chow, C.W.K., Chen, V., Drikas, M., 2007. Pre-treatments to reduce fouling of low pressure micro-filtration (MF) membranes. J. Membr. Sci. 289, 231–240.
Fearing, D.A., Banks, J., Guyetand, S., Monfort Eroles, C., Jefferson, B., Wilson, D., Hillis, P., Campbell, A.T., Parsons, S.A., 2004. Combination of ferric and MIEX® for the treatment of a humic rich water. Water Res. 38, 2551–2558.
Fettig, J., 1999. Removal of humic substances by adsorption/ion exchange. Water Sci. Technol. 40, 173–182.
Fonseca, A.C., Kennon, J., Booth, S., Labisi, B., Meyerhofer, J., White, P., 2005. Testing bromide and DBP precursors removal in California State Project Water using a magnetized ion exchange resin. Proceedings American Water Works Association Annual Conference.
Gibert, O., Lefèvre, B., Fernández, M., Bernat, X., Paraira, M., Calderer, M., Martínez-Lladó, X., 2013a. Characterizing biofilm development on granular activated carbon used for drinking water production. Water Res. 47, 1101–1110.
Gibert, O., Lefèvre, B., Fernández, M., Bernat, X., Paraira, M., Pons, M., 2013b. Fractionation and removal of dissolved organic carbon in a full-scale granular activated carbon filter used for drinking water production. Water Res. 47, 2821–2829.

Harland, C.E., 1994. Ion Exchange. Theory and Practice, second ed. The Royal Society of Chemistry, Bath.
Helfferich, F., 1995. Ion Exchange. Dover Publications, USA.
Hongve, D., Baann, J., Becher, G., Beckmann, O., 1999. Experiences from operation and regeneration of an anionic exchanger for natural organic matter (NOM) removal. Water Sci. Technol. 40, 215–221.
Hsu, S., Singer, P.C., 2009. Application of anion exchange to control NOM interference on lime softening. J. Am. Water Works Assoc. 101, 85–94.
Hsu, S., Singer, P.C., 2010. Removal of bromide and natural organic matter by anion exchange. Water Res. 44, 2133–2140.
Huang, H., Cho, H., Schwab, K.J., Jacangelo, J.G., 2012. Effects of magnetic ion exchange pretreatment on low pressure membrane filtration of natural surface water. Water Res. 46, 5483–5490.
Huang, H., Cho, H., Schwab, K., Jacangelo, J.G., 2011. Effects of feedwater pretreatment on the removal of organic microconstituents by a low fouling reverse osmosis membrane. Desalination 281, 446–454.
Humbert, H., Gallard, H., Jacquemet, V., Croué, J., 2007. Combination of coagulation and ion exchange for the reduction of UF fouling properties of a high DOC content surface water. Water Res. 41, 3803–3811.
Humbert, H., Gallard, H., Suty, H., Croué, J., 2008. Natural organic matter (NOM) and pesticides removal using a combination of ion exchange resin and powdered activated carbon (PAC). Water Res. 42, 1635–1643.
Humbert, H., Gallard, H., Suty, H., Croué, J., 2005. Performance of selected anion exchange resins for the treatment of a high DOC content surface water. Water Res. 39, 1699–1708.
Kaewsuk, J., Seo, G.T., 2011. Verification of NOM removal in MIEX-NF system for advanced water treatment. Sep. Purif. Technol. 80, 11–19.
Kennedy, M.D., Kamanyi, J., Heijman, B.G.J., Amy, G., 2008. Colloidal organic matter fouling of UF membranes: role of NOM composition and size. Desalination 220, 200–213.
Kim, H., Dempsey, B.A., 2013. Membrane fouling due to alginate, SMP, EfOM, humic acid, and NOM. J. Membr. Sci. 428, 190–197.
Kim, H., Dempsey, B.A., 2010. Removal of organic acids from EfOM using anion exchange resins and consequent reduction of fouling in UF and MF. J. Membr. Sci. 364, 325–330.
Kingsbury, R.S., Singer, P.C., 2013. Effect of magnetic ion exchange and ozonation on disinfection by-product formation. Water Res. 47, 1060–1072.
Kitis, M., İlker Harman, B., Yigit, N.O., Beyhan, M., Nguyen, H., Adams, B., 2007. The removal of natural organic matter from selected Turkish source waters using magnetic ion exchange resin (MIEX®). React. Funct. Polym. 67, 1495–1504.
Lee, N., Amy, G., Croué, J., Buisson, H., 2004. Identification and understanding of fouling in low-pressure membrane (MF/UF) filtration by natural organic matter (NOM). Water Res. 38, 4511–4523.
Malekian, R., Abedi-Koupai, J., Eslamian, S.S., Mousavi, S.F., Abbaspour, K.C., Afyuni, M., 2011. Ion-exchange process for ammonium removal and release using natural Iranian zeolite. Appl. Clay Sci. 51, 323–329.
Mergen, M.R.D., Jefferson, B., Parsons, S.A., Jarvis, P., 2008. Magnetic ion-exchange resin treatment: impact of water type and resin use. Water Res. 42, 1977–1988.
Morran, J.Y., Drikas, M., Cook, D., Bursill, D.B., 2004. Comparison of MIEX® treatment and coagulation on NOM character. Water Sci. Technol. Water Supply 4, 129–137.
Nkambule, T.I., Krause, R.W., Mamba, B.B., Haarhoff, J., 2009. Removal of natural organic matter from water using ion-exchange resins and cyclodextrin polyurethanes. Phys. Chem. Earth Parts A/B/C 34, 812–818.

Singer, P.C., Bilyk, K., 2002. Enhanced coagulation using a magnetic ion exchange resin. Water Res. 36, 4009–4022.
Teixeira, M.R., Sousa, V.S., 2013. Fouling of nanofiltration membrane: effects of NOM molecular weight and microcystins. Desalination 315, 149–155.
Tian, C., Liu, R., Liu, H., Qu, J., 2013. Disinfection by-products formation and precursors transformation during chlorination and chloramination of highly-polluted source water: significance of ammonia. Water Res. 47, 5901–5910.
Wachinski, A.M., Etzel, J.E., 1997. Environmental Ion Exchange: Principles and Design. Lewis Publishers, USA.
Walker, K.M., Boyer, T.H., 2011. Long-term performance of bicarbonate-form anion exchange: removal of dissolved organic matter and bromide from the St. Johns River, FL, USA. Water Res. 45, 2875–2886.
Yang, X., Guo, W., Lee, W., 2013. Formation of disinfection byproducts upon chlorine dioxide preoxidation followed by chlorination or chloramination of natural organic matter. Chemosphere 91, 1477–1485.
Zularisam, A.W., Ismail, A.F., Salim, M.R., Sakinah, M., Hiroaki, O., 2007. Fabrication, fouling and foulant analyses of asymmetric polysulfone (PSF) ultrafiltration membrane fouled with natural organic matter (NOM) source waters. J. Membr. Sci. 299, 97–113.

CHAPTER 9

Integrated Methods

Mika Sillanpää*, Sari Metsämuuronen, Anu Matilainen†, Mika Mänttäri****

*Lappeenranta University of Technology, LUT Faculty of Technology, LUT Chemtech, Laboratory of Green Chemistry, Sammonkatu 12, 50130 Mikkeli, Finland; **Lappeenranta University of Technology, LUT Faculty of Technology, LUT Chemtech, Skinnarilankatu 34, 53850 Lappeenranta, Finland; †Finnish Safety and Chemicals Agency, Kalevantie 2, 33100 Tampere, Finland

Abstract

Integrating conventional coagulation with novel treatment methods is expected to enhance natural organic matter (NOM) removal in water treatment. The great variability in NOM composition makes it difficult to completely remove from drinking water by any single technique. This chapter discusses NOM removal by integrating coagulation with other treatment processes (ion exchange, oxidation, adsorption, and membrane technology) and hybrid processes combining membrane techniques with other unit processes (coagulation, adsorption, oxidation, and membrane bioreactors).

Keywords: Adsorption; Coagulation; Coupling; Filtration; Ion exchange; Membrane technology; Natural organic matter (NOM); Oxidation; Process integration; Water treatment.

9.1 INTRODUCTION

The great variability in the composition of natural organic matter (NOM) makes it difficult to remove NOM from drinking water completely by conventional coagulation or by any single technique. This chapter discusses the integration of various processes for NOM removal (partially reviewed earlier in Matilainen et al. (2010) and Metsämuuronen et al. (2014)).

9.2 COUPLING COAGULATION WITH OTHER PROCESSES

9.2.1 Magnetic Ion Exchange Resin (MIEX®)

The magnetic ion exchange resin (MIEX®) technique is based on the traditional ion exchange (IE) process, but it uses microsized resin particles dispersed as slurry on a stirred contactor, thus allowing maximum surface area for adsorption. It removes both organic and inorganic contaminants such as sulfate, nitrate, or phosphate. The negatively charged NOM compounds are removed efficiently by this strong-base anion resin. A high density and the magnetic properties of the resin provide rapid clarification following the contact stage (Morran et al., 2004). MIEX® can be easily regenerated, and even after several regenerations, its organic matter removal efficiency is almost the same. Moreover, it is suitable for both small and large plants, and complements other water treatment technologies like coagulation or membrane filtration. MIEX® has been used in pretreatment prior to coagulation to enhance the efficiency of coagulation and reduce the coagulant dose (up to 60%), with reduced sludge formation and turbidity load (Shorrock and Drage, 2006; Humbert et al., 2007; Jarvis et al., 2008; Sani et al., 2008). In the MIEX® process, the NOM compounds with increasing charge density are removed, the most and the presence of functional groups has been noted to be more important than aromatic structures (Mergen et al., 2009). Moreover, MIEX® can also remove hydrophilic, low molecular mass (LMM) NOM, thus can remove a fraction of NOM recalcitrant to coagulation (Mergen et al., 2009, 2008). The flocs formed after pretreatment with MIEX® have been larger and more resistant to shearing effects than those formed by conventional processes (Jarvis et al., 2008; Sani et al., 2008). Hence, MIEX® in combination with coagulation has increased the dissolved organic carbon (DOC) and disinfection by-product (DBP) precursor reduction (Jarvis et al., 2008; Sani et al., 2008; Mergen et al., 2009), although this has not been evident in all cases (Bond et al., 2009). According to the literature, an addition of 8–63% and about 10–30% of DOC removal has been observed in jar (Mergen et al., 2008) and pilot tests (Shorrock and Drage, 2006), respectively.

9.2.2 Oxidation

When advanced oxidation processes (AOPs) are used as a pretreatment method, oxidation changes the structural properties of NOM constituents, affecting both the adsorption and coagulation characteristics of NOM (Uyguner et al., 2007b). Photocatalysis as a preoxidation method before coagulation has been found to result in approximately <15% decrease in removal efficiency, measured as $color_{436}$ and UV_{254}. Coagulation prior to oxidation has enhanced the removal of NOM about 5–32% and 8–33% as DOC and UV_{254}, respectively, compared to coagulation alone. Also, decreases in specific UV absorbance (SUVA) values were noted (Uyguner et al., 2007a).

Preoxidation has been traditionally used to improve the coagulation process. Prechlorination was widely utilized until the discovery of DBP formation during chlorination. Some other oxidants have since been in use and, for example, chloramine preoxidation has been noticed to enhance the coagulation of surface water (Yang et al., 2007). Ozonation has become a preferential preoxidation method prior the coagulation, and is used for elimination of taste, odor, and color, and inactivation of microorganisms (Yan et al., 2007). The influence of preozonation in coagulation has been closely related to NOM characteristics and basic properties of water to be treated (Liu et al., 2007, 2009). Moreover, the ozone dosage in preozonation has been observed to be highly important. At lower ozone dosage, ozone has produced some hydrophobic neutral and intermedial MM NOM compounds, which are removed efficiently in the following coagulation process, but at higher ozone doses the NOM has been oxidized further into more hydrophilic and LMM compounds, thus being more difficult to coagulate (Bose and Reckhow, 2007; Yan et al., 2007; Liu et al., 2007, 2009; Chinag et al., 2009). Accordingly, the ozonation process placement should be carefully determined, since postozonation might get better results than preozonation in combination with biological activated carbon (BAC) in some cases (Liu et al., 2009). The beneficial effect of ozonation towards better coagulation of nonhumic fraction has also been observed. Hence, to maximize DOC and SUVA removals, two-staged coagulation with intermediate ozonation has been proposed for waters containing both humic and nonhumic NOM (Bose and Reckhow, 2007). The composite polyaluminium chloride (HPAC) coagulant efficiency has not declined even after high ozone doses (Yan et al., 2007), and PACl has been found to be suitable in coagulation with preozonation (Liu et al., 2007).

Advanced oxidation processes are powerful techniques in disinfection and organic compounds (e.g., contaminants) degradation (Rämö and Sillanpää, 2001; Pirkanniemi and Sillanpää, 2002; Pirkanniemi et al., 2007). The photo-Fenton process, in which iron species can play a role as enhanced catalyst as well as coagulant, has been investigated for NOM removal with rather good results (Murray and Parson, 2004; Goslan et al., 2006). NOM removal improved compared to conventional coagulation with iron salts; in particular, LMM compounds were removed more efficiently (Murray and Parson, 2004). Photocatalytic oxidation prior to coagulation reduced the coagulation efficiency by 15%, most likely because the oxidation changes the characteristics of NOM and degrades NOM molecules towards LMM compounds (Uyguner et al., 2007b). Oxidation after coagulation, however, enhanced DOC and UV_{254} removal by about 5–32% and 8–33%, respectively (Uyguner et al., 2007a). Photocatalytic coagulation with nanoparticles of copper-doped titania has been observed to provide both better NOM removal and reduction of membrane fouling (Choi et al., 2007).

9.2.3 Activated Carbon Filtration

Activated carbon (AC) is an effective adsorbent used widely in drinking water treatment. Micropollutants such as pesticides, industrial chemicals, tastes, odors, and algal toxins are removed efficiently in the filtration process. NOM may also be removed in filtration, although it often competes for adsorption sites, decreasing the removal of other pollutants. If enhanced coagulation is insufficient for NOM reduction, additional treatment with granular or powdered AC (GAC/PAC) can maintain the required water quality in terms of, e.g., DBP formation (Szlachta and Adamski, 2009; Uyak et al., 2007; Garcia and Moreno, 2009). AC can also be colonized by a heterotropic biomass, creating BAC, thus reducing the proportion of easily assimilated and biodegradable elements. In addition, efficient removal of LMM and uncharged fractions of NOM is gained (Uyak et al., 2007). Yet, in most cases a part of LMM NOM is not removed, even after AC filtration (Szlachta and Adamski, 2009). The water can be filtered also only when necessary, e.g., during elevated NOM or alkalinity periods (Smith and Kamal, 2009). PAC-enhanced coagulation has also been noted to reduce the required coagulant dose, while increasing the process efficiency (Szlachta and Adamski, 2009).

9.2.4 Membrane Filtration

Membranes have become increasingly common in water treatment. Pressure-driven membrane processes include reverse osmosis (RO), nanofiltration (NF), ultrafiltration (UF) and microfiltration (MF), which all have different NOM and particle removal potentials (Jacangelo et al., 1995; Zularisam et al., 2006). The fouling of membranes and decline of flux are the major problems associated with membrane techniques (Zularisam et al., 2006). Several studies have indicated that NOM—especially the hydrophobic fraction and high molar mass (HMM) compounds—has a great impact on fouling and decline of flux (Zularisam et al., 2006; Chen et al., 2007; Dong et al., 2007; Howe et al., 2006; Jung et al., 2006). Pretreatment with coagulation is an effective method to prevent these adverse effects (Chen et al., 2007; Dong et al., 2007; Howe et al., 2006; Jung et al., 2006; Qiao et al., 2008), and a combination of coagulation and membrane filtration also reduces the amount of coagulant needed while improving turbidity and DOC reduction (Blankert et al., 2007). Moreover, coagulation has been observed to provide an effective hygienic barrier in combination with membranes (Leiknes, 2009), although contradictory results have been obtained (Barbot et al., 2008), and the removal of NOM has been significantly affected by the type of coagulant, coagulation conditions, type of membrane, filtration conditions, and characteristics of the water to be treated (Bond et al., 2009; Meyn et al., 2008; Park and Yoon, 2009; Barbot et al., 2008; Howe et al., 2006; Cai et al., 2008; Cho et al., 2006; Choo et al., 2008; Kim et al., 2006; Tran et al., 2006; Zularisam et al., 2009).

A combination of the coagulation with UF (Barbot et al., 2008; Blankert et al., 2007; Zularisam et al., 2009; Konieczny et al., 2009) or MF (Meyn et al., 2008; Cho et al., 2006; Loi-Brugger et al., 2006) is an in-line hybrid process. The coagulant dose and the duration of water treatment are reduced by batching the coagulant to the feed immediately before membrane (Konieczny et al., 2009). Moreover, a novel submerged UF membrane coagulation bioreactor (MCBR) process has been studied for treating slightly contaminated surface water to produce potable water (Tian et al., 2008). A proposed new way of reducing membrane fouling is the immobilization of TiO_2 photocatalyst on the membrane surface, thus creating a composite membrane (Ma et al., 2009). The use of nanoparticles (including alumina, silica, silver, and iron oxide) in membrane structures has gained a lot of interest, but their use in water treatment requires further research due to the possibility of nanoparticles being released into the water (Kim and Van der Bruggen, 2010).

9.3 COUPLING MEMBRANE TECHNOLOGY WITH OTHER PROCESSES

Coupling membrane techniques such as RO, NF, UF, and MF, with other unit processes has become a common way to comply with drinking water legislation. It is estimated that MF and UF plant capacity is around 20 million m^3 per day, 60% of which is for drinking water production (Schrotter and Bozkaya-Schrotter, 2010), and UF applications represent about 74% of the total capacity (Fane et al., 2008). MF and UF plants are primarily developed for turbidity, bacteria, and protozoa removal as standalone processes, as retrofits of existing conventional treatment plants, or as pretreatments to NF and RO (Fane et al., 2008).

An essential advantage of membrane filtration is its compact and modular design that allows for a high backing density and small footprint. The fouling of membranes and decline of flux are the major challenges associated with membrane techniques (Zularisam et al., 2006). These problems can be reduced by coupling membranes with other techniques.

9.3.1 Coupling with Filters

Coarse materials can be detrimental to membrane filtration performance. Hence, membranes are normally protected at least by an upstream screen, cartridge, or backed bed filters. An upstream screen with a mesh from 500 to 50 μm may be high enough protection for the MF membranes (Bonnélye et al., 2008), but the smaller pore size membranes need at least a 50-μm cartridge filter to ensure stable performance in NF of groundwater (Nederlof et al., 2000) or surface waters (Ødegaard et al., 2010).

A typical flow diagram of a small NF plant that has operated successfully for soft water characterized by high color and low turbidity is shown in Figure 9.1. The pretreatment consists only of a self-cleaning cartridge filter with 50-μm mesh (Ødegaard et al., 2010). A sand filter can be installed upstream of the cartridge when high turbidity raw water is treated. Approximately 150 plants of this type have been opened worldwide by 2005, about 70% of them in Norway. Most of them are small, with capacities in the range of 100–15,000 m^3 per day (Thorsen and Fløgstad, 2006).

Packed bed filters are effective in capturing particles of large size distribution and, depending on the properties of the downstream membrane, both the membrane surface and pore fouling may be reduced. Conventional packed bed filters can be effectively restored by backwashing, whereas a disadvantage of granular filters is that they may deteriorate and be difficult to recover. In principle, the removal mechanism for packed bed filters is either physical sieving or chemical adsorption/deposition (Huang et al., 2009).

Sand filtration-UF plants have been successfully operated for the production of drinking water (Botes et al., 1998; Pryor et al., 1998; Alborzfar et al., 1998; Mierzwa et al., 2008; Ødegaard et al., 2010; Mierzwa et al., 2012). A periodic relaxation and flushing procedure with a chlorine dosage has been observed as necessary to maintain stable operational conditions in a sand filtration-UF process (Mierzwa et al., 2012). Dual-media filters (sand and anthracite), followed by a 5-μm cartridge prior to membrane filtration

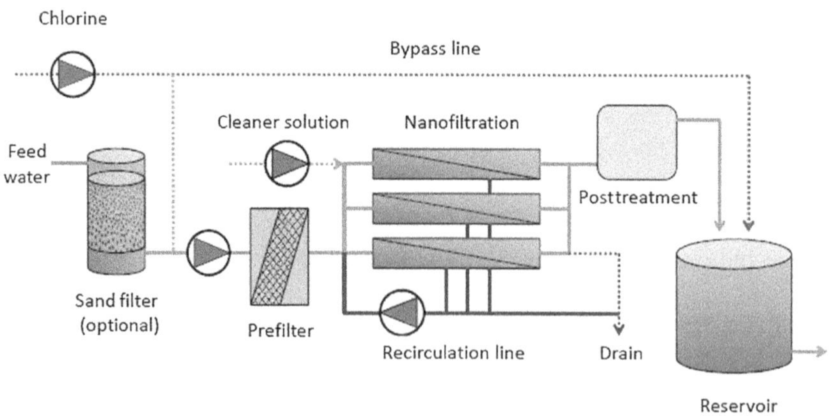

Figure 9.1 Flow diagram of a simple nanofiltration plant. *Redrawn from Ødegaard et al. (2010).*

with spiral-wound and tubular membranes, are applied in a number of small water treatment plants, producing high-quality drinking water for remote communities (Grose et al., 1998).

Ahn et al. (2004) investigated a coagulation/sedimentation-sand filtration-NF process (Figure 9.2), and found that the installation of MF between the sand filter and the NF process can significantly increase the permeability of the NF membrane. Several researchers have compared the coagulation/sedimentation-sand filtration pretreatment with the UF pretreatment for NF or RO (Glucina et al., 2000; Kim and Yu, 2005; Chang et al., 2009). It has been observed that the former reduced the NOM effectively with lower energy consumption than the latter, but in general the UF–NF process exhibited better removal efficiency for NOM and lower fouling (Chang et al., 2009). Also, Kim and Yu (2005) obtained the best pretreatment results with the UF-NF process for trihalomethane (THM) and haloacetic acid (HAA) formation potential removal (85% and 89%, respectively).

In RO for seawater and brackish water desalination plants, MF, UF, and NF pretreatments have shown great advantages over conventional pretreatment processes (Hilal et al., 2006). Yet, Glucina et al. (2000) found that both pretreatments were effective to control RO fouling and that both combinations produced water that exceeded the most stringent quality level proposed in the EU Directives. Altogether, MF and UF membranes have increasingly been applied as prefilters for NF and RO membranes. They remove suspended particles, colloids, *Giardia lamblia* and *Cryptosporidium* cysts, algae, some bacterial species, and a certain amount of HMM NOM (Kajitvichyanukul et al., 2008). Lamsal et al. (2012) reported that UF

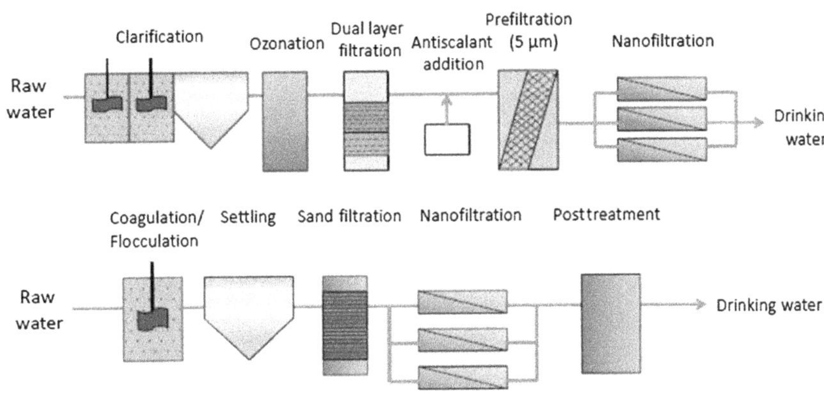

Figure 9.2 Flow chart of typical pretreatments for nanofiltration.

pretreatment removed 66% of DOC and 54% of THM formation potential from river water, and that NF was able to remove most of the remaining DOC with excellent performance.

Low-pressure membranes have been observed to be more effective than a cartridge MF filter for the pretreatment of NF membrane processes (Siddiqui et al., 2000). This is because these membranes are able to remove colloids, which is crucial for preventing NF fouling (Lee and Lee, 2007).

Lee and Lee (2006) found that UF can efficiently remove NF foulants, and that the efficiency of the NF membrane improves with decreasing prefilter pore size. Yet, pretreatment cost also increases with decreasing prefilter pore size.

9.3.2 Coupling with Coagulation

Coagulation has remained the most common process for water treatment, normally for the removal of turbidity and NOM (Liu et al., 2011). The conventional process incorporates several physicochemical processes including rapid mixing, slow mixing (flocculation), sedimentation, filtration, and disinfection. Coagulants are used to increase the size of aquatic substances, making them large enough to settle or to be removed by sand, dual layer filtration, or the membrane. Coagulation is based on charge effects: Particles suspended in water usually carry a negative charge due to the ionization of surface groups, adsorption of ions, or ion substitution, hence they can be coagulated by adding positively charged coagulants.

Integration of coagulation into membrane filtration has two main advantages: Enhanced removal of NOM molecules and reduction of membrane fouling. These two unit operations can be combined with or without settling (Figure 9.3). The most conventional way is to use a coagulation tank, which may be divided to rapid mixing and slow mixing/flocculation parts, and settling to remove flocs from the feed water. In this case, the supernatant is conveyed to membrane filtration. The advantage of the sedimentation step is that it smoothes out the sudden increase in solid loading before the membrane system (Moon et al., 2009). Another approach is to convey water directly from flocculation to the membrane unit without removal of flocs (direct filtration).

The most recent method of combining coagulant with MF (Meyn et al., 2008; Cho et al., 2006; Loi-Brugger et al., 2006) or UF (Barbot et al., 2008; Blankert et al., 2007; Zularisam et al., 2009; Konieczny et al., 2009) is to patch coagulant into the feed stream immediately prior to the membrane process without removing coagulated solids (in-line coagulation). These

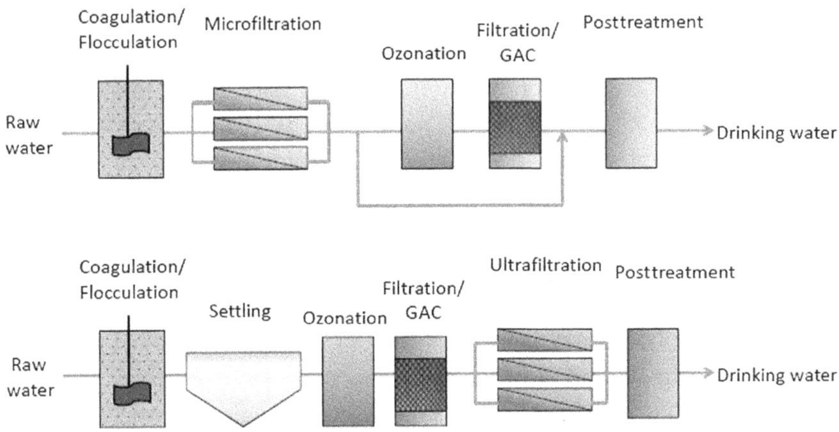

Figure 9.3 Flow chart of coagulation without settling before microfiltration and with settling before ultrafiltration.

integrated processes composed of MF or UF and in-line coagulation have become particularly attractive because they are expected to enhance operational efficiency (Bergamasco et al., 2011). The advantages of in-line coagulation are reduced treatment plant footprint, reduced water treatment time, and the possibility of lower coagulant dosing, since flocs do not need to settle and instead will be removed by the membrane (Jung and Kang, 2003; Konieczny et al., 2009). Furthermore, coagulant selection and dosing can be optimized specifically for NOM removal as particle removal is assured by the membrane (Vickers et al., 1995). Charge neutralization and sweep coagulation conditions have both been found to be effective for the removal of NOM and turbidity by in-line coagulation-UF (Choi and Dempsey, 2004).

In water treatment plants, MF membranes are generally positioned toward the front of a treatment chain after a light pretreatment such as coagulation/flocculation (Figure 9.3). Flocculated water is clarified typically with outside-in hollow-fiber MF membranes, which are able to treat a high solid loading. This process is adapted to treat karstic water and is mainly developed to remove turbidity and bacteria (Schrotter and Bozkaya-Schrotter, 2010). Tight inside-out hollow-fiber UF membranes are commonly situated at the end of the treatment chain because they require a more advanced pretreatment.

An example of a full-scale integrated water treatment plant is in Heemskerk, The Netherlands (Kamp et al., 2000). It is a direct surface water treatment plant based on the combined application of UF and RO from the Lake Ijssel (River Rhine) water. Its major quality control objectives are the removal of pathogenic microorganisms, organic micropollutants, and inorganics, as well as maintenance of biological stability. The treatment chain consists in coagulation, sedimentation, filtration, GAC (optional), UF, and RO. The permanent fouling of UF membrane by coagulant residuals was avoided by a temporary application of GAC filtration.

The removal of NOM has been significantly affected by the type and dosage of coagulant, coagulation conditions, type of membrane, filtration conditions, and characteristics of the water to be treated (Kabsch-Korbutowicz, 2005; Howe et al., 2006; Kim et al., 2006; Tran et al., 2006; Cho et al., 2006; Chooi et al., 2008; Cai et al., 2008; Meyn et al., 2008; Barbot et al., 2008; Bond et al., 2009; Park and Yoon, 2009; Zularisam et al., 2009). A sufficient dose of coagulant is required to produce large enough flocs to avoid pore blocking, but increased coagulation dose has been observed to lead to increased fouling (Tran et al., 2006). Small flocs result in rapid blocking of the membrane (Rojas et al., 2010). Therefore, coagulant dosing might be difficult if raw water quality varies significantly. It is also reported that floc strength is more important than floc size for membrane performance (Kim et al., 2013).

The most commonly used coagulants in practice are the hydrolyzing metal ions Al^{3+} and Fe^{3+} (Capodaglio et al., 2011). Polyaluminium chloride (PACl), aluminum sulfate (alum, $Al_2(SO_4)_3{\cdot}14H_2O$), and ferric chloride ($FeCl_3{\cdot}6H_2O$) have been widely adopted in drinking water plants. NOM removal with these coagulants is achieved through several mechanisms, which include double layer compression, charge neutralization, sweep coagulation, and interparticle bridging (Capodaglio et al., 2011).

Aluminum-based and iron-based coagulants are ideal for charge neutralization and sweep flocculation, and are known to preferentially remove hydrophobic, charged, and larger sized (>10 kDa) substances (Choi and Dempsey, 2004; Jung and Kang, 2003; Tran et al., 2006; Maartens et al., 1999; Carroll et al., 2000; Wang and Wang, 2006; Chen et al., 2007; Zhang et al., 2011a). Several studies have indicated that these compounds of NOM have great impact on fouling and decline of flux (Zularisam et al., 2006; Chen et al., 2007; Dong et al., 2007; Howe et al., 2006; Jung et al., 2006). Thus, pretreatment with coagulation is an effective method to prevent these adverse effects (Chen et al., 2007; Dong et al., 2007; Howe et al., 2006; Jung et al., 2006; Qiao et al., 2008).

Chen et al. (2007) reported that the flux was higher for coagulated water than for supernatant water when the coagulant doses were low, and vice versa at higher coagulant doses. The higher flux that was observed with coagulated water was supposed to be due to the formation of a permeable cake layer that prevented membrane fouling, leading to the deposition of hydrophilic neutral substances on the cake layer instead of on the membrane surface. This layer is then effectively removed by backwashing to maintain the flux at desired level (Tran et al., 2006; Chen et al., 2007; Liu et al., 2006). Other researchers have reported that filtration performance was enhanced when flocs are removed from membrane feed water. (Carroll et al., 2000; Chen et al., 2007).

Two-stage coagulant addition is believed to be an effective strategy to reduce membrane fouling in the coagulation-UF process (Liu et al., 2011). This is reported to be due to larger flocs with lower density compared to one-stage addition. Microflocs formed in the first stage might fill the cake layer and/or block the membrane pores, whereas floc aggregates formed in the second stage formed a permeable cake layer.

Coagulation has shown poor removal efficiency in the neutral hydrophilic and small MM fractions (Li et al., 2011; Chen et al., 2007). A coagulation-MF-NF chain has been reported to exhibit DOC removal of 75–87%, HAA formation potential removal of >72%, and final DOC and HAA formation potential concentrations of $<1\,mg\,L^{-1}$ and $<30\,mg\,L^{-1}$, respectively (Ahn et al., 2004). Another study of a coagulation-sedimentation-MF-NF process observed DOC, biodegradable dissolved organic carbon (BDOC), and assimilable organic carbon (AOC) removals of 68–80%, 33–80%, and 45–67%, respectively (Park et al., 2005).

Härmä (1999) performed filtration experiments using tight NF membranes in six water treatment plants where raw water was coagulated with alum, PACl, or ferric coagulants. All permeates contained total organic carbon (TOC) of less than $0.3\,mg\,L^{-1}$ and NOM removal efficiency (>90%) was independent of water recovery (50–80%), *trans*-membrane pressure (500–1000 kPa), and temperature (0–20 °C). The removal of AOC was lower, varying between 25% and 94% depending on the raw water. Tran et al. (2006) have compared different coagulants in a coagulation-UF process. They noticed that polysilicato-iron (PSI) coagulant was more effective in removing NOM and improving fouling conditions than aluminum-based coagulants: A DOC removal of 82–89% was obtained with PSI coagulant in comparison to the 67–86% removal with aluminum-based coagulants.

Novel electrocoagulation (Ben-Sasson et al., 2013; Dubrawski et al., 2013) and submerged coagulation membrane bioreactor (Tian et al., 2008; Li et al., 2011) processes for the production of drinking water have also been tested. A new proposal for reducing membrane fouling is to coat membranes by coagulants (Lin et al., 2012) and to immobilize TiO_2 photocatalyst on the membrane surface, thus creating a composite membrane (Ma et al., 2009). The use of nanoparticles (e.g., nano-alumina, silica, silver, or iron oxide) in membrane structures has raised a great deal of interest, but the risk is that nanoparticles may be released into the water, thus further research is needed (Kim and van der Bruggen, 2010).

9.3.3 Coupling with Adsorption

Another extensively used drinking water treatment method is the hybrid adsorption-membrane process. Suspended powders have been widely used to remove organic compounds that are scarcely removed by UF and MF membrane processes. Similarly to the hybrid coagulation process, membrane filtration can be performed with clarification as a pretreatment or without clarification as a combined adsorbent/membrane process. Adsorbents generally require long contact times or high concentrations to make the process effective. Therefore, regeneration of the adsorbent is required in many cases to make the process feasible. The membrane provides a physical barrier retaining the adsorbent and the organic compounds adsorbed on it. Two configurations exist: external loop and submerged membrane tank. In the latter, the membrane module is immersed into the adsorption tank, which significantly reduces the size and space of the equipment compared to the traditional configuration.

The performance of the hybrid adsorption-membrane process has been shown to depend on adsorbent dosing, membrane properties, backwashing frequency, reactor size and configuration, filtration mode, and NOM concentration and characteristics (Campos et al., 1998; Campinas and Rosa, 2010a; Mavrov et al., 1998; Lebeau et al., 1998; Oh et al., 2006).

The most widely studied adsorbent for MF and UF is PAC. It has been reported to remove small (>500 Da) and nonpolar NOM molecules (Zhang et al., 2011a). PAC/UF systems demonstrated about 90% humic acid and 100% phenol removal (Tomaszewska and Mozia, 2002), and very effective removal of fine contaminants such as trichloroethene, tetrachloroethane, and atrazine (Pianta et al., 1998). Yet, NOM removal from drinking water has not been as successful as removal of the above-mentioned small surrogate compounds (Xia et al., 2007). Enhanced NOM removal from micromolecular sizes up to 17 kDa has been observed when UF is combined with PAC (Kang and Choo, 2010; Lin et al., 1999; Lohwacharin et al., 2010).

PAC was also observed to have a lesser impact on the removal of algal organic matter. PAC enhanced the TOC rejection of algogenic organic matter in UF (35% vs 55%), but had no significant effect on rejections of the highly hydrophilic compounds of extracellular organic matter (Campinas and Rosa, 2010b). It was suggested that PAC was not able to adsorb the highly hydrophilic compounds of polysaccharidelike substances and, as such, did not lower the membrane reversible fouling. This is in accordance with the findings that the hydrophilic fraction exhibits the worst flux decline during UF, regardless of whether the PAC treatment was applied (Lin et al., 2000). In the latter study, both the PAC pretreated UF and combined PAC/UF system rejected 35% of DOC of the hydrophilic fraction. The hydrophobic fraction exhibited ~55% and ~40% DOC rejection with the PAC pretreatment and combined PAC/UF system, respectively.

In most of the above-mentioned studies, 100 kDa cut-off membranes were used. Yet, a comparison of different cut-off membranes in the range of 1–100 kDa has shown that the highest cut-off membrane exhibited the highest flux decline (60%) and the lowest UV_{254} removal (21%) of humic acid in the PAC/UF process (Li and Chen, 2004). The highest removal (89%) and the lowest flux decline (<5%) were observed when a 1 kDa cut-off membrane was used. These findings favor the use of finer UF membranes in integrated membrane systems based on adsorbents.

GAC has been reported to be effective in the removal of the lower MM and hydrophilic fractions of NOM (Yee et al., 2009). Other researchers have reported that the main removed compounds have been mostly hydrophobic (Kim et al., 2009a,b; Gur-Reznik et al., 2008; Tsujimoto et al., 1998).

The GAC filters that are not regenerated contain microbiological communities that make them effective in removing the biodegradable compounds, hence reduce BDOC and AOC. Biological GAC filtering is a high-performance, cost-effective and chemical-free method, which has been widely used for drinking water pretreatment (Hallé et al., 2009). It removes large fractions of polysaccharides and proteins (Hallé et al., 2009; Huang et al., 2011) and colloidal NOM (Kwon et al., 2005). Huang et al. (2011) found that the biodegradation of NOM not only depends on its chemical composition, but is also affected by molecule size and hydrophobicity. GAC adsorption is more suitable after pretreatment to remove large molecules or reduce them in size sufficiently to enter the pores of the adsorbent (e.g., coagulation or ozonation). As a biological process, it is influenced by ambient conditions and contact time, such that a higher temperature can enhance biopolymer removal (Hallé et al., 2009).

Other adsorbent particles that have been tested in hybrid systems include freshly precipitated iron and aluminum oxide particles, heated aluminum and iron oxide particles, ferrihydrite, and carbon black nanoparticles. Heated aluminum and iron oxide particles have consistently been reported to reduce fouling caused by NOM (Zhang et al., 2003; Shi and Benjamin, 2008; Kim et al., 2008). They adsorb foulants preferentially over nonfouling particles, whereas the opposite is true for PAC. As a result, higher foulant removal has been achieved with these oxide particles than with PAC, even though the highest NOM removal was achieved by PAC (Kim et al., 2008). The NOM rejection of 79–92%, 64–72%, and 46–65% were observed with PAC, heated aluminum, and iron oxide particles, respectively (Kim et al., 2008).

Higher NOM removal efficiencies have been observed when oxide particles were predeposited on the membrane surface than when adsorbent particles were suspended in raw water (Kim et al., 2008; Ha et al., 2004; Yao et al., 2009). It has been hypothesized that the formation of adsorbent cake layers at the membrane would contribute to further NOM rejection due to the sieving effect during UF. Heated aluminum oxide particles deposited on the membrane surface have been observed to effectively remove NOM and foulants like polysaccharides and proteins (Cai and Benjamin, 2011).

Ferrihydrite (Kang and Choo, 2010) and carbon black nanoparticles (Lohwacharin et al., 2010) have been reported to mitigate fouling more effectively than PAC. They remove more macromolecular NOM than PAC and form more porous cake layers that are more easily detachable by backwashing than the PAC cake layer.

9.3.4 Coupling with Ion Exchange

Combinations of fluidized IE and MIEX® with membrane filtration have been studied in the past few years. NOM can be removed by macroporous anion exchangers because of the negative charge of the humic substances. The MIEX® particles contain a high proportion of magnetic iron oxides that agglomerate into larger, fast-settling particles.

Cornelissen et al. (2009, 2010) applied fluidized IE treatment before UF and NF, and observed removals of 60% for NOM and >90% for humic substances, respectively. The fluidized IE treatment reduced fouling of a 100-kDa UF membrane, but not fouling of an NF membrane. Humbert et al. (2012) used a combined anion exchange resin and UF with a 100-kDa membrane. They observed that small resin particles fouled the membrane, but that irreversible fouling was significantly lowered.

Kaewsuk and Seo (2011) treated river water with a MIEX®-NF process. They observed that MIEX® could significantly remove NOM compounds with a low SUVA value (phenolic compounds, aliphatic hydrocarbon, and amino sugars), but it showed low efficiency for removing carboxylic compounds. Protein, polysaccharides, and amino sugars were completely retained by the NF membranes. They also noted that membrane roughness influences rejection.

Also, Humbert et al. (2007) and Kaewsuk and Seo (2011) observed that the MIEX® selectively adsorbs the low and medium MM fractions of NOM, but is not effective in removing the high MM compounds. The combination of MIEX®-UF with a 100-kDa membrane was observed to remove up to 80% of DOC (Humbert et al., 2012, 2007). The same removal rate was observed when MIEX® was combined with much tighter (5, 10 and 30 kDa) membranes (Kabsch-Korbutowicz et al., 2008).

Kaewsuk and Seo (2011) used a fluidized bed type MIEX® column as a pretreatment process for submerged NF membranes. The MIEX® process could remove the NOM MM above 500 Da and remove ~60% of polysaccharide compound (>43 kDa). A hydrophilic 350-Da membrane removed 80% and a hydrophobic 550-Da membrane 88% of DOC.

Bond et al. (2010) compared ferric sulfate coagulation, MIEX® anion exchange resin, and two NF membranes in the removal of DBP formation potential from NOM surrogates. MIEX® experiments were performed with a resin dose of $10\,mL\,L^{-1}$ and a bed volume of 1500. Coagulation and MIEX® preferentially removed hydrophobic organic acids, which are typically highly negatively charged, whereas neutral compounds were not significantly removed. Hydrophobic anionic tannic acid was well removed with all these methods: coagulation removed 89% and MIEX® and NF both 92%. With NF (150–430 Da) membranes, removal generally increased with molecular weight from glycine ($75\,g\,mol^{-1}$, 24% removal) to tannic acid ($1701\,g\,mol^{-1}$). The researchers concluded that full-scale MIEX® would not offer improved performance over coagulation.

Mergen et al. (2008) noted that MIEX® treatment is capable of removing both hydrophilic and hydrophobic NOM. Water dominated by hydrophilic acids showed consistently high levels (60%) of removal onto the resin. Yet, HMM NOM quickly saturated or blocked the resin and algal less charged NOM was poorly removed. To mitigate resin blockage, raw water was pretreated by a 50-μm microsieve or by a rapid sand filter when the turbidity was higher than 0.5–1.0 FTU (Ødegaard et al., 2010).

9.3.5 Coupling with Oxidation

Ozonation is the preferred oxidation method for humic water where the main goals are color removal and disinfection (Ødegaard et al., 2010).

Ozone can act directly or via an indirect reaction where it is decomposed to hydroxyl radicals. Direct oxidation targets mainly activated aromatics and double bonds, resulting in high color removal, but little TOC removal. Hydroxyl radicals are more powerful but less selective oxidants than ozone; they remove less color and more TOC.

The reaction of ozone and humic material produces a number of LMM polar compounds that are more biodegradable than the original compounds (Karnik et al., 2005; Malek et al., 2006; Song et al., 2010; Yan et al., 2010; Yang et al., 2010). Therefore, the UV_{254} removal of NOM can be much higher than the corresponding DOC removal. In an ozone-MF treatment, 71% UV_{254} removal, but only 10% DOC removal were observed (Song et al., 2010). Ozonation does not lead to full mineralization of most compounds, thus the initial DOC concentration declines only slightly. Karnik et al. (2005) noticed that an ozonation-UF process resulted in 50% reduction in humic substances and an increase of 20% in nonhumic substances. Furthermore, partially oxidized compounds that were less reactive with chlorine were formed, hence the total THMs and HAAs distributed in the water were reduced by 80% and 65%, respectively.

Byun et al. (2011) studied the effect of ozone dosage on DOC removal and fouling of a 200–400-Da NF membrane and a 30-kDa UF membrane. It was found that DOC removal by the NF membrane was >92% regardless of the ozone dosage, whereas DOC removal by the UF membrane increased slightly from 11% to 26% with an increase in the ozone dosage. The researchers reported that increased fouling due to ozonation was caused by the changes in the chemistry, not the size of the foulants.

Ribau Teixeira et al. (2011) compared conventional sequence, ozonation-NF and ozonation-coagulation/flocculation-sedimentation-NF for the removal of NOM and DBPs. They treated hard water with a low organic matter content (DOC $1.31\,mg\,L^{-1}$) consisting of mainly hydrophilic, less aromatic material of a lower molecular weight. All treatment sequences removed large molecules and reduced DOC content to below $1\,mg\,L^{-1}$. When NF was applied, smaller MM compounds were removed more than in the conventional sequence.

One disadvantage of preoxidation is the formation of small assimilable compounds that may block membrane pores and increase biofouling in the downstream membrane processes and distribution systems. The increased degree of fouling has been reported to be due to the increase in AOC and extracellular biopolymers released by algae upon ozonation (Her et al., 2007). Another disadvantage is the formation of harmful by-products, such as formaldehyde, glyoxal, and glyoxylic acid, which have been reported to be mutagenic (Koltuniewich and Drioli, 2008).

The above-mentioned problems can be avoided by combining oxidation with downstream biologically active filters, BAC or GAC (Figure 9.3) (Ødegaard et al., 2010; Yan et al., 2010; Yang et al., 2010). Biofiltration removes biodegradable organic matter produced by the oxidation step, improving the biostability of the water. Yang et al. (2010) reported that the conventional treatment followed by the O_3/BAC process exhibited DOC removal of 63%, which was further increased to 89% when UF-NF was added at the end of the treatment chain. Moreover, very high removals were observed for THM formation potential (84%) and HAA formation potential (98%).

A treatment facility combining UF with ozonation to treat surface water is one of the world's largest UF plants, located in Mississauga, Ontario, Canada (Byun et al., 2011). The UF membrane filtration unit at the Mississauga plant is combined with ozonation and biologically activated carbon contactors to meet water quality regulations and improve the water's aesthetic quality.

The combination of an AOP and membrane separation has been shown to have significant potential in mitigating both NOM fouling of membranes and the formation of DBPs (Le-Clech et al., 2006; Choo et al., 2008; Huang et al., 2008; Bai et al., 2009; Yao et al., 2009).

Among AOPs, TiO_2/UV oxidation has been observed to effectively remove HMM and hydrophobic organic compounds, and to reduce membrane fouling not only through NOM mineralization, but also by altering NOM characteristics and therefore their fouling potential. Thus, photocatalytic oxidation prior to membrane filtration has been considered quite an efficient method regarding membrane performance and controlling of NOM fouling (Choo et al., 2008; Huang et al., 2008; Bai et al., 2009), although contradictory results have also been observed (Syafei et al., 2008). When membranes are coated using iron oxide particles, the NOM removal performance of the photocatalysis/membrane system is further improved (Yao et al., 2009).

9.3.6 Membrane Bioreactor

The hybrid membrane bioreactor (MBR) process is quite a new technique in drinking water production. As indicated previously, different pretreatment technologies often preferentially remove certain components of NOM. In the MBR process, several removal mechanisms can be organized in a single step. Biomass contains microorganisms that can degrade biodegradable organic matter, adsorbents can be used to remove small-medium MM NOM, coagulants can be used to flocculate hydrophobic large MM NOM and, finally, the membrane provides a physical barrier to separate the entire sludge containing these substances and chemicals. Furthermore, an oxidation step can be added to a submerged MBR process to remove the residual nonadsorbable, nonbiodegradable fraction of DOC (Treguer et al., 2010).

MBR systems are implemented based on two main configurations (Figure 9.4). The submerged configuration consists of outer skin membranes that are situated inside the bioreactor. The driving force across the membrane is achieved by suction pumps or by pressurizing the bioreactor. It operates at a much lower *trans*-membrane pressure and at a lower liquid cross-flow velocity than the external MBR. Plate and frame modules or hollow-fiber membranes in bundles or curtains have been used in submerged membrane systems (Fane et al., 2008).

In the external configuration, mixed liquid is recirculated through membrane modules that are situated outside the bioreactor. MBRs operate at high concentration of suspended solids, so back transport of solids is necessary to sustain membrane permeability. MBRs can produce high-quality water from polluted raw water, but marginally higher capital and operating costs are considered disadvantages of MBRs compared to other biological processes (Fane et al., 2008).

Li and Chu (2003) used a laboratory MBR equipped with a submerged polyethylene hollow-fiber membrane module with a pore size of 0.4 μm

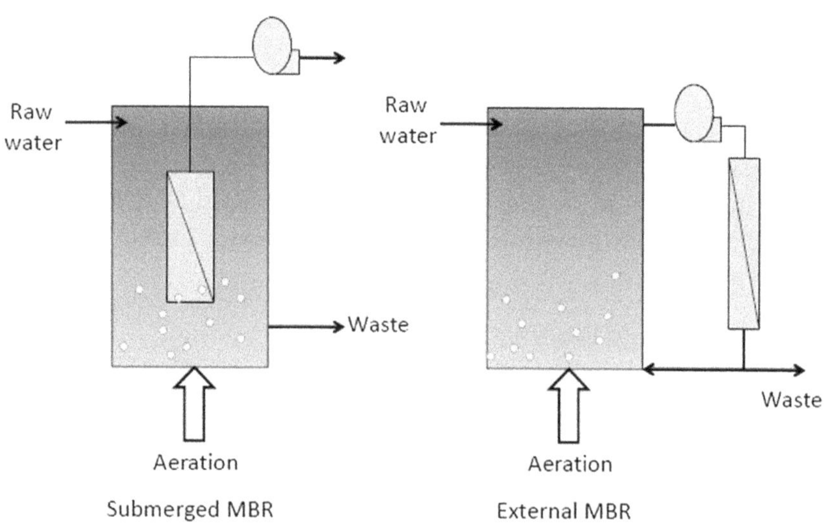

Figure 9.4 Flow charts for submerged and external membrane bioreactors.

and a total surface area of 0.2 m^2 for the treatment of raw water slightly polluted by domestic sewage (3–5 mgTOC L^{-1}). The MBR ran continuously for over 500 days with a ≤1-h hydraulic retention time. Reductions of 60% and 75% were achieved for TOC and THM formation potential, respectively. Furthermore, the biostability of the effluent improved considerably, as the AOC removal was 81%. Visvanathan et al. (2003) studied a membrane bioreactor as a pretreatment for seawater RO. They noted a 78% DOC removal and quasitotal biodegradable DOC removal when treating sodium acetate-spiked seawater (6–8 mgDOC L^{-1}).

MBR removal rates improved with the addition of coagulant (Tian et al., 2008a) or adsorbent (Tian et al., 2008b). Adding biomass did not improve removal as much as other aids and BDOC removal was even reduced compared to plain MBR (Tian et al., 2009). In this system, DOC was removed through a combination of adsorption and biodegradation mechanisms. The best results were achieved with the combination of coagulant-adsorbent-MBR (Tian et al., 2010). The retentions observed were: DOC 63%, UV254 76%, THM formation potential 55%, HAA formation potential 56%, BDOC 67% and AOC 76%.

Sagbo et al. (2008) compared MBR and MBR-PAC hybrid processes for drinking water production. They achieved a TOC removal of 25% for MBR and 50% for MBR-PAC (2 gL^{-1}) system when compared to the MBR. Ravindran et al. (2009) reported TOC removals of 63% when an MBR was operated without biomass and PAC, 70% with biomass alone, and 84% with a combination of PAC and biomass. In the latter case, the THM formation potential removal was 98%.

Zhang et al. (2011b) filtered algal-rich water with a submerged hollow-fiber PVC UF membrane. UF could achieve absolute removal of *Microcystis aeruginosa* cells but poor removal of algogenic organic matter that was released into the water. The addition of 4 gPAC L^{-1} to the reactor increased DOC removal from 54% to 65%. Yet, PAC had little effect on the rejection of hydrophilic high molecular weight assimilable organic matter such as carbohydrates and proteins. Williams and Pirbazari (2007) found that MBR combined with PAC and biomass was effective in the removal of AOC and aldehydes from preozonated water. Yet, Huang et al. (2011) noted that proteinlike substances are more readily eliminated by microorganisms than humiclike substances, indicating that the effect of adsorbent is less important.

It was suggested that an MBR with a short hydraulic retention time could be developed as an effective biological water treatment process for seriously contaminated surface water resources in developing countries (Wang and Menon, 2008).

9.4 CONCLUSIONS

Different treatment combinations with coagulation processes have been investigated. Preoxidation has been used to enhance the coagulation efficiency, with rather contradictory results (Liu et al., 2009). Also, pretreatment with MIEX® has been studied widely with promising results (Jarvis et al., 2008). After coagulation, the adsorption process with AC filtration has proved rather efficient in enhancing NOM reduction, particularly the better removal of LMM and NOM recalcitrant to coagulation (Uyak et al., 2007). Membrane filtrations have also become more common, and hybrid coagulation/membrane systems have been tested by several researchers. Postfiltration (NF) has reduced recalcitrant NOM and its corrosion aggressive constituents significantly (Bond et al., 2009; Shi and Taylor 2007). Accordingly, for high DOC waters, the combination of MIEX®, coagulation, and membrane filtration (UF) is highly efficient for NOM reduction (Humbert et al., 2007). Optimized coagulation may sometimes be sufficient for DBP control alone.

REFERENCES

Ahn, W., Yim, S., Son, B., Kim, G.T., Ahn, H., Lee, D., Lee, S., Kwon, B., Cho, J., 2004. Experimental evaluations of a pilot nanofiltration system with respect to NOM and BOM removals and stable permeability with appropriate pre-treatments: a case study in Korea. J. Water Supply: Res. Technol. – Aqua 53, 169–181.

Alborzfar, M., Jonsson, G., Grøn, C., 1998. Removal of natural organic matter from two types of humic ground waters by nanofiltration. Water Res. 32, 2983–2994.

Bai, H., Zhang, X., Pan, J., Sun, D., Shao, J., 2009. Combination of nano TiO_2 photocatalytic oxidation with microfiltration (MF) for natural organic matter removal. Water Sci. Technol. Water Supply 9, 31–37.

Barbot, E., Moustier, S., Bottero, J., Moulin, P., 2008. Coagulation and ultrafiltration: understanding of the key parameters of the hybrid process. J. Membr. Sci. 325, 520–527.

Ben-Sasson, M., Zidon, Y., Calvo, R., Adin, A., 2013. Enhanced removal of natural organic matter by hybrid process of electrocoagulation and dead-end microfiltration. Chem. Eng. J. 232, 338–345.

Bergamasco, R., Konradt-Moraes, L.C., Vieira, M.F., Fagundes-Klen, M.R., Vieira, A.M.S., 2011. Performance of a coagulation–ultrafiltration hybrid process for water supply treatment. Chem. Eng. J. 166, 483–489.

Blankert, B., Betlem, B., Roffel, B., 2007. Development of a control system for in-line coagulation in an ultrafiltration process. J. Membr. Sci. 301, 39–45.

Bond, T., Goslan, E., Parsons, S., Jefferson, B., 2010. Disinfection by-product formation of natural organic matter surrogates and treatment by coagulation, MIEX® and nanofiltration. Water Res. 44, 1645–1653.

Bonnélye, V., Guey, L., Del Castillo, J., 2008. UF/MF as RO pre-treatment: the real benefit. Desalination 222, 59–65.
Bose, P., Reckhow, D., 2007. The effect of ozonation on natural organic matter removal by alum coagulation. Water Res. 41, 1516–1524.
Botes, J.P., Jacobs, E.P., Bradshaw, S.M., 1998. Long-term evaluation of a UF pilot plant for potable water production. Desalination 115, 229–238.
Byun, S., Taurozzi, J.S., Alpatova, A.L., Wang, F., Tarabara, V.V., 2011. Performance of polymeric membranes treating ozonated surface water: effect of ozone dosage. Sep. Pur. Technol. 81, 270–278.
Cai, Z., Benjamin, M.M., 2011. NOM fractionation and fouling of low-pressure membranes in microgranular adsorptive filtration. Environ. Sci. Technol. 45, 8935–8940.
Cai, Z., Kim, J., Benjamin, M., 2008. NOM removal by adsorption and membrane filtration using heated aluminum oxide particles. Environ. Sci. Technol. 42, 619–623.
Campinas, M., Rosa, M.J., 2010a. Removal of microcystins by PAC/UF. Sep. Pur. Technol. 71, 114–120.
Campinas, M., Rosa, M.J., 2010b. Assessing PAC contribution to the NOM fouling control in PAC/UF systems. Water Res. 44, 1636–1644.
Campos, C., Mariñas, B.J., Snoeyink, V.L., Baudin, I., Laîné, J.M., 1998. Adsorption of trace organic compounds in CRISTAL(R) process. Desalination 117, 265–271.
Capodaglio, A.G., Callegari, A., Sauvignet, P., 2011. Advanced Separation Technology Application for NOM Removal from a Freshwater Supply. IWA Specialty Conference on Natural Organic Matter, July 27–29, 2011, Costa Mesa, California.
Carroll, T., King, S., Gray, S.R., Bolto, B.A., Booker, N.A., 2000. The fouling of microfiltration membranes by NOM after coagulation treatment. Water Res. 34, 2861–2868.
Chang, E.E., Chen, Y., Lin, Y., Chiang, P., 2009. Reduction of natural organic matter by nanofiltration process. Chemosphere 76, 1265–1272.
Chen, Y., Dong, B., Gao, N., Fan, J., 2007. Effect of coagulation pretreatment on fouling of an ultrafiltration membrane. Desalination 204, 181–188.
Chiang, P.-C., Chang, E.-E., Chang, P.-C., Huang, C.-P., 2009. Effects of pre-ozonation on the removal of THM precursors by coagulation. Sci. Total Environ. 407, 5735–5742.
Cho, M.-H., Lee, C.-H., Lee, S., 2006. Effect of flocculation conditions on membrane permeability in coagulation–microfiltration. Desalination 191, 386–396.
Choi, I.-H., Lee, S., Kim, I.-C., Min, B.-R., Lee, K.-H., 2007. Effect of new photocatalytic coagulant on NF membrane fouling. Ind. Eng. Chem. Res. 46, 2280–2285.
Choi, K.Y., Dempsey, B.A., 2004. In-line coagulation with low-pressure membrane filtration. Water Res. 38, 4271–4281.
Choo, K.-W., Tao, R., Kim, M.-J., 2008. Use of photocatalytic membrane reactor for the removal of natural organic matter in water: effect of photoinduced desorption and ferrihydrite adsorption. J. Membr. Sci. 322, 368–374.
Cornelissen, E.R., Beerendonk, E.F., Nederlof, M.N., van der Hoek, J.P., Wessels, L.P., 2009. Fluidized ion exchange (FIX) to control NOM fouling in ultrafiltration. Desalination 236, 334–341.
Cornelissen, E.R., Chasseriaud, D., Siegers, W.G., Beerendonk, E.F., van der Kooij, D., 2010. Effect of anionic fluidized ion exchange (FIX) pre-treatment on nanofiltration (NF) membrane fouling. Water Res. 44, 3283–3293.
Dong, B., Chen, Y., Gao, N., Fan, J., 2007. Effect of coagulation pretreatment on the fouling of ultrafiltration membrane. J. Environ. Sci. 19, 278–283.
Dubrawski, K.L., Fauvel, M., Mohseni, M., 2013. Metal type and natural organic matter source for direct filtration electrocoagulation of drinking water. J. Hazard. Mat. 244–245, 135–141.
Fane, A.G., Wang, R., Jia, Y., 2008. Membrane technology: past, present and future. In: Wang, L.K., Chen, J.P., Hung, Y., Shammas, N.K. (Eds.), Handbook of Environmental Engi-

neering. Membrane and Desalination Technologies, vol 13. Springer Science & Business Media, New York, pp. 1–45.
Garcia, I., Moreno, L., 2009. Use of GAC after enhanced coagulation for the removal of natural organic matter from water for purification. Water Sci. Technol. Water Supply 9, 173–180.
Glucina, K., Alvarez, A., Laîné, J.M., 2000. Assessment of an integrated membrane system for surface water treatment. Desalination 132, 73–82.
Goslan, E., Gurses, F., Banks, J., Parsons, S., 2006. An investigation into reservoir NOM reduction by UV photolysis and advanced oxidation processes. Chemosphere 65, 1113–1119.
Grose, A.B.F., Smith, A.J., Donn, A., O'Donnell, J., Welch, D., 1998. Supplying high quality drinking water to remote communities in Scotland. Desalination 117, 107–117.
Gur-Reznik, S., Katz, I., Dosoretz, C.G., 2008. Removal of dissolved organic matter by granular-activated carbon adsorption as a pretreatment to reverse osmosis of membrane bioreactor effluents. Water Res. 42, 1595–1605.
Ha, T., Choo, K., Choi, S., 2004. Effect of chlorine on adsorption/ultrafiltration treatment for removing natural organic matter in drinking water. J. Coll. Interf. Sci. 274, 587–593.
Hallé, C., Huck, P.M., Peldszus, S., Haberkamp, J., Jekel, M., 2009. Assessing the performance of biological filtration as pretreatment to low pressure membranes for drinking water. Environ. Sci. Technol. 43, 3878–3884.
Härmä, V., 1999. Tuloksia suomalaisilla pintavesilaitoksilla tehdyistä nanosuodatuskokeista. Vesitalous 6, 5–10.
Her, N., Amy, G., Plottu-Pecheux, A., Yoon, Y., 2007. Identification of nanofiltration membrane foulants. Water Res. 41, 3936–3947.
Hilal, N., Al-Abri, M., Al-Hinai, H., 2006. Enhanced membrane pre-treatment processes using macromolecular adsorption and coagulation in desalination plants: a Review. Sep. Sci. Technol. 41, 403–453.
Howe, K., Marwah, A., Chiu, K.-P., Adham, S., 2006. Effect of coagulation on the size of MF and UF membrane foulants. Environ. Sci. Technol. 40, 7908–7913.
Huang, G., Meng, F., Zheng, X., Wang, Y., Wang, Z., Liu, H., Jekel, M., 2011. Biodegradation behavior of natural organic matter (NOM) in a biological aerated filter (BAF) as a pretreatment for ultrafiltration (UF) of river water. Appl. Microbiol. Biotechnol. 90, 1795–1803.
Huang, H., Schwab, K., Jacangelo, J.G., 2009. Pretreatment for low pressure membranes in water treatment: a review. Environ. Sci. Technol. 43, 3011–3019.
Huang, X., Leal, M., Li, Q., 2008. Degradation of natural organic matter by TiO_2 photocatalytic oxidation and its effect on fouling of low-pressure membranes. Water Res. 42, 1142–1150.
Humbert, H., Gallard, H., Croué, J., 2012. A polishing hybrid AER/UF membrane process for the treatment of a high DOC content surface water. Water Res. 46, 1093–1100.
Humbert, H., Gallard, H., Jacquemet, V., Croué, J., 2007. Combination of coagulation and ion exchange for the reduction of UF fouling properties of a high DOC content surface water. Water Res. 41, 3803–3811.
Jacangelo, J., DeMarco, J., Owen, D., Randtke, S., 1995. Biofiltration performance: part 2, effect of backwashing. J. Amer. Water Works Assoc. 87, 64–70.
Jarvis, P., Mergen, M., Banks, J., McIntosh, B., Parsons, S., Jefferson, B., 2008. Pilot scale comparison of enhanced coagulation with magnetic resin plus coagulation systems. Environ. Sci. Technol. 42, 1276–1282.
Jung, C., Kang, L., 2003. Application of combined coagulation-ultrafiltration membrane process for water treatment. Kor. J. Chem. Eng. 20, 855–861.

Jung, C.-W., Son, H.-J., Kang, L.-S., 2006. Effects of membrane material and pretreatment coagulation on membrane fouling: fouling mechanism and NOM removal. Desalination 197, 154–164.
Kabsch-Korbutowicz, M., 2005. Effect of Al coagulant type on natural organic matter removal efficiency in coagulation/ultrafiltration process. Desalination 185, 327–333.
Kabsch-Korbutowicz, M., Majewska-Nowak, K., Winnicki, T., 2008. Water treatment using MIEX-DOC/ultrafiltration process. Desalination 221, 338–344.
Kaewsuk, J., Seo, G.T., 2011. Verification of NOM removal in MIEX-NF system for advanced water treatment. Sep. Pur. Technol. 80, 11–19.
Kajitvichyanukul, P., Shammas, N.K., Hung, Y., Wang, L.K., Ananpattarachai, J., 2008. Portable water biotechnology, membrane filtration and biofiltration. In: Wang, L.K., Chen, J.P., Hung, Y., Shammas, N.K. (Eds.), Handbook of Environmental Engineering. Membrane and Desalination Technologies, vol. 13. Springer Science & Business Media, New York, pp. 477–523.
Kamp, P.C., Kruithof, J.C., Folmer, H.C., 2000. UF/RO treatment plant Heemskerk: from challenge to full scale application. Desalination 131, 27–35.
Kang, S., Choo, K., 2010. Why does a mineral oxide adsorbent control fouling better than powdered activated carbon in hybrid ultrafiltration water treatment? J. Membr. Sci. 355, 69–77.
Karnik, B.S., Davies, S.H., Baumann, M.J., Masten, S.J., 2005. The effects of combined ozonation and filtration on disinfection by-product formation. Water Res. 39, 2839–2850.
Kim, H., Hong, J.-H., Lee, S., 2006. Fouling of microfiltration membranes by natural organic matter after coagulation treatment: a comparison of different initial mixing conditions. J. Membr. Sci. 283, 266–272.
Kim, J., Van der Bruggen, B., 2010. The use of nanoparticles in polymeric and ceramic membrane structures: review of manufacturing procedures and performance improvement for water treatment. Environ. Pollut. 158, 2335–2349.
Kim, J., Cai, Z., Benjamin, M.M., 2008. Effects of adsorbents on membrane fouling by natural organic matter. J. Membr. Sci. 310, 356–364.
Kim, J., Shan, W., Davies, S., Baumann, M., Masten, S., Tarabara, V., 2009a. Interactions of aqeuos NOM with nanoscale TiO_2: Implications for ceramic membrane filtration-ozonation hybrid process. Environ. Sci. Technol. 43, 5488–5494.
Kim, J.H., Lee, C.H., Lee, E.J., Lee, K.H., Kwon, S.B., Park, H., Kim, H.S., Jang, A., 2013. The effect of re-aggregated floc by additional coagulant on membrane. J. Taiwan Inst. Chem. Eng. 44, 802–807.
Kim, K., Kim, H., Kim, J., Nam, J., Kim, J., Son, S., 2009b. A hybrid microfiltration–granular activated carbon system for water purification and wastewater reclamation/reuse. Desalination 243, 132–144.
Kim, M.H., Yu, M.J., 2005. Characterization of NOM in the Han River and evaluation of treatability using UF–NF membrane. Environ. Res. 97, 116–123.
Koltuniewich, A.B., Drioli, E., 2008. Membranes in Clean Technologies, Theory and Practice. Wiley-VCH Verlag, Weinheim.
Konieczny, K., Sakol, D., Plonka, J., Rajca, M., Bodzek, M., 2009. Coagulation—ultrafiltration system for river water treatment. Desalination 240, 151–159.
Kwon, B., Lee, S., Cho, J., Ahn, H., Lee, D., Shin, H.S., 2005. Biodegradability, DBP fromation, and membrane fouling potential of natural organic matter: characterization and controllability. Environ. Sci. Technol. 39, 732–739.
Lamsal, R., Montreuil, K.R., Kent, F.C., Walsh, M.E., Gagnon, G.A., 2012. Characterization and removal of natural organic matter by an integrated membrane system. Desalination 303, 12–16.
Lebeau, T., Lelièvre, C., Buisson, H., Cléret, D., Van de Venter, L.W., Côté, P., 1998. Immersed membrane filtration for the production of drinking water: combination with PAC for NOM and SOCs removal. Desalination 117, 219–231.

Le-Clech, P., Lee, E.-K., Chen, V., 2006. Hybrid photocatalysis/membrane treatment for surface waters containing low concentrations of natural organic matters. Water Res. 40, 323–330.

Lee, S., Lee, C., 2006. Microfiltration and ultrafiltration as a pretreatment for nanofiltration of surface water. Sep. Sci. Technol. 41, 1–23.

Lee, S., Lee, C., 2007. Effect of membrane properties and pretreatment on flux and NOM rejection in surface water nanofiltration. Sep. Pur. Technol. 56, 1–8.

Leiknes, T., 2009. The effect of coupling coagulation and flocculation with membrane filtration in water treatment: a review. J. Environ. Sci. 21, 8–12.

Li, C., Chen, Y., 2004. Fouling of UF membrane by humic substance: effects of molecular weight and powder-activated carbon (PAC) pre-treatment. Desalination 170, 59–67.

Li, X., Chu, H.P., 2003. Membrane bioreactor for the drinking water treatment of polluted surface water supplies. Water Res. 37, 4781–4791.

Li, Y., Wang, J., Zhang, W., Zhang, X., Chen, C., 2011. Effects of coagulation on submerged ultrafiltration membrane fouling caused by particles and natural organic matter (NOM). Chin. Sci. Bull. 56, 584–590.

Lin, C., Huang, Y., Hao, O.J., 1999. Ultrafiltration processes for removing humic substances: effect of molecular weight fractions and PAC treatment. Water Res. 33, 1252–1264.

Lin, C., Lin, T., Hao, O.J., 2000. Effects of humic substance characteristics on UF performance. Water Res. 34, 1097–1106.

Lin, J.C., Cheng, A., Lin, Y., Lee, D., Guo, W., 2012. Dense membranes precoated with aluminium tridecamer (Al_{13}) for water treatment. Desalination 284, 349–352.

Liu, H., Cheng, F., Wang, D., 2009. Interaction of ozone and organic matter in coagulation with inorganic polymer flocculant-PACl: role of organic components. Desalination 249, 596–601.

Liu, H., Wang, D., Wang, M., Tang, H., Yang, M., 2007. Effect of pre-ozonation on coagulation with IPF–PACls: role of coagulant speciation. Coll. Surf. A: Physicochem. Eng. Asp. 294, 111–116.

Liu, Q., Seung-Hyun, K., Jong-Sup, Y., Seong-Yong, M., 2006. Impacts of coagulation pretreatment on MF filtration and a comparative study of different membrane module types. J. Ocean. Univ. China (English Ed. 5), 295–299.

Liu, T., Chen, Z., Yu, W., Shen, J., Gregory, J., 2011. Effect of two-stage coagulant addition on coagulation-ultrafiltration process for treatment of humic-rich water. Water Res. 45, 4260–4268.

Lohwacharin, J., Oguma, K., Takizawa, S., 2010. Use of carbon black nanoparticles to mitigate membrane fouling in ultrafiltration of river water. Sep. Pur. Technol. 72, 61–69.

Loi-Brugger, A., Panglisch, S., Buchta, P., Hattori, K., Yonekawa, H., Tomita, Y., Gimbel, R., 2006. Ceramic membranes for direct river water treatment applying coagulation and microfiltration. Water Sci. Technol. Water Supply 6, 89–98.

Ma, N., Quan, X., Zhang, Y., Chen, S., Zhao, H., 2009. Integration of separation and photocatalysis using an inorganic membrane modified with Si-doped TiO_2 for water purification. J. Membr. Sci. 335, 58–67.

Maartens, A., Swart, P., Jacobs, E.P., 1999. Feed-water pretreatment: methods to reduce membrane fouling by natural organic matter. J. Membr. Sci. 163, 51–62.

Malek, F., Harris, J.L., Roddick, F.A., 2006. Interrelationship of photooxidation and microfiltration in drinking water treatment. J. Membr. Sci. 281, 541–547.

Matilainen, A., Vepsäläinen, M., Sillanpää, M., 2010. Natural organic matter removal by coagulation during drinking water treatment: a review. Adv. Colloid Interface Sci. 159, 189–197.

Mavrov, V., Chmiel, H., Kluth, J., Meier, J., Heinrich, F., Ames, P., Backes, K., Usner, P., 1998. Comparative study of different MF and UF membranes for drinking water production. Desalination 117, 189–196.

Mergen, M., Adams, B., Vero, G., Price, T., Parsons, S., Jefferson, B., Jarvis, P., 2009. Characterization of natural organic matter (NOM) removed by magnetic ion exchange resin (MIEX® Resin). Water Sci. Technol. Water Supply 9, 199–205.
Mergen, M.R.D., Jefferson, B., Parsons, S.A., Jarvis, P., 2008. Magnetic ion-exchange resin treatment: Impact of water type and resin use. Water Res. 42, 1977–1988.
Metsämuuronen, S., Sillanpää, M., Bhatnagar, A., Mänttäri, M., 2014. Natural organic matter removal from drinking water by membrane technology. Sep. Pur. Rev. 43, 1–61.
Meyn, T., Bahn, A., Leiknes, T., 2008. Significance of flocculation for NOM removal by coagulation–ceramic membrane microfiltration. Water Sci. Technol. Water Supply 8, 691–700.
Mierzwa, J.C., da Silva, M.C.C., Veras, L.R.V., Subtil, E.L., Rodrigues, R., Li, T., Landenberger, K.R., 2012. Enhancing spiral-wound ultrafiltration performance for direct drinking water treatment through operational procedures improvement: a feasible option for the Sao Paulo metropolitan region. Desalination 307, 68–75.
Mierzwa, J.C., Hespanhol, I., da Silva, M.C.C., Rodrigues, L.D.B., Giorgi, C.F., 2008. Direct drinking water treatment by spiral-wound ultrafiltration membranes. Desalination 230, 41–50.
Moon, J., Kang, M., Lim, J., Kim, C., Park, H., 2009. Evaluation of a low-pressure membrane filtration for drinking water treatment: pretreatment by coagulation/sedimentation for the MF membrane. Desalination 247, 271–284.
Morran, J., Drikas, M., Cook, D., Bursill, D., 2004. Comparison of MIEX® treatment and coagulation on NOM character. Water Sci. Technol. Water Supply 4, 129–137.
Murray, C., Parsons, S., 2004. Removal of NOM from drinking water: Fenton's and photo-Fenton's processes. Chemosphere 54, 1017–1023.
Nederlof, M.M., Kruithof, J.C., Taylor, J.S., van der Kooij, D., Schippers, J.C., 2000. Comparison of NF/RO membrane performance in integrated membrane systems. Desalination 131, 257–269.
Ødegaard, H., Østerhus, S., Melin, E., Eikebrokk, B., 2010. NOM removal technologies – Norwegian experiences. Drink. Water Eng. Sci. 3, 1–9.
Oh, H., Yu, M., Takizawa, S., Ohgaki, S., 2006. Evaluation of PAC behavior and fouling formation in an integrated PAC–UF membrane for surface water treatment. Desalination 192, 54–62.
Park, N., Kwon, B., Sun, M., Ahn, H., Kim, C., Kwoak, C., Lee, D., Chae, S., Hyung, H., Cho, J., 2005. Application of various membranes to remove NOM typically occurring in Korea with respect to DBP, AOC and transport parameters. Desalination 178, 161–169.
Park, S., Yoon, T., 2009. Effects of iron species and inert minerals on coagulation and direct filtration for humic acid removal. Desalination 239, 146–158.
Pianta, R., Boller, M., Janex, M., Chappaz, A., Birou, B., Ponce, R., Walther, J., 1998. Micro- and ultrafiltration of karstic spring water. Desalination 117, 61–71.
Pirkanniemi, K., Sillanpää, M., 2002. Heterogeneous water phase catalysis as an environmental application: a review. Chemosphere 48, 1047–1060.
Pirkanniemi, K., Metsärinne, S., Sillanpää, M., 2007. Degradation of EDTA and novel complexing agents in pulp and paper mill process and waste waters by Fenton's reagent. J. Hazard. Mater. 147, 556–561.
Pryor, M.J., Jacobs, E.P., Botes, J.P., Pillay, V.L., 1998. A low pressure ultrafiltration membrane system for potable water supply to developing communities in South Africa. Desalination 119, 103–111.
Qiao, X., Zhang, Z., Wang, N., Wee, V., Low, M., Loh, C., Ng, T., 2008. Coagulation pretreatment for a large-scale ultrafiltration process treating water from the Taihu River. Desalination 230, 305–313.
Rämö, J., Sillanpää, M., 2001. Degradation of EDTA by hydrogen peroxide in alkaline conditions. J. Clean. Prod. 9, 191–195.

Ravindran, V., Tsai, H., Williams, M.D., Pirbazari, M., 2009. Hybrid membrane bioreactor technology for small water treatment utilities: process evaluation and primordial considerations. J. Memb. Sci. 344, 39–54.
Ribau Teixeira, M., Rosa, S.M., Sousa, V., 2011. Natural organic matter and disinfection by-products formation potential in water treatment. Water Res. Manag. 25, 3005–3015.
Rojas, J.C., Moreno, B., Garralon, G., Plaza, F., Perez, J., Gomez, M.A., 2010. Influence of velocity gradient in a hydraulic flocculator on NOM removal by aerated spiral-wound ultrafiltration membranes (ASWUF). J. Hazard. Mater. 178, 535–540.
Sagbo, O., Sun, Y., Hao, A., Gu, P., 2008. Effect of PAC addition on MBR process for drinking water treatment. Sep. Pur. Technol. 58, 320–327.
Sani, B., Basile, E., Rossi, L., Lubello, C., 2008. Effects of pre-treatment with magnetic ion exchange resins on coagulation/flocculation process. Water Sci. Technol. 57, 57–64.
Schrotter, J., Bozkaya-Schrotter, B., 2010. Current and emerging membrane processes for water treatment. In: Peinemann, K., Pereira Nunes, S. (Eds.), Membranes for Water Treatment. Wiley-VCH Verlag, Weinheim, pp. 53–91.
Shi, W., Benjamin, M.M., 2008. Membrane interactions with NOM and an adsorbent in a vibratory shear enhanced filtration process (VSEP) system. J. Memb. Sci. 312, 23–33.
Shorrock, K., Drage, B., 2006. A pilot plant evaluation of the magnetic ion exchange® process for the removal of dissolved organic carbon at Draycote water treatment works. Water Environ. J. 20, 65–70.
Siddiqui, M., Amy, G., Ryan, J., Odem, W., 2000. Membranes for the control of natural organic matter from surface waters. Water Res. 34, 3355–3370.
Smith, E., Kamal, Y., 2009. Optimizing treatment for reduction of disinfection by-product (DBP) formation. Water Sci. Technol. Water Supply 9, 191–198.
Song, Y., Dong, B., Gao, N., Xia, S., 2010. Huangpu River water treatment by microfiltration with ozone pretreatment. Desalination 250, 71–75.
Syafei, A., Lin, C.-F., Wu, C.-H., 2008. Removal of natural organic matter by ultrafiltration with TiO_2-coated membrane under UV irradiation. J. Coll. Interf. Sci. 323, 112–119.
Szlachta, M., Adamski, W., 2009. Effects of natural organic matter removal by integrated processes: alum coagulation and PAC-adsorption. Water Sci. Technol. 59, 1951–1957.
Thorsen, T., Fløgstad, H., 2006. Nanofiltration in Drinking Water Treatment, Literature Review. Research report for Techneau Project, December11, 2006.
Tian, J., Chen, Z., Nan, J., Liang, H., Li, G., 2010. Integrative membrane coagulation adsorption bioreactor (MCABR) for enhanced organic matter removal in drinking water treatment. J. Membr. Sci. 352, 205–212.
Tian, J., Liang, H., Li, X., You, S., Tian, S., Li, G., 2008a. Membrane coagulation bioreactor (MCBR) for drinking water treatment. Water Res. 42, 3910–3920.
Tian, J., Liang, H., Nan, J., Yang, Y., You, S., Li, G., 2009. Submerged membrane bioreactor (sMBR) for the treatment of contaminated raw water. Chem. Eng. J. 148, 296–305.
Tian, J., Liang, H., Yang, Y., Tian, S., Li, G., 2008b. Membrane adsorption bioreactor (MABR) for treating slightly polluted surface water supplies: as compared to membrane bioreactor (MBR). J. Membr. Sci. 325, 262–270.
Tomaszewska, M., Mozia, S., 2002. Removal of organic matter from water by PAC/UF system. Water Res. 36, 4137–4143.
Tran, T., Gray, S., Naughton, R., Bolto, B., 2006. Polysilicato-iron for improved NOM removal and membrane performance. J. Membr. Sci. 280, 560–571.
Treguer, R., Tatin, R., Couvert, A., Wolbert, D., Tazi-Pain, A., 2010. Ozonation effect on natural organic matter adsorption and biodegradation – application to a membrane bioreactor containing activated carbon for drinking water production. Water Res. 44, 781–788.
Tsujimoto, W., Kimura, H., Izu, T., Irie, T., 1998. Membrane filtration and pre-treatment by GAC. Desalination 119, 323–326.
Uyak, V., Yavuz, S., Toroz, I., Ozaydin, S., Genceli, E., 2007. Disinfection by-products precursors removal by enhanced coagulation and PAC adsorption. Desalination 216, 334–344.

Uyguner, C., Bekbolet, M., Selcuk, H., 2007a. A comparative approach to the application of a physico–chemical and advanced oxidation combined system to natural water samples. Sep. Sci. Technol. 42, 1405–1419.
Uyguner, C., Suphandag, S., Kerc, A., Bekbolet, M., 2007b. Evaluation of adsorption and coagulation characteristics of humic acids preceded by alternative advanced oxidation techniques. Desalination 210, 183–193.
Vickers, J.C., Thompson, M.A., Kelkar, U.G., 1995. The use of membrane filtration in conjunction with coagulation processes for improved NOM removal. Desalination 102, 57–61.
Visvanathan, C., Boonthanon, N., Sathasivan, A., Jegatheesan, V., 2003. Pretreatment of seawater for biodegradable organic content removal using membrane bioreactor. Desalination 153, 133–140.
Wang, J., Wang, X., 2006. Ultrafiltration with in-line coagulation for the removal of natural humic acid and membrane fouling mechanism. J. Environ. Sci. 18, 880–884.
Wang, L.K., Menon, R., 2008. Treatment of industrial effluents, municipal wastes, and portable water by membrane bioreactors. In: Wang, L.K., Chen, J.P., Hung, Y., Shammas, N.K. (Eds.), Handbook of Environmental Engineering. Membrane and Desalination Technologies, vol. 13. Springer Science & Business Media, New York, pp. 201–236.
Williams, M.D., Pirbazari, M., 2007. Membrane bioreactor process for removing biodegradable organic matter from water. Water Res. 41, 3880–3893.
Xia, S., Liu, Y., Li, X., Yao, J., 2007. Drinking water production by ultrafiltration of Songhuajiang River with PAC adsorption. J. Environ. Sci. 19, 536–539.
Yan, M., Wang, D., Ma, X., Ni, J., Zhang, H., 2010. THMs precursor removal by an integrated process of ozonation and biological granular activated carbon for typical Northern China water. Sep. Pur. Technol. 72, 263–268.
Yan, M., Wang, D., Shi, B., Wang, M., Yan, Y., 2007. Effect of pre-ozonation on optimized coagulation of a typical North-China source water. Chemosphere 69, 1695–1702.
Yang, J., Yuan, D., Weng, T., 2010. Pilot study of drinking water treatment with GAC, O_3/BAC and membrane processes in Kinmen Island, Taiwan. Desalination 263, 271–278.
Yang, W., Chen, J., Li, X., Liang, H., He, W., Li, G., 2007. Using chloramine as a coagulant aid in enhancing coagulation of Yellow River water in China. J. Zhejiang Univ. – Sci. A 8, 1475–1481.
Yao, P., Choo, K., Kim, M., 2009. A hybridized photocatalysis–microfiltration system with iron oxide-coated membranes for the removal of natural organic matter in water treatment: effects of iron oxide layers and colloids. Water Res. 43, 4238–4248.
Yee, L.F., Abdullah, P.M., Abdullah, A., Ishak, B., Abidin, K.N.Z., 2009. Hydrophobicity characteristics of natural organic matter and the formation of THM. Malay. J. Anal. Sci. 13, 94–99.
Zhang, M., Li, C., Benjamin, M.M., Chang, Y., 2003. Fouling and natural organic matter removal in adsorbent/membrane systems for drinking water treatment. Environ. Sci. Technol. 37, 1663–1669.
Zhang, W., Zhang, X., Li, Y., Wang, J., Chen, C., 2011a. Membrane flux dynamics in the submerged ultrafiltration hybrid treatment process during particle and natural organic matter removal. J. Environ. Sci. 23, 1970–1976.
Zhang, Y., Tian, J., Nan, J., Gao, S., Liang, H., Wang, M., Li, G., 2011b. Effect of PAC addition on immersed ultrafiltration for the treatment of algal-rich water. J. Hazard. Mat. 186, 1415–1424.
Zularisam, A., Ismail, A., Salim, R., 2006. Behaviours of natural organic matter in membrane filtration for surface water treatment – a review. Desalination 194, 211–231.
Zularisam, A., Ismail, A., Salim, M., Sakinah, M., Matsuura, T., 2009. Application of coagulation–ultrafiltration hybrid process for drinking water treatment: optimization of operating conditions using experimental design. Sep. Pur. Technol. 65, 193–210.

BIBLIOGRAPHY

Abbaszadegan, M., Mayer, B.K., Ryu, H., Nwachuku, N., 2007. Efficacy of removal of CCL viruses under enhanced coagulation conditions. Environ. Sci. Technol. 41, 971–977.

Abu-Lail, L., Liu, Y., Atabek, A., Camesano, T., 2007. Quantifying the adhesion and interaction forces between *Pseudomonas aeruginosa* and natural organic matter. Environ. Sci. Technol. 41, 8031–8037.

Afcharian, A., Levi, Y., Kiene, L., Scribe, P., 1997. Fractionation of dissolved organic matter from surface waters using macroporous resins. Water Res. 31, 2989–2996.

Agustina, T., Ang, H., Vareek, V., 2005. A review of synergistic effect of photocatalysis and ozonation on wastewater treatment. J. Photochem. Photobiol. Photochem. Rev. 6, 264–273.

Ahn, W., Yim, S., Son, B., Kim, G.T., Ahn, H., Lee, D., Lee, S., Kwon, B., Cho, J., 2004. Experimental evaluations of a pilot nanofiltration system with respect to NOM and BOM removals and stable permeability with appropriate pre-treatments: a case study in Korea. J. Water Supply Res. Technol. Aqua 53, 169–181.

Aiken, G., 1985. Isolation and concentration techniques for aquatic humic substances. In: Aiken, G., McKnight, D., Wershaw, R., MacCarthy, P. (Eds.), Humic Substances in Soil, Sediment and Water. John Wiley & Sons, pp. 363–385.

Aiken, G., McKnight, D., Thorn, K., Thurman, E., 1992. Isolation of hydrophilic organic acids from water using non-ionic macroporous resins. Org. Geochem. 18 (4), 567–573.

Aiken, G., 1984. Evaluation of ultrafiltration for determining molecular weight of fulvic acid. Environ. Sci. Technol. 18, 978–981.

Akyol, A., Bayramoglu, M., 2010. Preparation and characterization of supported ZnO photocatalyst by zincate method. J. Hazard. Mater. 175, 484–491.

Al-Amoudi, A.S., 2010. Factors affecting natural organic matter (NOM) and scaling fouling in NF membranes: a review. Desalination 259, 1–10.

Alasonati, E., Slaveykova, V., Gallard, H., Croué, J.-P., Benedetti, M., 2010. Characterization of the colloidal organic matter from Amazonian basin by asymmetrical flow field-flow fractionation and size exclusion chromatography. Water Res. 44, 223–231.

Alborzfar, M., Jonsson, G., Grøn, C., 1998. Removal of natural organic matter from two types of humic ground waters by nanofiltration. Water Res. 32, 2983–2994.

Alkan, U., Teksoy, A., Ateply, A., Bapkaya, H., 2007. Efficiency of the UV/H_2O_2 process for the disinfection of humic surface waters. J. Environ. Sci. Health A Tox. Hazard. Subst. Environ. Eng. 42, 497–506.

Allpike, B., Heitz, A., Joll, C., Kagi, R., 2007. A new organic carbon detector for size exclusion chromatography. J. Chromatogr. A 1157, 472–476.

Allpike, B.P., Heitz, A., Joll, C.A., Kagi, R.I., Abbt-Braun, G., Frimmel, F.H., Brinkmann, T., Her, N., Amy, G., 2005. Size exclusion chromatography to characterize DOC removal in drinking water treatment. Environ. Sci. Technol. 39, 2334–2342.

Al-Naseri, S.K., Abbas, T.R., 2009. Predicting NOM removal by fixed-bed GAC adsorbers. Jordan J. Civ. Eng. 3, 172–183.

Alsheyab, M., Muñoz, A., 2006. Reducing the formation of trihalomethanes (THMs) by ozone combined with hydrogen peroxide (H_2O_2/O_3). Desalination 194, 121–126.

Alsheyab, M., Muñoz, A., 2007. Comparative study of ozone and MnO_2/O_3 effects on the elimination of TOC and COD of raw water at Valmayor station. Desalination 207, 179–183.

Álvarez-Uriarte, J.I., Iriarte-Velasco, U., Chimeno-Alanís, N., González-Velasco, J.R., 2010. The effect of mixed oxidants and powdered activated carbon on the removal of natural organic matter. J. Hazard. Mater. 181, 426–431.

Amin, M., Faraji, M., Momenbeik, F., Hasanzadeh, A., Sadani, M., 2012. Performance evaluation of an Anaerobic Migrating Blanket Reactor in the biodegradation of perchloroethylene from industrial wastewaters. Int. J. Environ. Health Eng. 1, 1–8.

Amy, G., 2008. Fundamental understanding of organic matter fouling of membranes. Desalination 231, 44–51.

Amy, G., Cho, J., 1999. Interactions between natural organic matter (NOM) and membranes: rejection and fouling. In: Ødegaard, H. (Ed.), International IAWQ–IWSA Joint Specialist Conference on Removal of Humic Substances from Water. June 24–26, 1999, p. 141.

Amy, G., Collins, M., Kuo, C., King, P., 1987. Comparing gel permeation chromatography and ultrafiltration for the molecular weight characterization of aquatic organic matter. J. Am. Water Works Assoc. 79 (1), 43–49.

Anglada, A., Urtiaga, A., Ortiz, I., 2009. Pilot scale performance of the electro-oxidation of landfill leachate at boron-doped diamond anodes. Environ. Sci. Technol. 43, 2035–2040.

Anglada, A., Urtiaga, A., Ortiz, I., Mantzavinos, D., Diamadopoulos, E., 2011. Boron-doped diamond anodic treatment of landfill leachate: evaluation of operating variables and formation of oxidation by-products. Water Res. 45, 828–838.

Anirudhan, T.S., Ramachandran, M., 2007. Surfactant-modified bentonite as adsorbent for the removal of humic acid from wastewaters. Appl. Clay Sci. 35, 276–281.

Anopsitakis, G., Dionysiou, D., 2004. Transition metal/UV-based advanced oxidation technologies for water decontamination. Appl. Catal. B Environ. 54, 155–163.

Aoustin, E., Schäfer, A.I., Fane, A.G., Waite, T.D., 2001. Ultrafiltration of natural organic matter. Sep. Purif. Technol. 22–23, 63–78.

Apell, J.N., Boyer, T.H., 2010. Combined ion exchange treatment for removal of dissolved organic matter and hardness. Water Res. 44, 2419–2430.

Archer, A., Singer, P., 2006. An evaluation of the relationship between SUVA and NOM coagulation using the ICR database (PDF). J. Am. Water Works Assoc. 98 (7), 110–123.

Assemi, S., Newcombe, G., Hepplewhite, C., Beckett, R., 2004. Characterization of natural organic matter fractions separated by ultrafiltration using flow field-flow fractionation. Water Res. 38, 1467–1476.

Ates, N., Kitis, M., Yetis, U., 2007. Formation of chlorination by-products in waters with low SUVA – correlations with SUVA and differential UV spectroscopy. Water Res. 41, 4139–4148.

Ates, N., Yilmaz, L., Kitis, M., Yetis, U., 2009. Removal of disinfection by-product precursors by UF and NF membranes in low-SUVA waters. J. Membr. Sci. 328, 104–112.

Awad, H.S., Abo Galwa, N., 2005. Electrochemical degradation of Acid Blue and Basic Brown dyes on Pb/PbO_2 electrode in the presence of different conductive electrolyte and effect of various operating factors. Chemosphere 61, 1327–1335.

Ayu Sari, M., Chellam, S., 2013. Surface water nanofiltration incorporating (electro) coagulation–microfiltration pretreatment: fouling control and membrane characterization. J. Membr. Sci. 437, 249–256.

Baalousha, M., Motelica-Heino, M., Le Coustumer, P., 2006. Conformation and size of humic substances: effects of major cation concentration and type, pH, salinity, and residence time. Colloids Surf. A 272, 48–55.

Baalousha, M., Stolpe, B., Lead, J.R., 2011. Flow field-flow fractionation for the analysis and characterization of natural colloids and manufactured nanoparticles in environmental systems: a critical review. J. Chromatogr. A 1218, 4078–4103.

Bagastyo, A.Y., Radjenovic, J., Mu, Y., Rozendal, R.A., Batstone, D.J., Rabaey, K., 2011. Electrochemical oxidation of reverse osmosis concentrate on mixed metal oxide (MMO) titanium coated electrodes. Water Res. 45, 4951–4959.

Bagga, A., Chellam, S., Clifford, D.A., 2008. Evaluation of iron chemical coagulation and electrocoagulation pretreatment for surface water microfiltration. J. Membr. Sci. 309, 82–93.

Baghoth, S., Dignum, M., Grefte, A., Kroesbergen, J., Amy, G., 2009. Characterization of NOM in a drinking water treatment process train with no disinfectant residual. Water Sci. Technol. 9 (1), 379–386.

Baghoth, S., Sharma, S., Amy, G., 2011. Tracking natural organic matter (NOM) in a drinking water treatment plant using fluorescence excitation–emission matrices and PARAFAC. Water Res. 45, 797–809.

Bai, H., Zhang, X., Pan, J., Sun, D., Shao, J., 2009. Combination of nano TiO_2 photocatalytic oxidation with microfiltration (MF) for natural organic matter removal. Water Sci. Technol. Water Supply 9, 31–37.

Bai, R., Zhang, X., 2001. Polypyrrole-coated granules for humic acid removal. J. Colloid Interface Sci. 243, 52–60.

Baker, A., Tipping, E., Thacker, S., Gondar, D., 2008. Relating dissolved organic matter fluorescence and functional properties. Chemosphere 73, 1765–1772.

Bansal, R.C., Goyal, M., 2005. Activated Carbon Adsorption. CRC Press, Boca Raton.

Barber, L., Leenher, J., Noyes, T., Stiles, E., 2001. Nature and transformation of dissolved organic matter in treatment wetlands. Environ. Sci. Technol. 35, 4805–4816.

Barbot, E., Moustier, S., Bottero, J., Moulin, P., 2008. Coagulation and ultrafiltration: understanding of the key parameters of the hybrid process. J. Membr. Sci. 325, 520–527.

Barron, P., Wilson, M., 1981. Humic soil and coal structure study with magic-angle spinning ^{13}C CP NMR. Nature 289, 275–276.

Beckett, R., Jue, Z., Giddings, J.C., 1987. Determination of molecular weight distributions of fulvic and humic acids using flow field-flow fractionation. Environ. Sci. Technol. 21, 289–295.

Bekbölet, M., Özkösemen, G., 1996. A preliminary investigation on the photocatalytic degradation of a model humic acid. Water Sci. Technol. 33, 189–194.

Bell-Ajy, K., Abbaszadegan, M., Ibrahim, E., Verges LeChevallier, D., 2000. Conventional and optimized coagulation for NOM removal. J. Am. Water Works Assoc. 92, 44–58.

Benotti, M., Stanford, B., Wert, E., Snyder, S., 2009. Evaluation of a photocatalytic reactor membrane pilot system for the removal of pharmaceuticals and endocrine disrupting compounds from water. Water Res. 43, 1513–1522.

Ben-Sasson, M., Adin, A., 2010a. Fouling mechanisms and energy appraisal in microfiltration pretreated by aluminum-based electroflocculation. J. Membr. Sci. 352, 86–94.

Ben-Sasson, M., Adin, A., 2010b. Fouling mitigation by iron-based electroflocculation in microfiltration: mechanisms and energy minimization. Water Res. 44, 3973–3981.

Ben-Sasson, M., Calmano, W., Adin, A., 2009. Iron-oxidation processes in an electroflocculation (electrocoagulation) cell. J. Hazard. Mater. 171, 704–709.

Ben-Sasson, M., Lin, Y.M., Adin, A., 2011. Electrocoagulation-membrane filtration hybrid system for colloidal fouling mitigation of secondary-effluent. Sep. Purif. Technol. 82, 63–70.

Ben-Sasson, M., Zidon, Y., Calvo, R., Adin, A., 2013. Enhanced removal of natural organic matter by hybrid process of electrocoagulation and dead-end microfiltration. Chem. Eng. J. 232, 338–345.

Bergamasco, R., Konradt-Moraes, L.C., Vieira, M.F., Fagundes-Klen, M.R., Vieira, A.M.S., 2011. Performance of a coagulation–ultrafiltration hybrid process for water supply treatment. Chem. Eng. J. 166, 483–489.

Bergmann, M., Rollin, J., Iourtchouk, T., 2009. The occurrence of perchlorate during drinking water electrolysis using BDD anodes. Electrochim. Acta 54, 2102–2107.

Bergmann, M.E.H., Koparal, A.S., 2005. Studies on electrochemical disinfectant production using anodes containing RuO_2. J. Appl. Electrochem. 35, 1321–1329.

Bérubê, D., Dorea, C., 2008. Optimizing alum coagulation for turbidity, organics, and residual Al reductions. Water Sci. Technol. Water Supply 8, 505–511.

Berwick, L., Greenwood, P., Smernik, R., 2010. The use of MSSV pyrolysis to assist the molecular characterization of aquatic natural organic matter. Water Res. 44, 3039–3054.

Berwick, L., Greenwood, P., Kagi, R., Croué, J.-P., 2007. Thermal release of nitrogen organics from natural organic matter using micro scale sealed vessel pyrolysis. Org. Geochem. 38, 1073–1090.

Bessiere, Y., Jefferson, B., Goslan, E., Bacchin, P., 2009. Effect of hydrophilic/hydrophobic fractions of natural organic matter on irreversible fouling of membranes. Desalination 249, 182–187.

Bhatnagar, A., Hogland, W., Marques, M., Sillanpää, M., 2013. An overview of the modification methods of activated carbon for its water treatment applications. Chem. Eng. J. 219, 499–511.

Bhattacharya, P., Conroy, N., Rao, A.M., Powell, B.A., Ladner, D.A., Ke, P.C., 2012. PAMAM dendrimer for mitigating humic foulant. RSC Adv. 2, 7997–8001.

Bieroza, M., Baker, A., Bridgeman, J., 2009. Relating freshwater organic matter fluorescence to organic carbon removal efficiency in drinking water treatment. Sci. Total Environ. 407, 1765–1774.

Bjelopavlic, M., Newcombe, G., Hayes, R., 1999. Adsorption of NOM onto activated carbon: effect of surface charge, ionic strength, and pore volume distribution. J. Colloid Interface Sci. 210, 271–280.

Blankert, B., Betlem, B., Roffel, B., 2007. Development of a control system for in-line coagulation in an ultrafiltration process. J. Membr. Sci. 301, 39–45.

Blodau, C., Bauer, M., Regenspurg, S., Macalady, D., 2009. Electron accepting capacity of dissolved organic matter as determined by reaction with metallic zinc. Chem. Geol. 260, 186–195.

Bo, X., Gao, B., Peng, N., Wang, Y., Yue, Q., Zhao, Y., 2012. Effect of dosing sequence and solution pH on floc properties of the compound bioflocculant–aluminum sulfate dual-coagulant in kaolin–humic acid solution treatment. Bioresour. Technol. 113, 89–96.

Bodzek, M., Konieczny, K., Kwiecińska, A., 2011. Application of membrane processes in drinking water treatment – state of art. Desalin. Water Treat. 35, 164–184.

Bokhorst, S., Huiskes, A., Convey, P., Aerts, R., 2007. Climate change effects on organic matter decomposition rates in ecosystems from the Maritime Antarctic and Falkland Islands. Global Change Biol. 13, 2642–2653.

Bolto, B., Gregory, J., 2007. Organic polyelectrolytes in water treatment. Water Res. 41, 2301–2324.

Bolto, B., Dixon, D., Eldridge, R., 2004. Ion exchange for the removal of natural organic matter. React. Funct. Polym. 60, 171–182.

Bolto, B., Dixon, D., Eldridge, R., King, S., 2001. Cationic polymer and clay or metal oxide combinations for natural organic matter removal. Water Res. 35, 2669–2676.

Bolto, B., Dixon, D., Eldridge, R., King, S., Linge, K., 2002. Removal of natural organic matter by ion exchange. Water Res. 36, 5057–5065.

Bond, T., Goslan, E., Jefferson, B., Roddick, F., Fan, L., Parsons, S., 2009a. Chemical and biological oxidation of NOM surrogates and effect on HAA formation. Water Res. 43, 2615–2622.

Bond, T., Goslan, E., Parsons, S., Jefferson, B., 2010. Disinfection by-product formation of natural organic matter surrogates and treatment by coagulation, MIEX® and nanofiltration. Water Res. 44, 1645–1653.

Bond, T., Goslan, E.H., Parsons, S.A., Jefferson, B., 2009b. Disinfection by-product formation of natural organic matter surrogates and treatment by coagulation, MIEX and nanofiltration. Water Res. 44, 1645–1653.

Bond, T., Goslan, E.H., Parsons, S.A., Jefferson, B., 2011. Treatment of disinfection by-product precursors. Environ. Technol. 32, 1–25.

Bonnélye, V., Guey, L., Del Castillo, J., 2008. UF/MF as RO pre-treatment: the real benefit. Desalination 222, 59–65.

Bose, P., Reckhow, D., 2007. The effect of ozonation on natural organic matter removal by alum coagulation. Water Res. 41, 1516–1524.
Botes, J.P., Jacobs, E.P., Bradshaw, S.M., 1998. Long-term evaluation of a UF pilot plant for potable water production. Desalination 115, 229–238.
Boyer, T.H., Singer, P.C., 2005. Bench-scale testing of a magnetic ion exchange resin for removal of disinfection by-product precursors. Water Res. 39, 1265–1276.
Boyer, T.H., Singer, P.C., 2006. A pilot-scale evaluation of magnetic ion exchange treatment for removal of natural organic material and inorganic anions. Water Res. 40, 2865–2876.
Braeken, L., Ramaekers, R., Zhang, Y., Maes, G., Bruggen, B.V.d., Vandecasteele, C., 2005. Influence of hydrophobicity on retention in nanofiltration of aqueous solutions containing organic compounds. J. Membr. Sci. 252, 195–203.
Brattebø, H., Ødegaard, H., Halle, O., 1987. Ion exchange for the removal of humic acids in water treatment. Water Res. 21, 1045–1052.
Brezonik, P., Arnold, W., 2012. Water chemistry: fifty years of change and progress. Environ. Sci. Technol. 46, 5650–5657.
Bruchet, A., Rousseau, C., Mallevialle, J., 1990. Pyrolysis-GC-MS for investigating high-molecular-weight THM precursors and other refractory organics. J. Am. Water Works Assoc. 82, 66–74.
van der Bruggen, B., Vandecasteele, C., 2003. Removal of pollutants from surface water and groundwater by nanofiltration: overview of possible applications in the drinking water industry. Environ. Pollut. 122, 435–445.
van der Bruggen, B., Schaep, J., Wilms, D., Vandecasteele, C., 1999. Influence of molecular size, polarity and charge on the retention of organic molecules by nanofiltration. J. Membr. Sci. 156, 29–41.
Brunauer, S., Emmett, P.H., Teller, E., 1938. Adsorption of gases in multimolecular layers. J. Am. Chem. Soc. 60, 309–319.
Buchanan, W., Roddick, F., Porter, N., 2006. Formation of hazardous by-products resulting from irradiation of natural organic matter: comparison between UV and VUV irradiation. Chemosphere 63, 1130–1141.
Buchanan, W., Roddick, F., Porter, N., Drikas, M., 2005. Fractionation of UV and VUV pretreated natural organic matter from drinking water. Environ. Sci. Technol. 39, 4647–4654.
Budd, G., Hess, A., Shorney-Darby, H., Neemann, J., Spencer, C., Bellamy, J., Hargette, P., 2004. Coagulation applications for new treatment goals. J. Am. Water Works Assoc. 92, 102–113.
Budd, G.C., Long, B., Edwards, J.C., Singer, P.C., Meisch, M., 2005. Evaluation of MIEX process impacts on different source waters. American Water Works Association Research Foundation, 91067F.
Buffle, J., Wilkinson, K.J., Stoll, S., Filella, M., Zhang, J., 1998. A generalized description of aquatic colloidal interactions: the three-colloidal component approach. Environ. Sci. Technol. 32, 2887–2899.
Bunting, L., Leavitt, P., Weidman, P., Vinebrooke, R., 2010. Regulation of the nitrogen biogeochemistry of mountain lakes by subsidies of terrestrial dissolved organic matter and the implications for climate studies. Limnol. Oceanogr. 55 (1), 333–345.
Byun, S., Taurozzi, J.S., Alpatova, A.L., Wang, F., Tarabara, V.V., 2011. Performance of polymeric membranes treating ozonated surface water: effect of ozone dosage. Sep. Purif. Technol. 81, 270–278.
Cai, Z., Benjamin, M.M., 2011. NOM fractionation and fouling of low-pressure membranes in microgranular adsorptive filtration. Environ. Sci. Technol. 45, 8935–8940.
Cai, Z., Kim, J., Benjamin, M., 2008. NOM removal by adsorption and membrane filtration using heated aluminum oxide particles. Environ. Sci. Technol. 42, 619–623.
Campinas, M., Rosa, M.J., 2010a. Removal of microcystins by PAC/UF. Sep. Purif. Technol. 71, 114–120.

Campinas, M., Rosa, M.J., 2010b. Assessing PAC contribution to the NOM fouling control in PAC/UF systems. Water Res. 44, 1636–1644.

Campos, C., Mariñas, B.J., Snoeyink, V.L., Baudin, I., Laîné, J.M., 1998. Adsorption of trace organic compounds in CRISTAL(R) process. Desalination 117, 265–271.

Cañizares, P., Jiménez, C., Martínez, F., Rodrigo, M.A., Sáez, C., 2009. The pH as a key parameter in the choice between coagulation and electrocoagulation for the treatment of wastewaters. J. Hazard. Mater. 163, 158–164.

Cao, B., Gao, B., Liu, X., Wang, M., Yang, Z., Yue, Q., 2011. The impact of pH on floc structure characteristic of polyferric chloride in a low DOC and high alkalinity surface water treatment. Water Res. 45, 6181–6188.

Capar, G., Yetis, U., 2001. Removal of natural organic matter and trihalomethanes from the drinking water of Ankara City. Turk. J. Eng. Environ. Sci. 25, 527–535.

Capodaglio, A.G., Callegari, A., Sauvignet, P., 2011. Advanced Separation Technology Application for NOM Removal from a Freshwater Supply. IWA Specialty Conference on Natural Organic Matter, July 27–29, 2011, Costa Mesa, California.

Carp, O., Huisman, C., Reller, A., 2004. Photoinduced reactivity of titanium dioxide. Prog. Solid State Chem. 32, 33–177.

Carroll, T., King, S., Gray, S.R., Bolto, B.A., Booker, N.A., 2000. The fouling of microfiltration membranes by NOM after coagulation treatment. Water Res. 34, 2861–2868.

Chae, S., Yamamura, H., Choi, B., Watanabe, Y., 2009. Fouling characteristics of pressurized and submerged PVDF (polyvinylidene fluoride) microfiltration membranes in a pilot-scale drinking water treatment system under low and high turbidity conditions. Desalination 244, 215–226.

Chamoli, U., 2013. Disinfection and Self-Sensitized Degradation of Natural Organic Matter (NOM) by TiO_2 Photocatalysis with Visible Light (M.Sc. thesis). University of Calgary, Calgary, Canada. 129 pp.

Chang, E.E., Chen, Y., Lin, Y., Chiang, P., 2009a. Reduction of natural organic matter by nanofiltration process. Chemosphere 76, 1265–1272.

Chang, P.-Y., Hsin, C.-H., Doong, R., 2009b. Characterization and photocatalytic activity of vanadium-doped titanium dioxide nanocatalyst. Water Sci. Technol. 59, 523–530.

Chatzisymeon, E., Xekoukoulotakis, N.P., Coz, A., Kalogerakis, N., Mantzavinos, D., 2006. Electrochemical treatment of textile dyes and dyehouse effluents. J. Hazard. Mater. B137, 998–1007.

Chen, C., Zhang, X., He, W., Lu, W., Han, H., 2007a. Comparison of seven kinds of drinking water treatment processes to enhance organic material removal: a pilot test. Sci. Total Environ. 382, 93–102.

Chen, C., Zhang, X., Zhu, L., He, W., Han, H., 2011. Changes in different organic matter fractions during conventional treatment and advanced treatment. J. Environ. Sci. 23, 582–586.

Chen, C., Zhang, X., Zhu, L., Liu, J., He, W., Han, H., 2008. Disinfection by-products and their precursors in a water treatment plant in North China: seasonal changes and fraction analysis. Sci. Total Environ. 397, 140–147.

Chen, D., Weavers, L., Walker, H., Lenhart, J., 2006. Ultrasonic control of ceramic membrane fouling caused by natural organic matter and silica particles. J. Membr. Sci. 276, 135–144.

Chen, G., 2004. Electrochemical technologies in wastewater treatment. Sep. Purif. Technol. 38, 11–41.

Chen, J., Gu, B., LeBoef, E.J., Pan, H., Dai, S., 2002. Spectroscopic characterization of the structural and functional properties of natural organic matter fractions. Chemosphere 48, 59–68.

Chen, P., Rosenfeldt, E., Kullman, S., Hinton, D., Linden, K., 2007b. Biological assessments of a mixture of endocrine disruptors at environmentally relevant concentrations in water following UV/H_2O_2 oxidation. Sci. Total Environ. 376, 18–26.

Chen, V., Mansouri, J., Charlton, T., 2010. Biofouling in membrane systems. In: Peinemann, K., Pereira Nunes, S. (Eds.), Membranes for Water Treatment. Wiley-VCH Verlag, Weinheim, pp. 25–49.

Chen, W., Westerhoff, P., Leenheer, J., Booksh, K., 2003. Fluoresence exitation-emission matrix regional integration to quantify spectra for dissolved organic matter. Environ. Sci. Technol. 37, 5701–5710.

Chen, X., Chen, G., 2006. Anodic oxidation of Orange II on Ti/BDD electrode: variable effects. Sep. Purif. Technol. 48, 45–49.

Chen, X., Chen, G., 2011. Fabrication and application of Ti/BDD for wastewater treatment. In: Brillas, E., Martinez-Huitle, C.A. (Eds.), Synthetic Diamond Films: Preparation, Electrochemistry, Characterization, and Applications. John Wiley & Sons, New Jersey, USA, pp. 353–371.

Chen, X., Deng, H., 2012. Removal of humic acids from water by hybrid titanium-based electrocoagulation with ultrafiltration membrane processes. Desalination 300, 51–57.

Chen, X., Chen, G., Yue, P.L., 2000. Separation of pollutants from restaurant wastewater by electrocoagulation. Sep. Purif. Technol. 19, 65–76.

Chen, Y., Dong, B., Gao, N., Fan, J., 2007c. Effect of coagulation pretreatment on fouling of an ultrafiltration membrane. Desalination 204, 181–188.

Chen, Y.-H., Chen, L.-L., Shang, N.-C., 2009. Photocatalytic degradation of dimethyl phthalate in an aqueous solution with Pt-doped TiO_2-coated magnetic PMMA micropheres. J. Hazard. Mater. 172, 20–29.

Chena, X., Chuaia, X., Yanga, L., Zhao, H., 2012. Climatic warming and overgrazing induced the high concentration of organic matter in Lake Hulun, a large shallow eutrophic steppe lake in northern China. Sci. Total Environ. 431, 332–338.

Cheng, W., Chi, F., Li, C., Yu, R., 2008. A study on the removal of organic substances from low-turbidity and low-alkalinity water with metal-polysilicate coagulants. Colloids Surf. A Physicochem. Eng. Aspects 312, 238–244.

Cheng, W., Dastgheib, S.A., Karanfil, T., 2005. Adsorption of dissolved natural organic matter by modified activated carbons. Water Res. 39 (2005), 2281–2290.

Cheng, Y.-L., Wong, R.-J., Lin, J., Huang, C., Lee, D.-J., Mujumdar, A., 2010. Water coagulation using electrostatic patch coagulation (EPC) mechanism. Drying Technol. 28, 850–857.

Cheryan, M., 1998. Ultrafiltration and Microfiltration Handbook. Technomic Publishing, Lancaster, USA.

Chiang, P.-C., Chang, E.-E., Chang, P.-C., Huang, C.-P., 2009. Effects of pre-ozonation on the removal of THM precursors by coagulation. Sci. Total Environ. 407, 5735–5742.

Chin, A., Bérubé, P., 2005. Removal of disinfection by-product precursors with ozone-UV advanced oxidation process. Water Res. 39, 2136–2144.

Chin, Y., Aiken, G., O'Loughlin, E., 1994. Molecular weight, polydispersity, and spectroscopic properties of aquatic humic substances. Environ. Sci. Technol. 28, 1853–1858.

Cho, J., Amy, G., Pellegrino, J., 1999. Membrane filtration of natural organic matter: initial comparison of rejection and flux decline characteristics with ultrafiltration and nanofiltration membranes. Water Res. 33, 2517–2526.

Cho, J., Amy, G., Pellegrino, J., 2000b. Membrane filtration of natural organic matter: factors and mechanisms affecting rejection and flux decline with charged ultrafiltration (UF) membrane. J. Membr. Sci. 164, 89–110.

Cho, J., Amy, G., Pellegrino, J., 2000a. Membrane filtration of natural organic matter: comparison of flux decline, NOM rejection, and foulants during filtration with three UF membranes. Desalination 127, 283–298.

Cho, M.-H., Lee, C.-H., Lee, S., 2006. Effect of flocculation conditions on membrane permeability in coagulation–microfiltration. Desalination 191, 386–396.

Choi, H., Al-Abed, S.R., Dionysiou, D.D., Stathatos, E., Lianos, P., 2010. TiO_2-based advanced oxidation nanotechnologies for water purification and reuse. In: Abraham, M.A., Sheldon, R. (Eds.), Sustainability Science and Engineering, Book Series, vol. 2. In: Escobar, I.C., Schäfer, A.I. (Eds.), Sustainable Water for the Future: Water Recycling versus Desalination. Elsevier, pp. 229–254.

Choi, I.-H., Lee, S., Kim, I.-C., Min, B.-R., Lee, K.-H., 2007. Effect of new photocatalytic coagulant on NF membrane fouling. Ind. Eng. Chem. Res. 46, 2280–2285.

Choi, K.Y., Dempsey, B.A., 2004. In-line coagulation with low-pressure membrane filtration. Water Res. 38, 4271–4281.

Choi, Y., Kim, H., Kweon, J., 2008. Role of hydrophobic natural organic matter flocs on the fouling in coagulation-membrane processes. Sep. Purif. Technol. 62, 529–534.

Choo, K.-H., Kang, S.-K., 2003. Removal of residual organic matter from secondary effluent by iron oxides adsorption. Desalination 154, 139–146.

Choo, K.-W., Tao, R., Kim, M.-J., 2008. Use of photocatalytic membrane reactor for the removal of natural organic matter in water: effect of photoinduced desorption and ferrihydrite adsorption. J. Membr. Sci. 322, 368–374.

Chow, A.T., Dahlgren, R.A., Gao, S., 2005. Physical and chemical fractionation of dissolved organic matter and trihalomethane precursors: a review. J. Water Supply Res. Technol. 54, 475–507.

Chow, C., Fabris, R., Drikas, M., 2004. A rapid fractionation technique to characterize natural organic matter for the optimization of water treatment processes. J. Water Supply Res. Technol. Aqua 53, 85–92.

Chow, C., Fabris, R., van Leeuwen, J., Wang, D., Drikas, M., 2008. Assessing natural organic matter treatability using high performance size exclusion chromatography. Environ. Sci. Technol. 42, 6683–6689.

Chow, C., Kuntke, P., Fabris, R., Drikas, M., 2009a. Organic characterization tools for distribution system management. Water Sci. Technol. 9, 1–8.

Chow, C., van Leeuwen, J., Fabris, R., Drikas, M., 2009b. Optimized coagulation using aluminium sulfate for the removal of dissolved organic carbon. Desalination 245, 120–134.

Chow, C.W.K., Majewski, P., Bauer, S., Fabris, R., Drikas, M., 2009c. Removal of natural organic matter using self-assembled monolayer technology. Desalin. Water Treat. 12, 344–351.

Christy, A., Bruchet, A., Rybacki, D., 1999. Characterization of natural organic matter by pyrolysis/GC-MS. Environ. Int. 25, 181–189.

Chu, Y.Y., Qian, Y., Wang, W.J., Deng, X.L., 2012. A dual-cathode electro-Fenton oxidation coupled with anodic oxidation system used for 4-nitrophenol degradation. J. Hazard. Mater. 199–200, 179–185.

Clark, M.M., Lucas, P., 1998. Diffusion and partitioning of humic acid in a porous ultrafiltration membrane. J. Membr. Sci. 143, 13–25.

Clemets, W., Brooks, M., Kashian, D., Zuellig, R., 2008. Changes in dissolved organic material determine exposure of stream benthic communities to UV-B radiation and heavy metals: implications for climate change. Global Change Biol. 14, 2201–2214.

Coble, P.G., 1996. Characterization of marine and terrestrial DOM in seawater using excitation–emission matrix spectroscopy. Mar. Chem. 51, 325–346.

Collin, B., Oostveen, E., Tsyusko, O.V., Unrine, J.M., 2014. Influence of natural organic matter and surface charge on the toxicity and bioaccumulation of functionalized ceria nanoparticles in *Caenorhabditis elegans*. Environ. Sci. Technol. 48, 1280–1289.

Comninellis, C., Kapalka, A., Malato, S., Parsons, S., Poulios, I., Mantzavinos, D., 2008. Advanced oxidation processes for water treatment: advances and trends for R&D. J. Chem. Technol. Biotechnol. 83, 769–776.

Comstock, S.E.H., Boyer, T.H., 2014. Combined magnetic ion exchange and cation exchange for removal of DOC and hardness. Chem. Eng. J. 241, 366–375.

Comstock, S.E.H., Boyer, T.H., Graf, K.C., 2011. Treatment of nanofiltration and reverse osmosis concentrates: comparison of precipitative softening, coagulation, and anion exchange. Water Res. 45, 4855–4865.

Cook, R.L., 2004. Coupling NMR to NOM. Anal. Bioanal. Chem. 376, 1484–1503.

Cooper, W., Song, W., Gonsior, M., Kalnina, D., Peake, B., Mezyk, S., 2008. Recent advances in structure and reactivity of dissolved organic matter in natural waters. Water Sci. Technol. 8, 615–623.

Cornelissen, E.R., Beerendonk, E.F., Nederlof, M.N., van der Hoek, J.P., Wessels, L.P., 2009. Fluidized ion exchange (FIX) to control NO NOM fouling in ultrafiltration. Desalination 236, 334–341.

Cornelissen, E.R., Chasseriaud, D., Siegers, W.G., Beerendonk, E.F., van der Kooij, D., 2010. Effect of anionic fluidized ion exchange (FIX) pre-treatment on nanofiltration (NF) membrane fouling. Water Res. 44, 3283–3293.

Cornelissen, E.R., Moreau, N., Siegers, W.G., Abrahamse, A.J., Rietveld, L.C., Grefte, A., Dignum, M., Amy, G., Wessels, L.P., 2008. Selection of anionic exchange resins for removal of natural organic matter (NOM) fractions. Water Res. 42, 413–423.

Cortés-Francisco, N., Caixach, J., 2013. Molecular characterization of dissolved organic matter through a desalination process by high resolution mass spectrometry. Environ. Sci. Technol. 47, 9619–9627.

Costa, A.R., de Pinho, M.N., 2006a. Comparison of the performance of ultrafiltration and nanofiltration in surface water treatment. Desalination 199, 73–75.

Costa, A.R., de Pinho, M.N., 2006b. Performance and cost estimation of nanofiltration for surface water treatment in drinking water production. Desalination 196, 55–65.

Couture, S., Houle, D., Gagnon, C., 2012. Increases of dissolved organic carbon in temperate and boreal lakes in Quebec, Canada. Environ. Sci. Pollut. Res. 19, 361–371.

Criquet, J., Vel Leitner, K., 2009. Degradation of acetic acid with sulfate radical generated by persulfate ions photolysis. Chemosphere 77, 194–200.

Croue, J.-P., 2004. Isolation of humic and non-humic NOM fractions: structural characterization. Environ. Monit. Assess. 92, 193–207.

Croue, J.-P., Korshin, G.V., Benjamin, M.M., 2000. Characterization of Natural Organic Matter in Drinking Water, American Water Works Association, p. 324.

Crozes, G., White, P., Marshall, M., 1995. Enhanced coagulation: its effect on NOM removal and chemical costs. J. Am. Water Works Assoc. 87, 78–89.

Culea, M., Cozar, O., Ristoiu, D., 2006. Methods validation for the determination of trihalomethanes in drinking water. J. Mass Spectrom. 41, 1594–1597.

Dąbrowski, A., 2001. Adsorption — from theory to practice. Adv. Colloid Interface Sci. 93, 135–224.

Danielsson, L., 1982. On the use of filters for distinguishing between dissolved and particulate fractions in natural waters. Water Res. 16, 179–182.

Dastgheib, S.A., Karanfil, T., Cheng, W., 2004. Tailoring activated carbons for enhanced removal of natural organic matter from natural waters. Carbon 42, 547–557.

Datta, C., Ghosh, K., Mukherjee, S., 1971. Fluorescence excitation spectra of different fractions of humus. J. Ind. Chem. Soc. 48, 279–287.

Davis, J.A., 1982. Adsorption of natural dissolved organic matter at the oxide/water interface. Geochim. Cosmochim. Acta 46, 2381–2393.

Day, G.M., Hart, B.T., McKelvie, I.D., Beckett, R., 1994. Adsorption of natural organic matter onto goethite. Colloids Surf. A Physicochem. Eng. Aspects 89, 1–13.

Delpla, I., Jung, A.-V., Baures, E., Clement, M., Thomas, O., 2009. Impacts of climate change on surface water quality in relation to drinking water production. Environ. Int. 35, 1225–1233.

Deng, S., 2006. Sorbent technology. In: Encyclopedia of Chemical Processing. Taylor & Francis, New York.

Deng, S., Bai, R.B., 2003. Aminated polyacrylonitrile fibers for humic acid adsorption: behaviors and mechanisms. Environ. Sci. Technol. 37, 5799–5805.

Deshmukh, A., Pacheco, C., Hay, M., Myneni, S., 2007. Structural environments of carboxyl groups in natural organic molecules from terrestrial systems. Part 2: 2D NMR spectroscopy. Geochim. Cosmochim. Acta 71, 3533–3544.

Dhaouadi, A., Monser, L., Adhoum, N., 2009. Anodic oxidation and electro-Fenton treatment of rotenone. Electrochim. Acta 54, 4473–4480.

Diem, S., von Rohra, M., Heringa, J., Kohlera, H.-P., Schirmera, M., von Gunten, U., 2013. NOM degradation during river infiltration: effects of the climate variables temperature and discharge. Water Res. 47, 6585–6595.

Ding, C., Shang, C., 2010. Mechanisms controlling adsorption of natural organic matter on surfactant-modified iron oxide-coated sand. Water Res. 44, 3651–3658.

Ding, C., Yang, X., Liu, W., Chang, Y., Shang, C., 2010. Removal of natural organic matter using surfactant-modified iron oxide-coated sand. J. Hazard. Mater. 174, 567–572.

Dobrovic, S., Juretic, H., Ruzinski, N., 2007. Photodegradation of natural organic matter in water with UV irradation at 185 and 254 nm: importance of hydrodynamic conditions on the decomposition rate. Sep. Sci. Technol. 42, 1421–1432.

Dong, B., Chen, Y., Gao, N., Fan, J., 2007. Effect of coagulation pretreatment on the fouling of ultrafiltration membrane. J. Environ. Sci. 19, 278–283.

Dong, B.Z., Chen, Y., Gao, N.Y., Fan, J.C., 2006. Effect of pH on UF membrane fouling. Desalination 195, 201–208.

Dong, C., Chen, W., Liu, C., 2014. Preparation of novel magnetic chitosan nanoparticle and its application for removal of humic acid from aqueous solution. Appl. Surf. Sci. 292, 1067–1076.

Dongsheng, W., Hong, L., Chunhua, L., Hongxiao, T., 2006. Removal of humic acid by coagulation with nano-Al_{13}. Water Sci. Technol. Water Supply 6, 59–67.

Doong, R.-A., Chiang, L.-F., 2008. Coupled removal of organic compounds and heavy metals by titanate/carbon nanotube composites. Water Sci. Technol. 58, 1958–1992.

Doulia, D., Leodopoulos, C., Gimouhopoulos, K., Rigas, F., 2009. Adsorption of humic acid on acid-activated Greek bentonite. J. Colloid Interface Sci. 340, 131–141.

Drikas, M., Chow, C.W.K., Cook, D., 2003. The impact of recalcitrant organic character on disinfection stability, trihalomethane formation and bacterial regrowth: an evaluation of magnetic ion exchange resin (MIEX®) and alum coagulation. J. Water Supply Res. Technol. Aqua 52, 475–487.

Drikas, M., Dixon, M., Morran, J., 2009. Removal of MIB and geosmin using granular activated carbon with and without MIEX® pre-treatment. Water Res. 43, 5151–5159.

Drikas, M., Dixon, M., Morran, J., 2011. Long term case study of MIEX® pre-treatment in drinking water; understanding NOM removal. Water Res. 45, 1539–1548.

Drikas, M., Morran, J.Y., Pelekani, C., Hepplewhite, C., Bursill, D.B., 2002. Removal of natural organic matter – a fresh approach. Water Sci. Technol. Water Supply 2, 71–79.

Duan, J., Gregory, J., 2003. Coagulation by hydrolysing metal salts. Adv. Colloid Interface Sci. 100–102, 475–502.

Dubrawski, K.L., Mohseni, M., 2013a. Standardizing electrocoagulation reactor design: Iron electrodes for NOM removal. Chemosphere 91, 55–60.

Dubrawski, K.L., Mohseni, M., 2013b. In-situ identification of iron electrocoagulation speciation and application for natural organic matter (NOM) removal. Water Res. 47, 5371–5380.

Dubrawski, K.L., Fauvel, M., Mohseni, M., 2013. Metal type and natural organic matter source for direct filtration electrocoagulation of drinking water. J. Hazard. Mater. 244–245, 135–141.

Duguet, J., Brodard, E., Dussert, B., Mallevialle, J., 1985. Improvement in the effectiveness of ozonation of drinking water through the use of hydrogen peroxide. Ozone Sci. Eng. 7, 241–253.

Eckenfelder, W., 2000. Industrial Water Pollution Control, third ed. McGraw-Hill Series in Water Resources and Environmental Engineering.
Edwards, M., Benjamin, M.M., Ryan, J.N., 1996. Role of organic acidity in sorption of natural organic matter (NOM) to oxide surfaces. Colloids Surf. A Physicochem. Eng. Aspects 107, 297–307.
Edzwald, J., Tobiason, J., 1999. Enhanced coagulation: US requirements and a broader view. Water Sci. Technol. 40 (9), 63–70.
Egeberg, P., Christy, A., Eikenes, M., 2002. The molecular size of natural organic matter (NOM) determined by diffusivimetry and seven other methods. Water Res. 36, 925–932.
Eggins, B., Palme, F., Byrne, J., 1997. Photocatalytic treatment of humic substances in drinking water. Water Res. 31, 1223–1226.
Eikebrokk, B., Fabris, R., Drikas, M., Chow, C., 2007. NOM characteristics and treatability by coagulation. In: Hahn, H.H., Hoffman, E., Ødegaard, H. (Eds.), Chemical Water and Wastewater Treatment IX. IWA Publishing, London.
Eikebrokk, B., Vogt, R., Liltved, H., 2004. NOM increase in Northern European source waters: discussion of possible causes and impacts on coagulation/contact filtration processes. Water Sci. Technol. Water Supply 4 (4), 47–54.
El Gough, R., Holliman, P., Willis, N., Freeman, C., 2014. Dissolved organic carbon and trihalomethane precursor removal at a UK upland water treatment works. Sci. Total Environ. 468–469, 228–239.
El-Ashtoukhy, E.-S.Z., Amin, N.K., Abdelwahab, O., 2009. Treatment of paper mill effluents in a batch-stirred electrochemical tank reactor. Chem. Eng. J. 146, 205–210.
El-Masry, M.H., Sadek, O.M., Mekhemer, W.K., 2004. Purification of raw surface water using electro-coagulation method. Water Air Soil. Pollut. 158, 373–385.
Ericsson, B., Trägårdh, G., 1997. Treatment of surface water rich in humus — membrane filtration vs. conventional treatment. Desalination 108, 117–128.
Ericsson, B., Hallberg, M., Wachenfeldt, J., 1997. Nanofiltration of highly colored raw water for drinking water production. Desalination 108, 129–141.
Escobar, I., Randall, A., 2001. Assimilable organic carbon (AOC) and biodegradable dissolved organic carbon (BDOC): complementary measurements. Water Res. 35, 4444–4454.
Escobar, I.C., Hong, S., Randall, A.A., 2000. Removal of assimilable organic carbon and biodegradable dissolved organic carbon by reverse osmosis and nanofiltration membranes. J. Membr. Sci. 175, 1–17.
Evans, C., Monteith, D., Cooper, M., 2005. Long-term increases in surface water dissolved organic carbon: observations, possible causes and environmental impacts. Environ. Pollut. 137, 55–71.
Fabbri, D., Mongardi, M., Mintanari, L., Galletti, G., Chiavari, G., Scotti, R., 1998. Comparison between CP/MS ^{13}C-NMR and pyrolysis-GC/MS in the structural characterization of humins and humic acids of soil and sediments. J. Anal. Chem. 362, 299–306.
Fabris, R., Chow, C., Drikas, M., Eikebrokk, B., 2008. Comparison of NOM character in selected Australian and Norwegian drinking waters. Water Res. 42, 4188–4196.
Fabris, R., Chow, C.W.K., Drikas, M., 2004. Practical application of a combined treatment process for removal of recalcitrant NOM – alum and PAC. Water Sci. Technol. Water Supply 4, 89–94.
Fabris, R., Lee, E.K., Chow, C.W.K., Chen, V., Drikas, M., 2007. Pre-treatments to reduce fouling of low pressure micro-filtration (MF) membranes. J. Membr. Sci. 289, 231–240.
Fan, L., Harris, J.L., Roddick, F.A., Booker, N.A., 2001. Influence of the characteristics of natural organic matter on the fouling of microfiltration membranes. Water Res. 35, 4455–4463.

Fane, A.G., Chang, S., 2009. Techniques to enhance performance of membrane process. In: Paddy, A.K., Rizvi, S.S.H., Sastre, A.M. (Eds.), Handbook of Membrane Separations, Chemical, Pharmaceutical, Food, and Biotechnological Applications. CRC Press, Taylor & Francis Group, Boca Raton, USA, pp. 193–232.

Fane, A.G., Wang, R., Jia, Y., 2008. Membrane technology: past, present and future. In: Wang, L.K., Chen, J.P., Hung, Y., Shammas, N.K. (Eds.), Handbook of Environmental Engineering. Membrane and Desalination Technologies, vol. 13. Springer Science & Business Media, New York, pp. 1–45.

Faouzi, A.M., Nası, B., Abdellatif, G., 2007. Electrochemical degradation of anthraquinone dye Alizarin Red S by anodic oxidation on boron-doped diamond. Dyes Pigm. 73, 86–89.

Fearing, D.A., Banks, J., Guyetand, S., Monfort Eroles, C., Jefferson, B., Wilson, D., Hillis, P., Campbell, A.T., Parsons, S.A., 2004. Combination of ferric and MIEX® for the treatment of a humic rich water. Water Res. 38, 2551–2558.

Feng, Q., Li, X., Cheng, Y., Meng, L., Meng, Q., 2007. Removal of humic acid from groundwater by electrocoagulation. J. China Univ. Min. Technol. 17, 513–520.

Feng, X., Simpson, M., 2011. Molecular-level methods for monitoring soil organic matter responses to global climate change. J. Environ. Monit. 13, 1246–1254.

Fettig, J., 1999. Removal of humic substances by adsorption/ion exchange. Water Sci. Technol. 40, 173–182.

Fettig, J., 2005. Modelling the uptake of natural organic matter (NOM) by different granular sorbent media. J. Water Supply Res. Technol. Aqua 54, 83–93.

Filella, M., 2009. Freshwaters: which NOM matters? Environ. Chem. Lett. 7, 21–35.

Flaten, T., 2001. Aluminium as a risk factor in Alzheimer's disease, with emphasis on drinking water. Brain Res. Bull. 55, 187–196.

Flox, C., Arias, C., Brillas, E., Savall, A., Groenen-Serrano, K., 2009. Electrochemical incineration of cresols: a comparative study between PbO_2 and boron-doped diamond anodes. Chemosphere 74, 1340–1347.

Fonseca, A.C., Kennon, J., Booth, S., Labisi, B., Meyerhofer, J., White, P., 2005. Testing bromide and DBP precursors removal in California State Project Water using a magnetized ion exchange resin. Proceedings American Water Works Association Annual Conference.

Forsberg, C., Petersen Jr., R., 1990. A darkening of Swedish lakes due to increased humus inputs during the last 15 years. Verh. Int. Ver. Limnol. 24, 289–292.

Francaviglia, R., Coleman, K., Whitmore, A., Doro, L., Urracci, G., Rubino, M., Ledda, L., 2012. Changes in soil organic carbon and climate change – application of the RothC model in agro-silvo-pastoral Mediterranean systems. Agric. Syst. 112, 48–54.

Freundlich, H.M.F., 1906. Over the adsorption in solution. J. Phys. Chem. 57, 385–471.

Frimmel, F., 1994. Photochemical aspects related to humic substances. Environ. Int. 20, 373–385.

Frimmel, F., 1998. Impact of light on the properties of aquatic natural organic matter. Environ. Int. 24, 559–571.

Fu, J., Ji, M., Zhao, Y., Wang, L., 2006a. Kinetics of aqueous photocatalytic oxidation of fulvic acids in a photocatalysis-ultrafiltration reactor (PUR). Sep. Purif. Technol. 50, 107–113.

Fu, J., Min, J., Wang, Z., Jin, L., An, D., 2006b. A new submerged membrane photocatalysis reactor (SMPR) for fulvic acid removal using nano-structured photocatalyst. J. Hazard. Mater. 131, 238–242.

Fu, J., Zhao, Y., Wu, Q., 2007. Optimising photoelectrocatalytic oxidation of fulvic acid response surface methodology. J. Hazard. Mater. 144, 499–505.

Fu, Y., Yu, S., 2007. Characterization and coagulation performance of solid poly-silicic–ferric (PSF) coagulant. J. Non-Cryst. Solids 353, 2206–2213.

Fu, Y., Yu, S., Han, C., 2009. Morphology and coagulation performance during preparation of poly-silicic-ferric (PSF) coagulant. Chem. Eng. J. 149, 1–10.

Fujishima, A., 1971. Electrochemical photolysis of water at a semiconductor electrode. Nature 238, 37–38.
Fukano, K., Komiya, K., Sasaki, H., Hashimoto, T., 1978. Evaluation of new supports for high-pressure aqueous gel permeation chromatography: TSK-GEL SW type columns. J. Chromatogr. 166, 47–54.
Fukushima, M., Tatsumi, K., Nagao, S., 2001. Degradation characteristics of humic acid during photo-Fenton processes. Environ. Sci. Technol. 35, 3683–3690.
Galeano, L., Bravo, P., Luna, C., Vicente, M., Gil, A., 2012. Removal of natural organic matter for drinking water production by Al/Fe-PILC-catalyzed wet peroxide oxidation: effect of the catalyst preparation from concentrated precursors. Appl. Catal. B Environ. 111–112, 527–535.
Gamage, N.P., Chellam, S., 2011. Aluminum electrocoagulation pretreatment reduces fouling during surface water microfiltration. J. Membr. Sci. 379, 97–105.
Gamage, N.P., Rimer, J.D., Chellam, S., 2012. Improvements in permeate flux by aluminum electroflotation pretreatment during microfiltration of surface water. J. Membr. Sci. 411–412, 45–53.
Gao, B.-Y., Wang, Y., Yue, Q.-Y., Wei, J.-C., Li, Q., 2008. The size and coagulation behaviour of a novel composite inorganic–organic coagulant. Sep. Purif. Technol. 62, 544–550.
Garcia, I., Moreno, L., 2009. Use of GAC after enhanced coagulation for the removal of natural organic matter from water for purification. Water Sci. Technol. Water Supply 9, 173–180.
Genz, A., Baumgarten, B., Goernitz, M., Jekel, M., 2008. NOM removal by adsorption onto granular ferric hydroxide: equilibrium, kinetics, filter and regeneration studies. Water Res. 42, 238–248.
Gerrity, D., Mayer, B., Ryu, H., Crittenden, J., Abbaszadegan, M., 2009. A comparison of pilot-scale photocatalysis and enhanced coagulation for disinfection byproduct mitigation. Water Res. 43, 1597–1610.
Gerrity, D., Ryu, H., Crittenden, J., Abbaszadegan, M., 2008. Photocatalytic inactivation of viruses using titanium dioxide and low-pressure UV light. J. Environ. Sci. Health A Tox. Hazard. Subst. Environ. Eng. 43, 1261–1270.
Ghernaout, D., 2014. The hydrophilic/hydrophobic ratio *vs.* dissolved organics removal by coagulation – a review. J. King Saud Univ. Sci. 26, 169–180.
Ghernaout, D., Badis, A., Kellil, A., Ghernaout, B., 2008. Application of electrocoagulation in *Escherichia coli* culture and two surface waters. Desalination 219, 118–125.
Ghernaout, D., Ghernaout, B., Saiba, A., Boucherit, A., Kellil, A., 2009a. Removal of humic acids by continuous electromagnetic treatment followed by electrocoagulation in batch using aluminium electrodes. Desalination 239, 295–308.
Ghernaout, D., Ghernaout, B., Boucherit, A., Naceur, M.W., Khelifa, A., Kellil, A., 2009b. Study on mechanism of electrocoagulation with iron electrodes in idealized conditions and electrocoagulation of humic acids solution in batch using aluminium electrodes. Desalin. Water Treat. 8, 91–99.
Ghernaout, D., Irki, S., Boucherit, A., 2014. Removal of Cu^{2+} and Cd^{2+}, and humic acid and phenol by electrocoagulation using iron electrodes. Desalin. Water Treat. 52, 3256–3270.
Ghernaout, D., Mariche, A., Ghernaout, B., Kellil, A., 2010. Electromagnetic treatment doubled electrocoagulation of humic acid in continuous mode using response surface method for its optimization and application on two surface waters. Desalin. Water Treat. 22, 311–329.
Ghosh, K., Schnitzer, M., 1980. Macromolecular structures of humic substances. Soil Sci. 129, 266–276.
Gibert, O., Lefèvre, B., Fernández, M., Bernat, X., Paraira, M., Pons, M., 2013b. Fractionation and removal of dissolved organic carbon in a full-scale granular activated carbon filter used for drinking water production. Water Res. 47, 2821–2829.

Gibert, O., Lefèvre, B., Fernández, M., Bernat, X., Paraira, M., Calderer, M., Martínez-Lladó, X., 2013a. Characterizing biofilm development on granular activated carbon used for drinking water production. Water Res. 47, 1101–1110.

Giri, R., Ozaki, H., Takanami, R., Taniguchi, S., 2008. Heterogenous photocatalytic ozonation of 2,4-D in dilute aqeuous solution with TiO_2 fiber. Water Sci. Technol. 58, 207–216.

Gjessing, E., 1965. Use of Sephadex gels for estimation of humic substances in natural water. Nature 208, 1091–1092.

Gjessing, E., 1970. Ultrafiltration of aquatic humus. Environ. Sci. Technol. 4, 437–438.

Gjessing, E., Källqvist, T., 1991. Algicidal and chemical effect of u.v.-radiation of water containing humic substances. Water Res. 25, 491–494.

Gjessing, E., Alberts, J., Bruchet, A., Egeberg, P., Lydersen, E., McGown, L., Mobed, J., Munster, U., Pempkowiak, J., Perdue, M., Ratnawerra, H., Rybacki, D., Takacs, M., Abbt-Braun, G., 1998. Multi-method characterization of natural organic matter isolated from water: characterization of reverse osmosis-isolates from water of two semi-identical dystrophic lakes basins in Norway. Water Res. 32, 3108–3124.

Gjessing, E., Steiro, C., Becher, G., Christy, A., 2007. Reduced analytical availability of polychlorinated biphenyls (PCB's) in coloured surface water. Chemosphere 66, 644–649.

Gjessing, E.T., 1973. Gel- and ultramembrane filtration of aquatic humus: a comparison of the two methods. Schweiz. Z. Hydrol. 35, 286–294.

Gjessing, E., 1976. Physical and Chemical Characteristics of Aquatic Humus. Ann Arbor Science Publishers, Michigan.

Glaze, W., Kang, J., Chapin, D., 1987. The chemistry of water treatment processes involving ozone, hydrogen peroxide and ultraviolet radiation. Ozone Sci. Eng. 9, 335–352.

Glaze, W.H., Beltran, F., Tuhkanen, T., Kang, J.-W., 1992. Chemical models of advanced oxidation processes. Water Pollut. Res. J. Can. 27, 23–42.

Glaze, W., Peyton, G., Lin, S., Huang, R., Burleson, J., 1982. Destruction of pollutants in water with ozone in combination with ultraviolet radiation. 2. Natural trihalomethane precursors. Environ. Sci. Technol. 16, 454–458.

Glucina, K., Alvarez, A., Laîné, J.M., 2000. Assessment of an integrated membrane system for surface water treatment. Desalination 132, 73–82.

Golder, A., Chanda, A., Samanta, A., Ray, S., 2007. Removal of Cr(VI) from aqueous solution: electrocoagulation vs chemical coagulation. Sep. Sci. Technol. 42, 2177–2193.

Gomes, J.A.G., Daida, P., Kesmez, M., Weir, M., Moreno, H., Parga, J.R., Irwin, G., McWhinney, H., Grady, T., Peterson, E., Cocke, D.L., 2007. Arsenic removal by electrocoagulation using combined Al–Fe electrode system and characterization of products. J. Hazard. Mater. 139, 220–231.

Goncharuk, V., Malyarenko, V., Yaremenko, V., 2008. Use of ultrasound in water treatment. J. Water Chem. Technol. 30, 137–150.

Gonzalez, M., Oliveros, E., Wörner, M., Braun, A., 2004. Vacuum-ultraviolet photolysis of aqueous reaction systems. J. Photochem. Photobiol. C. Photochem. Rev. 5, 225–246.

Gorenflo, A., Velázquez-Padrón, D., Frimmel, F.H., 2003. Nanofiltration of a German groundwater of high hardness and NOM content: performance and costs. Desalination 151, 253–265.

Goslan, E., Gurses, F., Banks, J., Parsons, S., 2006. An investigation into reservoir NOM reduction by UV photolysis and advanced oxidation processes. Chemosphere 65, 1113–1119.

Gotsi, M., Kalogerakis, N., Psillakis, E., Samaras, P., Mantzavinos, D., 2005. Electrochemical oxidation of olive oil mill wastewaters. Water Res. 39, 4177–4187.

Gottfried, A., Shepard, A., Hardiman, K., Walsh, M., 2008. Impact of recycling filter backwash water on organic removal in coagulation–sedimentation processes. Water Res. 42, 4683–4691.

Granskog, M., Macdonald, R., Mundy, C.-J., Barber, D., 2007. Distribution, characteristics and potential impacts of chromophoric dissolved organic matter (CDOM) in Hudson Strait and Hudson Bay, Canada. Cont. Shelf Res. 27, 2032–2050.

Greenwood, P., Leenher, J., McIntyre, C., Berwick, L., Franzmann, P., 2006. Bacterial biomarkers thermally released from dissolved organic matter. Org. Geochem. 37, 597–609.
Grieco, S.A., Ramarao, B.V., 2013. Removal of TCEP from aqueous solutions by adsorption with zeolites. Colloids Surf. A Physicochem. Eng. Aspects 434, 329–338.
Grose, A.B.F., Smith, A.J., Donn, A., O'Donnell, J., Welch, D., 1998. Supplying high quality drinking water to remote communities in Scotland. Desalination 117, 107–117.
Gu, B., Schmitt, J., Chen, Z., Liang, L., McCarthy, J.F., 1994. Adsorption and desorption of natural organic matter on iron oxide: mechanisms and models. Environ. Sci. Technol. 28, 38–46.
von Gunten, U., 2003. Ozonation of drinking water: part I. Oxidation kinetics and product formation. Water Res. 37, 1443–1467.
Guinea, E., Garrido, J.A., Rodriguez, R.M., Cabot, P.-L., Arias, C., Centellas, F., Brillas, E., 2010. Degradation of the fluoroquinolone enrofloxacin by electrochemical advanced oxidation processes based on hydrogen peroxide electrogeneration. Electrochim. Acta 55, 2101–2115.
Guizard, C., Amblard, P., 2009. Current status and prospects for ceramic membrane applictions. In: Paddy, A.K., Rizvi, S.S.H., Sastre, A.M. (Eds.), Handbook of Membrane Separations, Chemical, Pharmaceutical, Food, and Biotechnological Applications. CRC Press, Taylor & Francis Group, Boca Raton, USA, pp. 139–179.
Gumy, A., Rincon, A.G., Hajdu, R., Pulgarin, C., 2006. Solar photocatalysis for detoxification and disinfection of water: different types of suspended and fixed TiO_2 catalysts study. Sol. Energy 80, 1376–1381.
Gur-Reznik, S., Katz, I., Dosoretz, C.G., 2008. Removal of dissolved organic matter by granular-activated carbon adsorption as a pretreatment to reverse osmosis of membrane bioreactor effluents. Water Res. 42, 1595–1605.
Ha, T., Choo, K., Choi, S., 2004. Effect of chlorine on adsorption/ultrafiltration treatment for removing natural organic matter in drinking water. J. Colloid Interface Sci. 274, 587–593.
Haarhoff, J., Kubare, M., Mamba, B., Krause, R., Nkambule, T., Matsebula, B., Menge, J., 2010. NOM characterization and removal at six Southern African water treatment plants. Drink. Water Eng. Sci. 3, 53–61.
Hallé, C., Huck, P.M., Peldszus, S., Haberkamp, J., Jekel, M., 2009. Assessing the performance of biological filtration as pretreatment to low pressure membranes for drinking water. Environ. Sci. Technol. 43, 3878–3884.
Hamza, M., Ammar, S., Abdelhedi, R., 2011. Electrochemical oxidation of 1,3, 5-trimethoxybenzene in aqueous solutions at gold oxide and lead dioxide electrodes. Electrochim. Acta 56, 3785–3789.
Hand, D., Perram, D., Crittenden, J., 1995. Destruction of DBP precursors with catalytic oxidation. J. Am. Water Works Assoc. 87, 84–96.
Harif, T., Adin, A., 2011. Size and structure evolution of kaolin-$Al(OH)_3$ flocs in the electroflocculation process: a study using static light scattering. Water Res. 45, 6195–6206.
Harland, C.E., 1994. Ion Exchange. Theory and Practice, second ed. The Royal Society of Chemistry, Bath.
Härmä, V., 1999. Tuloksia suomalaisilla pintavesilaitoksilla tehdyistä nanosuodatuskokeista. Vesitalous 6, 5–10.
Harman, B.I., Koseoglu, H., Yigit, N.O., Beyhan, M., Kitis, M., 2010. The use of iron oxide-coated ceramic membranes in removing natural organic matter and phenol from waters. Desalination 261, 27–33.
Hay, M., Myneni, S., 2007. Structural environments of carboxyl groups in natural organic molecules from terrestrial systems. Part 1: Infrared spectroscopy. Geochim. Cosmochim. Acta 71, 3518–3532.
van Hees, P., Jones, D., Finlay, R., Godbold, D., Lundstrom, U., 2005. The carbon we do not see – the impact of low molecular weight compounds on carbon dynamics and respiration in forest soils: a review. Soil Biol. Biochem. 37, 1–13.

Heijman, S.G.J., v. Paassen, A.M., v.d. Meer, W.G.J., Hopman, R., 1999. Adsorptive removal of natural organic matter during drinking water treatment. Water Sci. Technol. 40, 183–190.

Hejzlara, J., Dubrovskýb, M., Buchtelec, J., Růžičkac, M., 2003. The apparent and potential effects of climate change on the inferred concentration of dissolved organic matter in a temperate stream (the Malše River, South Bohemia). Sci. Total Environ. 310, 143–152.

Helfferich, F., 1995. Ion Exchange. Dover Publications, USA.

Hem, L.J., Efraimsen, H., 2001. Assimilable organic carbon in molecular weight fractions of natural organic matter. Water Res. 35, 1106–1110.

Henderson, R., Sharp, E., Jarvis, P., Parsons, S., Jefferson, B., 2006. Identifying the linkage between particle characteristics and understanding coagulation performance. Water Sci. Technol. Water Supply 6, 31–38.

Her, N., Amy, G., Chung, J., Yoon, J., Yoon, Y., 2008a. Characterizing dissolved organic matter and evaluating associated nanofiltration membrane fouling. Chemosphere 70, 495–502.

Her, N., Amy, G., Foss, D., Cho, J., Yoon, Y., Kosenka, P., 2002. Optimization of method for detecting and characterizing NOM by HPLC-size exclusion chromatography with UV and on-line DOC detection. Environ. Sci. Technol. 36, 1069–1076.

Her, N., Amy, G., Plottu-Pecheux, A., Yoon, Y., 2007. Identification of nanofiltration membrane foulants. Water Res. 41, 3936–3947.

Her, N., Amy, G., Sohn, J., Gunten von, U., 2008b. UV absorbance ratio index with size exclusion chromatography (URI-SEC) as an NOM property indicator. J. Water Supply Res. Technol. Aqua 57, 35–44.

Hering, J.G., Morel, F.M.M., 1988. Humic acid complexation of calcium and copper. Environ. Sci. Technol. 22, 1234–1237.

Hilal, N., Al-Abri, M., Al-Hinai, H., 2006. Enhanced membrane pre-treatment processes using macromolecular adsorption and coagulation in desalination plants: a review. Sep. Sci. Technol. 41, 403–453.

Hoffmann, M., Martin, S., Choi, W., Bahnemann, D., 1995. Environmental applications of semiconductor photocatalysis. Chem. Rev. 95, 69–96.

Holt, P.K., Barton, G.W., Mitchell, C.A., 2005. The future for electrocoagulation as a localized water treatment technology. Chemosphere 59, 355–367.

Hong, S., Elimelech, M., 1997. Chemical and physical aspects of natural organic matter (NOM) fouling of nanofiltration membranes. J. Membr. Sci. 132, 159–181.

Hongve, D., Baann, J., Becher, G., Beckmann, O., 1999. Experiences from operation and regeneration of an anionic exchanger for natural organic matter (NOM) removal. Water Sci. Technol. 40, 215–221.

Howe, K., Ishida, K., Clark, M., 2002. Use of ATR/FTIR spectrometry to study fouling of microfiltration membranes by natural waters. Desalination 147, 251–255.

Howe, K., Marwah, A., Chiu, K.-P., Adham, S., 2006. Effect of coagulation on the size of MF and UF membrane foulants. Environ. Sci. Technol. 40, 7908–7913.

Hsu, S., Singer, P.C., 2009. Application of anion exchange to control NOM interference on lime softening. J. Am. Water Works Assoc. 101, 85–94.

Hsu, S., Singer, P.C., 2010. Removal of bromide and natural organic matter by anion exchange. Water Res. 44, 2133–2140.

Hu, J., Liu, H., Qu, J., Wang, D., Ru, J., 2006. Coagulation behaviour of aluminum salts in eutrophic water: significance of Al_{13} species and pH control. Environ. Sci. Technol. 40, 325–331.

Hu, Q., Zhang, C., Wang, Z., Chen, Y., Mao, K., Zhang, X., Xiong, Y., Zhu, M., 2008. Photodegradation of methyl *tert*-butyl ether (MTBE) by UV/H_2O_2 and UV/TiO_2. J. Hazard. Mater. 154, 795–803.

Hua, G., Reckhow, D., 2007. Characterization of disinfection byproduct precursors based on hydrophobicity and molecular size. Environ. Sci. Technol. 41, 3309–3315.

Huang, G., Meng, F., Zheng, X., Wang, Y., Wang, Z., Liu, H., Jekel, M., 2011a. Biodegradation behavior of natural organic matter (NOM) in a biological aerated filter (BAF) as a pretreatment for ultrafiltration (UF) of river water. Appl. Microbiol. Biotechnol. 90, 1795–1803.

Huang, H., Cho, H., Schwab, K., Jacangelo, J.G., 2011b. Effects of feedwater pretreatment on the removal of organic microconstituents by a low fouling reverse osmosis membrane. Desalination 281, 446–454.

Huang, H., Cho, H., Schwab, K.J., Jacangelo, J.G., 2012. Effects of magnetic ion exchange pretreatment on low pressure membrane filtration of natural surface water. Water Res. 46, 5483–5490.

Huang, H., Schwab, K., Jacangelo, J.G., 2009. Pretreatment for low pressure membranes in water treatment: a review. Environ. Sci. Technol. 43, 3011–3019.

Huang, X., Leal, M., Li, Q., 2008. Degradation of natural organic matter by TiO_2 photocatalytic oxidation and its effect on fouling of low-pressure membranes. Water Res. 42, 1142–1150.

Huang, X., Gao, B., Wang, Y., Yue, Q., Li, Q., Zhang, Y., 2014. Coagulation performance and flocs properties of a new composite coagulant: polytitanium-silicate-sulfate. Chem. Eng. J. 245, 173–179.

Huber, S., Frimmel, F., 1992a. A liquid chromatographic system with multi-detection for direct analysis of hydrophilic organic compounds in natural waters. J. Anal. Chem. 342, 198–200.

Huber, S., Frimmel, F., 1992b. A new method for the characterization of organic carbon in aquatic systems. Int. J. Environ. Anal. Chem. 49, 49–57.

Huber, S.A., Balz, A., Abert, M., Pronk, W., 2011. Characterization of aquatic humic and non-humic matter with size-exclusion chromatography – organic carbon detection – organic nitrogen detection (LC-OCD-OND). Water Res. 45, 879–885.

Huber, S., Frimmel, F., 1994. Direct gel chromatic characterization and quantification of marine dissolved organic carbon using high-sensitivity DOC detection. Environ. Sci. Technol. 28, 1194–1197.

Hudson, N., Baker, A., Reynolds, D., 2007. Fluorescence analysis of dissolved organic matter in natural, waste and polluted waters—a review. River Res. Appl. 23, 631–649.

Hudson, N., Baker, A., Ward, D., Reynolds, D., Brunsdon, C., Carliell-Marquet, C., Browning, S., 2008. Can fluorescence spectrometry be used as a surrogate for the biochemical oxygen demand (BOD) test in water quality assessment? An example from South West England. Sci. Total Environ. 391, 149–158.

Humbert, H., Gallard, H., Croué, J., 2012. A polishing hybrid AER/UF membrane process for the treatment of a high DOC content surface water. Water Res. 46, 1093–1100.

Humbert, H., Gallard, H., Jacquemet, V., Croué, J., 2007. Combination of coagulation and ion exchange for the reduction of UF fouling properties of a high DOC content surface water. Water Res. 41, 3803–3811.

Humbert, H., Gallard, H., Suty, H., Croué, J., 2005. Performance of selected anion exchange resins for the treatment of a high DOC content surface water. Water Res. 39, 1699–1708.

Humbert, H., Gallard, H., Suty, H., Croué, J., 2008. Natural organic matter (NOM) and pesticides removal using a combination of ion exchange resin and powdered activated carbon (PAC). Water Res. 42, 1635–1643.

Hur, J., Williams, M., Schlautman, M., 2006. Evaluating spectroscopic and chromatographic techniques to resolve dissolved organic matter via end member mixing analysis. Chemosphere 63, 387–402.

Hyung, H., Kim, J.-H., 2008. Natural organic matter (NOM) adsorption to multi-walled carbon nanotubes: effect of NOM characteristics and water quality parameters. Environ. Sci. Technol. 42, 4416–4421.

Ikehata, K., El-Din, M., 2005. Aqeuous pesticide degradation by ozonation and ozone-based advanced oxidation processes: a review (part II). Ozone Sci. Eng. 27, 173–202.

Imoberdorf, G., Mohseni, M., 2011. Degradation of natural organic matter in surface water using vacuum-UV irradiation. J. Hazard. Mater. 186, 240–246.
Imyim, A., Prapalimrungsi, E., 2010. Humic acids removal from water by aminopropyl functionalized rice husk ash. J. Hazard. Mater. 184, 775–781.
Irabelli, A., Jasim, S., Biswas, N., 2008. Pilot-scale evaluation of ozone vs. peroxone for trihalomethane formation. Ozone Sci. Eng. 30, 356–366.
Iriarte-Velasco, U., Àlvarez-Uriarte, J., Gonzáles-Velasco, J., 2007. Enhanced coagulation under changing alkalinity-hardness conditions and its implications on trihalomethane precursors removal and relationship with UV absorbance. Sep. Purif. Technol. 55, 368–380.
Iriarte-Velasco, U., Álvarez-Uriarte, J.I., Chimeno-Alanís, N., González-Velasco, J.R., 2008. Natural organic matter adsorption onto granular activated carbons: implications in the molecular weight and disinfection byproducts formation. Ind. Eng. Chem. Res. 47, 7868–7876.
Jacangelo, J., DeMarco, J., Owen, D., Randtke, S., 1995a. Biofiltration performance: part 2, effect of backwashing. J. Am. Water Works Assoc. 87, 64–70.
Jacangelo, J., DeMarco, J., Owen, D., Randtke, S., 1995b. Selected processes for removing NOM: an overview. J. Am. Water Works Assoc. 87 (1), 64–77.
Jacangelo, J.G., Rhodes Trussell, R., Watson, M., 1997. Role of membrane technology in drinking water treatment in the United States. Desalination 113, 119–127.
Jaouadi, M., Amdouni, N., Duclaux, L., 2012. Characteristics of natural organic matter extracted from the waters of Medjerda dam (Tunisia). Desalination 305, 64–71.
Jarusutthirak, C., Mattaraj, S., Jiraratananon, R., 2007. Factors affecting nanofiltration performances in natural organic matter rejection and flux decline. Sep. Purif. Technol. 58, 68–75.
Jarvis, P., Banks, J., Molinder, R., Stephenson, T., Parsons, S., Jefferson, B., 2008a. Processes for enhanced NOM removal: beyond Fe and Al coagulation. Water Sci. Technol. Water Supply 8, 709–716.
Jarvis, P., Jefferson, B., Parsons, S., 2004. Characterizing natural organic matter flocs. Water Sci. Technol. Water Supply 4, 79–87.
Jarvis, P., Jefferson, B., Parsons, S., 2005. Breakage, regrowth, and fractal nature of natural organic matter flocs. Environ. Sci. Technol. 39, 2307–2314.
Jarvis, P., Jefferson, B., Parsons, S., 2006. Floc structural characteristics using conventional coagulation for a high doc, low alkalinity surface water source. Water Res. 40, 2727–2737.
Jarvis, P., Mergen, M., Banks, J., McIntosh, B., Parsons, S., Jefferson, B., 2008b. Pilot scale comparison of enhanced coagulation with magnetic resin plus coagulation systems. Environ. Sci. Technol. 42, 1276–1282.
Jarvis, P., Sharp, E., Pidou, M., Molinder, R., Parsons, S., Jefferson, B., 2012. Comparison of coagulation performance and floc properties using a novel zirconium coagulant against traditional ferric and alum coagulants. Water Res. 46, 4179–4187.
Jeong, J., Kim, C., Yoon, J., 2009. The effect of electrode material on the generation of oxidants and microbial inactivation in the electrochemical disinfection processes. Water Res. 43, 895–901.
Jeong, S., Kim, S., Min Kim, C., Vigneswaran, S., Vinh Nguyen, T., Shon, H., Kandasamy, J., Kim, I.S., 2013. A detailed organic matter characterization of pretreated seawater using low pressure microfiltration hybrid systems. J. Membr. Sci. 428, 290–300.
Jermann, D., Pronk, W., Kägi, R., Halbeisen, M., Boller, M., 2008. Influence of interactions between NOM and particles on UF fouling mechanisms. Water Res. 42, 3870–3878.
Jermann, D., Pronk, W., Meylan, S., Boller, M., 2007. Interplay of different NOM fouling mechanisms during ultrafiltration for drinking water production. Water Res. 41, 1713–1722.
Ji, Q., Liu, H., Hu, C., Qu, J., Wang, D., Li, J., 2008. Studies on the treatment of brilliant green solution by combination microwave induced oxidation with $CoFe_2SO_4$. Sep. Purif. Technol. 62, 464–469.

Jiang, C., Zhang, J., 2007. Progress and prospect in electro-Fenton process for wastewater treatment. J. Zhejiang Univ. Sci. A 8, 1118–1125.
Joseph, L., Flora, J.R.V., Park, Y.-G., Badawy, M., Saleh, H., Yoon, Y., 2012. Removal of natural organic matter from potential drinking water sources by combined coagulation and adsorption using carbon nanomaterials. Sep. Purif. Technol. 95, 64–72.
Jucker, C., Clark, M.M., 1994. Adsorption of aquatic humic substances on hydrophobic ultrafiltration membranes. J. Membr. Sci. 97, 37–52.
Jung, C., Kang, L., 2003. Application of combined coagulation-ultrafiltration membrane process for water treatment. Korean J. Chem. Eng. 20, 855–861.
Jung, C., Son, H., 2008. The relationship between disinfection by-products formation and characteristics of natural organic matter in raw water. Korean J. Chem. Eng. 25, 714–720.
Jung, C.-W., Son, H.-J., Kang, L.-S., 2006. Effects of membrane material and pretreatment coagulation on membrane fouling: fouling mechanism and NOM removal. Desalination 197, 154–164.
Jung, H.R., Seo, S.D., Choi, H.C., 2007. Preparation of mesostructured carbon (MSC) for removal of natural organic matter (NOM) in aqueous solution. Mater. Sci. Forum 544–545, 123–126.
Kabdaslı, I., Vardar, B., Arslan-Alaton, I., Tunay, O., 2009. Effect of dye auxiliaries on color and COD removal from simulated reactive dye bath effluent by electrocoagulation. Chem. Eng. J. 148, 89.
Kabsch-Korbutowicz, M., 2005. Effect of Al coagulant type on natural organic matter removal efficiency in coagulation/ultrafiltration process. Desalination 185, 327–333.
Kabsch-Korbutowicz, M., Majewska-Nowak, K., Winnicki, T., 2008. Water treatment using MIEX-DOC/ultrafiltration process. Desalination 221, 338–344.
Kaewsuk, J., Seo, G.T., 2011. Verification of NOM removal in MIEX-NF system for advanced water treatment. Sep. Purif. Technol. 80, 11–19.
Kajitvichyanukul, P., Shammas, N.K., Hung, Y., Wang, L.K., Ananpattarachai, J., 2008. Portable water biotechnology, membrane filtration and biofiltration. In: Wang, L.K., Chen, J.P., Hung, Y., Shammas, N.K. (Eds.), Handbook of Environmental Engineering. Membrane and Desalination Technologies, vol. 13. Springer Science & Business Media, New York, pp. 477–523.
Kamp, P.C., Kruithof, J.C., Folmer, H.C., 2000. UF/RO treatment plant Heemskerk: from challenge to full scale application. Desalination 131, 27–35.
Kaneco, S., Itoh, K., Katsumata, H., Suzuki, T., Masuyama, K., Funasaka, K., Hatano, K., Ohta, K., 2003. Removal of natural organic polyelectrolytes by adsorption onto tobermorite. Environ. Sci. Technol. 37, 1448–1451.
Kang, M.-G., Ku, Y.-H., Cho, Y.-K., Yu, M.-J., 2006. Variation of dissolved organic matter and microbial regrowth potential through drinking water treatment processes. Water Sci. Technol. 6, 57–66.
Kang, S., Choo, K., 2010. Why does a mineral oxide adsorbent control fouling better than powdered activated carbon in hybrid ultrafiltration water treatment? J. Membr. Sci. 355, 69–77.
Kanokkantapong, V., Marhaba, T., Panyapinyophol, B., Pavasant, P., 2006a. FTIR evaluation of functional groups involved in the formation of haloacetic acids during the chlorination of raw water. J. Hazard. Mater. B136, 188–196.
Kanokkantapong, V., Marhaba, T., Pavasant, P., Panyapinyophol, B., 2006b. Characterization of haloacetic acid precursors in source water. J. Environ. Manage. 80, 214–221.
Kapalka, A., Foti, G., Comninellis, C., 2010. Basic principles of the electrochemical mineralization of organic pollutants for wastewater treatment. In: Comninellis, C., Chen, G. (Eds.), Electrochemistry for the Environment. Springer, New York, USA, pp. 1–23.
Karanfil, T., Kitis, M., Kilduff, J.E., Wigton, A., 1999. Role of granular activated carbon surface chemistry on the adsorption of organic compounds. 2. Natural organic matter. Environ. Sci. Technol. 33, 3225–3233.
Karnik, B.S., Davies, S.H., Baumann, M.J., Masten, S.J., 2005. The effects of combined ozonation and filtration on disinfection by-product formation. Water Res. 39, 2839–2850.

Kasprzyk-Hordern, B., 2004. Chemistry of alumina, reactions in aqueous solution and its application in water treatment. Adv. Colloid Interface Sci. 110, 19–48.
Kasprzyk-Hordern, B., Raczyk-Stanislawiak, U., Swietlik, J., Nawrocki, J., 2005. Catalytic ozonation of natural organic matter on alumina. Appl. Catal. B Environ. 62, 345–358.
Katsumata, H., Sada, M., Kaneco, S., Suzuki, T., Ohta, K., Yobiko, Y., 2008. Humic acid degradation in aqeuous solution by the photo-Fenton process. Chem. Eng. J. 137, 225–230.
Kazpard, V., Lartiges, B., Frochot, C., d'Espinose de la Caillerie, J., Viriot, M., Portal, J., Görner, T., Bersillon, J., 2006. Fate of coagulant species and conformational effects during the aggregation of a model of a humic substance with Al_{13} polycations. Water Res. 40, 1965–1974.
Kennedy, M.D., Chun, H.K., Quintanilla Yangali, V.A., Heijman, B.G.J., Schippers, J.C., 2005. Natural organic matter (NOM) fouling of ultrafiltration membranes: fractionation of NOM in surface water and characterization by LC-OCD. Desalination 178, 73–83.
Kennedy, M.D., Kamanyi, J., Heijman, B.G.J., Amy, G., 2008a. Colloidal organic matter fouling of UF membranes: role of NOM composition & size. Desalination 220, 200–213.
Kennedy, M.D., Kamanyi, J., Salinas Rodríguez, S.G., Lee, N.H., Schippers, J.C., Amy, G., 2008b. Water treatment by microfiltration and ultrafiltration. In: Li, N.N., Fane, A.G., Ho, W.S.W., Matsuura, T. (Eds.), Advanced Membrane Technology and Applications. John Wiley & Sons, Hoboken, USA, pp. 131–170.
Kilduff, J.E., Mattaraj, S., Pieracci, J.P., Belfort, G., 2000. Photochemical modification of poly(ether sulfone) and sulfonated poly(sulfone) nanofiltration membranes for control of fouling by natural organic matter. Desalination 132, 133–142.
Kiliç, M.G., Hosten, Ç, Demirci, S., 2009. A parametric comparative study of electrocoagulation and coagulation using ultrafine quartz suspensions. J. Hazard. Mater. 171, 247–252.
Kim, D.H., Shon, H.K., Sharma, G., Cho, J., 2011. Charge effect of natural organic matter for ultrafiltration and nanofiltration membranes. J. Ind. Eng. Chem. 17, 109–113.
Kim, H., Dempsey, B.A., 2010. Removal of organic acids from EfOM using anion exchange resins and consequent reduction of fouling in UF and MF. J. Membr. Sci. 364, 325–330.
Kim, H., Dempsey, B.A., 2013. Membrane fouling due to alginate, SMP, EfOM, humic acid, and NOM. J. Membr. Sci. 428, 190–197.
Kim, H., Hong, J.-H., Lee, S., 2006a. Fouling of microfiltration membranes by natural organic matter after coagulation treatment: a comparison of different initial mixing conditions. J. Membr. Sci. 283, 266–272.
Kim, H., Lee, S., Byun, S., Yu, M., 2006b. Application of improved rapid mixing for enhanced removal of dissolved organic matter and DBPFP (disinfection by-product formation potential) control. Water Sci. Technol. Water Supply 6, 49–57.
Kim, H.-C., Yu, M.-J., 2007. Characterization of aquatic humic substances to DBPs formation in advanced treatment processes for conventionally treated water. J. Hazard. Mater. 143, 486–493.
Kim, H.-C., Yu, M.-J., Han, I., 2006c. Multi-method study of the characteristic chemical nature of aquatic humic substances isolated from the Han River, Korea. Appl. Geochem. 21, 1226–1239.
Kim, J., Van der Bruggen, B., 2010. The use of nanoparticles in polymeric and ceramic membrane structures: review of manufacturing procedures and performance improvement for water treatment. Environ. Pollut. 158, 2335–2349.
Kim, J., Cai, Z., Benjamin, M., 2010. NOM fouling mechanisms in a hybrid adsorption/membrane system. J. Membr. Sci. 349, 35–43.
Kim, J., Cai, Z., Benjamin, M.M., 2008. Effects of adsorbents on membrane fouling by natural organic matter. J. Membr. Sci. 310, 356–364.
Kim, J., Shan, W., Davies, S., Baumann, M., Masten, S., Tarabara, V., 2009a. Interactions of aqeuos NOM with nanoscale TiO_2: implications for ceramic membrane filtration-ozonation hybrid process. Environ. Sci. Technol. 43, 5488–5494.

Kim, J.H., Lee, C.H., Lee, E.J., Lee, K.H., Kwon, S.B., Park, H., Kim, H.S., Jang, A., 2013a. The effect of re-aggregated floc by additional coagulant on membrane. J. Taiwan Inst. Chem. Eng. 44, 802–807.

Kim, K., Kim, H., Kim, J., Nam, J., Kim, J., Son, S., 2009b. A hybrid microfiltration–granular activated carbon system for water purification and wastewater reclamation/reuse. Desalination 243, 132–144.

Kim, M.H., Yu, M.J., 2005. Characterization of NOM in the Han River and evaluation of treatability using UF–NF membrane. Environ. Res. 97, 116–123.

Kim, S., Kim, J., Seo, G., 2013b. Iron oxide nanoparticle-impregnated powder-activated carbon (IPAC) for NOM removal in MF membrane water treatment system. Desalin. Water Treat. 51, 1–9.

Kingsbury, R.S., Singer, P.C., 2013. Effect of magnetic ion exchange and ozonation on disinfection by-product formation. Water Res. 47, 1060–1072.

Kitis, M., Kaplan, S., 2007. Advanced oxidation of natural organic matter using hydrogen peroxide and iron-coated pumice particles. Chemosphere 68, 1846–1853.

Kitis, M., İlker Harman, B., Yigit, N.O., Beyhan, M., Nguyen, H., Adams, B., 2007a. The removal of natural organic matter from selected Turkish source waters using magnetic ion exchange resin (MIEX®). React. Funct. Polym. 67, 1495–1504.

Kitis, M., Kaplan, S.S., Karakaya, E., Yigit, N.O., Civelekoglu, G., 2007b. Adsorption of natural organic matter from waters by iron coated pumice. Chemosphere 66, 130–138.

Klavarioti, M., Mantzavinos, D., Kassinos, D., 2009. Removal of residual pharmaceuticals from aqueous systems by advanced oxidation processes. Environ. Int. 35, 402–417.

Kleiser, G., Frimmel, F., 2000. Removal of precursors for disinfection by-products (DBPs) – differences between ozone- and OH-radical-induced oxidation. Sci. Total Environ. 256, 1–9.

Kliaugaite, D., Yasadi, K., Euverink, G.-J., Bijmans, M.F.M., Racys, V., 2013. Electrochemical removal and recovery of humic-like substances from wastewater. Sep. Purif. Technol. 108, 37–44.

Kobya, M., Gebologlu, U., Ulu, F., Oncel, S., Demirbas, E., 2011. Removal of arsenic from drinking water by the electrocoagulation using Fe and Al electrodes. Electrochim. Acta 56, 5060–5070.

Kobya, M., Hiz, H., Senturk, E., Aydiner, C., Demirbas, E., 2006. Treatment of potato chips manufacturing wastewater by electrocoagulation. Desalination 190, 2201–2211.

Koltuniewich, A.B., Drioli, E., 2008. Membranes in Clean Technologies, Theory and Practice. Wiley-VCH Verlag, Weinheim.

Konieczny, K., Sakol, D., Plonka, J., Rajca, M., Bodzek, M., 2009. Coagulation—ultrafiltration system for river water treatment. Desalination 240, 151–159.

Koparal, A.S., Yildiz, Y., Ş, Keskinler B., Demircioglu, N., 2008. Effect of initial pH on the removal of humic substances from wastewater by electrocoagulation. Sep. Purif. Technol. 59, 175–182.

Koprivnjak, J., Perdue, E., Pfromm, P., 2006. Coupling reverse osmosis with electrodialysis to isolate natural organic matter from fresh waters. Water Res. 40, 3385–3392.

Korshin, G., Benjamin, M., Li, C.-W., 1999. Use of differential spectroscopy to evaluate the structure and reactivity of humics. Water Sci. Technol. 40, 9–16.

Korshin, G., Chow, C., Fabris, R., Drikas, M., 2009. Absorbance spectroscopy-based examination of effects of coagulation on the reactivity of fractions of natural organic matter with varying apparent molecular weights. Water Res. 43, 1541–1548.

Korshin, G.V., Benjamin, M.M., Sletten, R.S., 1997. Adsorption of natural organic matter (NOM) on iron oxide: effects on NOM composition and formation of organo-halide compounds during chlorination. Water Res. 31, 1643–1650.

Korth, A., Fiebiger, C., Bornmann, K., Schmidt, W., 2004. NOM increase in drinking water reservoirs – relevance for drinking water production. Water Sci. Technol. Water Supply 4 (4), 55–60.

Kothawala, D.N., Stedmon, C.A., Müller, R.A., Weyhenmeyer, G.A., Köhler, S.J., Tranvik, L.J., 2014. Controls of dissolved organic matter quality: evidence from a large-scale boreal lake survey. Global Change Biol. 20, 1101–1114.

Kourdali, S., Badis, A., Saiba, A., Boucherit, A., Boutoumi, H., 2014. Humic acid removal by electrocoagulation using aluminium sacrificial anode under influencing operational parameters. Desalin. Water Treat. 52, 5442–5453

Kraft, A., Stadelmann, M., Blaschke, M., 2003. Anodic oxidation with doped diamond electrodes: a new advanced oxidation process. J. Hazard. Mater. B103, 247–261.

Kraft, A., Stadelmann, M., Blaschke, M., Kreysig, D., Sandt, B., Schröder, F., Rennau, J., 1999. Electrochemical water disinfection part I: hypochlorite production from very dilute chloride solutions. J. Appl. Electrochem. 29, 861–868.

Krasner, S., Weinberg, H., Richardson, S., Pastor, S., Chinn, R., Sclimenti, M., Onstad, G., Thruston Jr., A., 2006. Occurrence of a new generation of disinfection by-products. Environ. Sci. Technol. 40, 7175–7185.

Kristiana, I., Allpike, B., Joll, C., Heitz, A., Trolio, R., 2010. Understanding the behaviour of molecular weight fractions of natural organic matter to improve water treatment processes. Water Sci. Technol. Water Supply 10, 59–68.

Kruithof, J., Kamp, P., Martijn, B., 2007. UV/H_2O_2 treatment: a practical solution for organic contaminant control and primary disinfection. Ozone Sci. Eng. 29, 273–280.

Kruithof, J.C., Schippers, J.C., Kamp, P.C., Folmer, H.C., Hofman, J.A.M.H., 1998. Integrated multi-objective membrane systems for surface water treatment: pretreatment of reverse osmosis by conventional treatment and ultrafiltration. Desalination 117, 37–48.

Küchler, I.L., Miekeley, N., 1994. Ultrafiltration of humic compounds through low molecular mass cut-off level membranes. Sci. Total Environ. 154, 23–28.

Kumar, E., Bhatnagar, A., Hogland, W., Marques, M., Sillanpää, M., 2014. Interaction of anionic pollutants with Al-based adsorbents in aqueous media – a review. Chem. Eng. J. 241, 443–456.

Kumar, P., Manivel, A., Anandan, S., 2009. Syntesis of Ag-ZnO nanoparticles for enhanced photocatalytic degradation of acid red 88 in aqeuous environment. Water Sci. Technol. 59, 1423–1430.

Kunenkov, E., Kononikhin, A., Perminova, I., Hertkorn, N., Gaspar, A., Scmitt-Kopplin, P., Popov, I., Garmash, A., Nikolaev, E., 2009. Total mass difference statistics algorithm: a new approach to identification of high-mass building blocks in electrospray ionization fourier transform ion cyclotron mass spectrometry data of natural organic matter. Anal. Chem. 81, 10106–10115.

Kutschera, K., Börnick, H., Worch, E., 2009. Photoinitiated oxidation of geosmin and 2-methylisoborneol by irradiation with 254 nm and 185 nm UV light. Water Res. 43, 2224–2232.

Kwon, B., Lee, S., Cho, J., Ahn, H., Lee, D., Shin, H.S., 2005. Biodegradability, DBP fromation, and membrane fouling potential of natural organic matter: characterization and controllability. Environ. Sci. Technol. 39, 732–739.

Laborda, F., Ruiz-Beguería, S., Bolea, E., Castillo, J., 2009. Functional speciation of metal-dissolved organic matter complexes by size exclusion chromatography coupled to inductively coupled plasma mass spectrometry and deconvolution analysis. Spectrochim. Acta B 392, 392–398.

Lahoussine-Turcaud, V., Wiesner, M.R., Bottero, J., 1990. Fouling in tangential-flow ultrafiltration: the effect of colloid size and coagulation pretreatment. J. Membr. Sci. 52, 173–190.

Laîné, J.M., Vial, D., Moulart, P., 2000. Status after 10 years of operation — overview of UF technology today. Desalination 131, 17–25.

Lambert, S.D., Graham, N.J.D., 1995. Removal of non-specific dissolved organic matter from upland potable water supplies—I. Adsorption. Water Res. 29, 2421–2426.

Lamsal, R., Montreuil, K.R., Kent, F.C., Walsh, M.E., Gagnon, G.A., 2012. Characterization and removal of natural organic matter by an integrated membrane system. Desalination 303, 12–16.

Lamsal, R., Walsh, M., Gagnon, G., 2011. Comparison of advanced oxidation processes for the removal of natural organic matter. Water Res. 45, 3263–3269.

Lanao, M., Ormad, M., Ibarz, C., Miguel, N., Ovelleiro, J., 2008. Bactericidal effectiveness of O_3, O_3/H_2O and O_3/TiO_2 on *Clostridium perfringens*. Ozone Sci. Eng. 30, 431–438.

Langlais, B., Reckhow, D.A., Brink, D.R., 1991. Ozone in Water Treatment: Application and Engineering. Lewis Publishers, Chelsea, Michigan, USA.

Langmuir, I., 1916. The adsorption of gases on plane surface of glass, mica and platinum. J. Am. Chem. Soc. 40, 1361–1368.

Lankes, U., Lüdemann, H.D., Frimmel, F.H., 2008. Search for basic relationships between "molecular size" and "chemical structure" of aquatic natural organic matter-Answers from ^{13}C and ^{15}N CPMAS NMR spectroscopy. Water Res. 42, 1051–1060.

Lankes, U., Müller, M., Weber, M., Frimmel, F., 2009. Reconsidering the quantitative analysis of organic carbon concentrations in size exclusion chromatography. Water Res. 43, 915–924.

Lead, J.R., Wilkinson, K.J., Balnois, E., Cutak, B.J., Larive, C.K., Assemi, S., Beckett, R., 2000. Diffusion coefficients and polydispersities of the Suwannee river fulvic acid: comparison of fluorescence correlation spectroscopy, pulsed-field gradient nuclear magnetic resonance, and flow field-flow fractionation. Environ. Sci. Technol. 34, 3508–3513.

Lebeau, T., Lelièvre, C., Buisson, H., Cléret, D., Van de Venter, L.W., Côté, P., 1998. Immersed membrane filtration for the production of drinking water: combination with PAC for NOM and SOCs removal. Desalination 117, 219–231.

Le-Clech, P., Lee, E.-K., Chen, V., 2006. Hybrid photocatalysis/membrane treatment for surface waters containing low concentrations of natural organic matters. Water Res. 40, 323–330.

Lee, C., Sedlak, D., 2009. A novel homogenous Fenton-like system with Fe(III)-phosphotungstate for oxidation of organic compounds at neutral pH values. J. Mol. Catal. A Chem. 311, 1–6.

Lee, D.-J., Cheng, Y.-L., Wong, R.-J., 2013. Characteristics of natural organic matters in raw and treated drinking water: comparison study. J. Taiwan Inst. Chem. Eng. 44, 103–110.

Lee, N., Amy, G., Croué, J., 2006. Low-pressure membrane (MF/UF) fouling associated with allochthonous versus autochthonous natural organic matter. Water Res. 40, 2357–2368.

Lee, N., Amy, G., Croué, J., Buisson, H., 2004a. Identification and understanding of fouling in low-pressure membrane (MF/UF) filtration by natural organic matter (NOM). Water Res. 38, 4511–4523.

Lee, S., Cho, J., 2004. Comparison of ceramic and polymeric membranes for natural organic matter (NOM) removal. Desalination 160, 223–232.

Lee, S., Lee, C., 2006. Microfiltration and ultrafiltration as a pretreatment for nanofiltration of surface water. Sep. Sci. Technol. 41, 1–23.

Lee, S., Lee, C., 2007. Effect of membrane properties and pretreatment on flux and NOM rejection in surface water nanofiltration. Sep. Purif. Technol. 56, 1–8.

Lee, S., Amy, G., Cho, J., 2004b. Applicability of Sherwood correlations for natural organic matter (NOM) transport in nanofiltration (NF) membranes. J. Membr. Sci. 240, 49–65.

Lee, S., Kwon, B., Sun, M., Cho, J., 2005. Characterizations of NOM included in NF and UF membrane permeates. Desalination 173, 131–142.

Leenher, J., 1981. Comprehensive approach to preparative isolation and fractionation of dissolved organic carbon from natural waters and wastewaters. Environ. Sci. Technol. 15, 578–587.

Leenher, J., 2004. Comprehensive assesment of precursors, diagenesis, and reactivity to water treatment of dissolved and colloidal organic matter. Water Sci. Technol. Water Supply 4 (4), 1–9.

Leenher, J.A., 1985. Fractionation techniques for aquatic humic substances. In: Aiken, G., McKnight, D., Wershaw, R., MacCarthy, P. (Eds.), Humic Substances in Soil, Sediment and Water. John Wiley & Sons, pp. 409–429.
Leenheer, J.A., Wershaw, R.L., Reddy, M.M., 1995. Strong-acid, carboxyl-group structures in fulvic acid from the Suwannee river, Georgia. 1. Minor structures. Environ. Sci. Technol. 29, 393–398.
Leenheer, J., Croué, J.-P., 2003a. Peer reviewed: characterizing aquatic dissolved organic matter. Environ. Sci. Technol. 37, 18A–26A.
Leenher, J., Croue, J.-P., 2003b. Characterizing dissolved aquatic organic matter. Environ. Sci. Technol. 37, 18A–26A.
Legrini, O., Oliveros, E., Braun, A., 1993. Photochemical processes for water treatment. Chem. Rev. 83, 671–698.
Legube, B., Vel Leitner, N.K., 1999. Catalytic ozonation: a promising advanced oxidation technology for water treatment. Catal. Today 53, 61–72.
Lei, G., Ma, X., Guan, X., Song, A., Cui, Y., 2009. Effect of basicity on coagulation performance of polyferric chloride applied in eutrophicated raw water. Desalination 247, 518–529.
Leiknes, T., 2009. The effect of coupling coagulation and flocculation with membrane filtration in water treatment: a review. J. Environ. Sci. 21, 8–12.
Li, A., Hu, J., Li, W., Zhang, W., Wang, X., 2009a. Polarity based fractionation of fulvic acids. Chemosphere 77, 1419–1426.
Li, C., Chen, Y., 2004. Fouling of UF membrane by humic substance: effects of molecular weight and powder-activated carbon (PAC) pre-treatment. Desalination 170, 59–67.
Li, F., Akira, Y., Yuka, A., 2008a. Characterization of micro-flocs of NOM coagulated by PACl, alum and polysilicate-iron in terms of molecular weight and floc size. Water Sci. Technol. 57, 83–90.
Li, K., Hokanson, D., Crittenden, J., Trussell, R., Minakata, D., 2008b. Evaluating UV/H_2O_2 processes for methyl *tert*-butyl ether and tertiary butyl alcohol removal: effect of pre-treatment options and light sources. Water Res. 42, 5045–5053.
Li, T., Zhu, Z., Wang, D., Yao, C., Tang, H., 2006. Characterization of floc size, strength and structure under various coagulation mechanisms. Powder Technol. 168, 104–110.
Li, X., Chu, H.P., 2003. Membrane bioreactor for the drinking water treatment of polluted surface water supplies. Water Res. 37, 4781–4791.
Li, Y., Wang, J., Zhang, W., Zhang, X., Chen, C., 2011. Effects of coagulation on submerged ultrafiltration membrane fouling caused by particles and natural organic matter (NOM). Chin. Sci. Bull. 56, 584–590.
Li, Z., Zhong, S., Lei, H.-Y., Chen, R.-W., Yu, Q., Li, H.-L., 2009b. Production of a novel bioflocculant by *Bacillus licheniformis* X14 and its application to low temperature drinking water treatment. Bioresour. Technol. 100, 3650–3656.
Liao, A.A., Spitzer, M., Motheo, A.J., Bertazzoli, R., 2008. Electrocombustion of humic acid and removal of algae from aqueous solutions. J. Appl. Electrochem. 38, 721–727.
Lin, C., Huang, Y., Hao, O.J., 1999. Ultrafiltration processes for removing humic substances: effect of molecular weight fractions and PAC treatment. Water Res. 33, 1252–1264.
Lin, C., Lin, T., Hao, O.J., 2000. Effects of humic substance characteristics on UF performance. Water Res. 34, 1097–1106.
Lin, C., Liu, S., Hao, O.J., 2001. Effect of functional groups of humic substances on uf performance. Water Res. 35, 2395–2402.
Lin, H.-C., Wang, G.-S., 2011. Effects of UV/H_2O_2 on NOM fractionation and corresponding DBPs formation. Desalination 270, 221–226.
Lin, J., Zhan, Y., 2012. Adsorption of humic acid from aqueous solution onto unmodified and surfactant-modified chitosan/zeolite composites. Chem. Eng. J. 200–202, 202–213.

Lin, J.C., Cheng, A., Lin, Y., Lee, D., Guo, W., 2012. Dense membranes precoated with aluminium tridecamer (Al_{13}) for water treatment. Desalination 284, 349–352.
Lin, L., Lei, Z., Wang, L., Liu, X., Zhang, Y., Wan, C., Lee, D.-J., Tay, J.H., 2013. Adsorption mechanisms of high-levels of ammonium onto natural and NaCl-modified zeolites. Sep. Purif. Technol. 103, 15–20.
Lindqvist, N., Jokela, J., Tuhkanen, T., 2004. In: Proceedings of the 11th International Gothenburg Symposium on Chemical Treatment of Water and Wastewater, November 8–11, Orlando, Florida, USA.
Lindqvist, N., Korhonen, S., Jokela, J., Tuhkanen, T., 2002. In: Hahn, H., Hofmann, E., Odegaard, H. (Eds.), Chemical Water and Wastewater Treatment VIII. IWA Publishing, London, p. 133.
Liu, F., Xu, Z., Wan, H., Wan, Y., Zheng, S., Zhu, D., 2011a. Enhanced adsorption of humic acids on ordered mesoporous carbon compared with microporous activated carbon. Environ. Toxicol. Chem. 30, 793–800.
Liu, F.-F., Fan, J.-L., Wang, S.-G., Ma, G.-H., 2013. Adsorption of natural organic matter analogues by multi-walled carbon nanotubes: comparison with powdered activated carbon. Chem. Eng. J. 219, 450–458.
Liu, H., Cheng, F., Wang, D., 2009a. Interaction of ozone and organic matter in coagulation with inorganic polymer flocculant-PACl: role of organic components. Desalination 249, 596–601.
Liu, H., Wang, D., Wang, M., Tang, H., Yang, M., 2007a. Effect of pre-ozonation on coagulation with IPF–PACls: role of coagulant speciation. Colloids Surf. A Physicochem. Eng. Aspects 294, 111–116.
Liu, J., Liu, L., Gao, B., Yang, F., 2012a. Cathode membrane fouling reduction and sludge property in membrane bioreactor integrating electrocoagulation and electrostatic repulsion. Sep. Purif. Technol. 100, 44–50.
Liu, L., Liu, J., Gao, B., Yang, F., 2012b. Minute electric field reduced membrane fouling and improved performance of membrane bioreactor. Sep. Purif. Technol. 86, 106–112.
Liu, L., Liu, J., Gao, B., Yang, F., Chellam, S., 2012c. Fouling reductions in a membrane bioreactor using an intermittent electric field and cathodic membrane modified by vapor phase polymerized pyrrole. J. Membr. Sci. 394–395, 202–208.
Liu, Q., Seung-Hyun, K., Jong-Sup, Y., Seong-Yong, M., 2006. Impacts of coagulation pretreatment on MF filtration and a comparative study of different membrane module types. J. Ocean Univ. China (Engl. Ed.) 5, 295–299.
Liu, R., Lead, J., Baker, A., 2007b. Fluorescence characterization of cross flow ultrafiltration derived freshwater colloidal and dissolved organic matter. Chemosphere 68, 1304–1311.
Liu, S., Lim, M., Chiang, K., Amal, R., Fabris, R., Chow, C., Drikas, M., 2007c. A study on the removal of humic acid using advanced oxidation process. Sep. Sci. Technol. 42, 1391–1404.
Liu, S., Lim, M., Fabris, R., Chow, C., Chiang, K., Drikas, M., Amal, R., 2008a. Removal of humic acid using TiO_2 photocatalytic process – fractionation and molecular weight characterization studies. Chemosphere 72, 263–271.
Liu, S., Lim, M., Fabris, R., Chow, C., Drikas, M., Amal, R., 2008b. TiO_2 photocatalysis of natural organic matter in surface water: Impact on trihalomethane and haloacetic acid formation potential. Environ. Sci. Technol. 42, 6218–6223.
Liu, S., Lim, M., Fabris, R., Chow, C., Drikas, M., Amal, R., 2010. Comparison of photocatalytic degradation of natural organic matter in two Australian surface waters using multiple analytical techniques. Org. Geochem. 41, 124–129.
Liu, T., Chen, Z., Yu, W., Shen, J., Gregory, J., 2011c. Effect of two-stage coagulant addition on coagulation-ultrafiltration process for treatment of humic-rich water. Water Res. 45, 4260–4268.

Liu, T.-K., Chin, C., 2009. Improved coagulation performance using preformed polymeric iron chloride (PICl). Colloids Surf. A Physicochem. Eng. Aspects 339, 192–198.
Liu, W., Wang, K., Li, B., Yuan, H., Yang, J., 2009b. Production and characterization of an intracellular bioflocculant by *Chryseobacterium daeguense* W6 cultured in low nutrition medium. Bioresour. Technol. 101, 1044–1048.
Liu, Z., Zu, Y., Meng, R., Xing, Z., Tan, S., Zhao, L., Sun, T., Zhou, Z., 2011b. Adsorption of humic acid onto carbonaceous surfaces: atomic force microscopy study. Microsc. Microanal. 17, 1015–1021.
Ljubas, D., 2005. Solar photocatalysis-a possible step in drinking water treatment. Energy 30, 1699–1710.
Lohwacharin, J., Oguma, K., Takizawa, S., 2010. Use of carbon black nanoparticles to mitigate membrane fouling in ultrafiltration of river water. Sep. Purif. Technol. 72, 61–69.
Loi-Brugger, A., Panglisch, S., Buchta, P., Hattori, K., Yonekawa, H., Tomita, Y., Gimbel, R., 2006. Ceramic membranes for direct river water treatment applying coagulation and microfiltration. Water Sci. Technol. Water Supply 6, 89–98.
van Loon, G.W., Duffy, S.J., 2005. Environmental Chemistry: A Global Perspective, second ed. Oxford University Press, USA.
Lopes, M.P., Matos, C.T., Pereira, V.J., Benoliel, M.J., Valério, M.E., Bucha, L.B., Rodrigues, A., Penetra, A.I., Ferreira, E., Cardoso, V.V., Reis, M.A.M., Crespo, J.G., 2013. Production of drinking water using a multi-barrier approach integrating nanofiltration: a pilot scale study. Sep. Purif. Technol. 119, 112–122.
Lou, I., Gong, S., Huang, X., Liu, Y., 2012. Coagulation optimization for low temperature and low turbidity source water using combined coagulants: a case study. Desalin. Water Treat. 46, 107–114.
Louie, S., Tilton, R., Lowry, G., 2013. Effects of molecular weight distribution and chemical properties of natural organic matter on gold nanoparticle aggregation. Environ. Sci. Technol. 47, 4245–4254.
Lu, C., Su, F., 2007. Adsorption of natural organic matter by carbon nanotubes. Sep. Purif. Technol. 58, 113–121.
Lu, J., Huang, Q., 2009. Removal of acetaminophen using enzyme-mediated oxidative coupling processes: II. Cross-coupling with natural organic matter. Environ. Sci. Technol. 43, 7068–7073.
Ma, H., Allen, H., Yin, Y., 2001. Characterization of isolated fractions of dissolved organic matter from natural waters and a wastewater effluent. Water Res. 35, 985–996.
Ma, H., Wang, B., Wang, Y., 2007. Application of molybdenum and phosphate modified kaolin in electrochemical treatment of paper mill wastewater. J. Hazard. Mater. 145, 417–423.
Ma, N., Quan, X., Zhang, Y., Chen, S., Zhao, H., 2009. Integration of separation and photocatalysis using an inorganic membrane modified with Si-doped TiO_2 for water purification. J. Membr. Sci. 335, 58–67.
Maartens, A., Swart, P., Jacobs, E.P., 1999. Feed-water pretreatment: methods to reduce membrane fouling by natural organic matter. J. Membr. Sci. 163, 51–62.
Mahvi, A.H., Malakootian, M., Heidari, M.R., 2011. Comparison of polyaluminum silicate chloride and electrocoagulation process, in natural organic matter removal from surface water in Ghochan, Iran. J. Water Chem. Technol. 33, 377–385.
Malato, S., Fernández-Ibáñez, P., Maldonado, M., Blanco, J., Gernjak, W., 2009. Decontamination and disinfection of water by solar photocatalysis: recent overview and trends. Catal. Today 147, 1–59.
Malato, S., Blanco, J., Alarcón, D., Maldonado, M., Fernández-Ibáñez, P., Gernjak, W., 2007. Photocatalytic decontamination and disinfection of water with solar collectors. Catal. Today 122, 137–149.

Malcolm, R., MacCarthy, P., 1992. Quantitative evaluation of XAD-8 and XAD-4 resins used in tandem for removing organic solutes from water. Environ. Int. 18, 597–607.
Malek, F., Harris, J.L., Roddick, F.A., 2006. Interrelationship of photooxidation and microfiltration in drinking water treatment. J. Membr. Sci. 281, 541–547.
Malekian, R., Abedi-Koupai, J., Eslamian, S.S., Mousavi, S.F., Abbaspour, K.C., Afyuni, M., 2011. Ion-exchange process for ammonium removal and release using natural Iranian zeolite. Appl. Clay Sci. 51, 323–329.
Mansouri, K., Hannachi, A., Abdel-Wahab, A., Bensalah, N., 2012. Electrochemically dissolved aluminum coagulants for the removal of natural organic matter from synthetic and real industrial wastewaters. Ind. Eng. Chem. Res. 51, 2428–2437.
Mantell, C.L., 1951. Adsorption. McGraw-Hill Book Company, New York.
Mao, J., Cory, R., McKnight, D., Schmidt-Rohr, K., 2007a. Characterization of a nitrogen-rich fulvic acid and its precursor algae from solid state NMR. Org. Geochem. 38, 1277–1292.
Mao, J., Tremblay, L., Gagné, J.-P., Kohl, S., Rice, J., Schmidt-Rohr, K., 2007b. Humic acids from particulate organic matter in the Saguenay Fjord and St. Lawrence Estuary investigated by advanced solid-state NMR. Geochim. Cosmochim. Acta 71, 5483–5499.
Marhaba, T., 2000. Fluorescence technique for rapid identification of DOM fractions. J. Environ. Eng. Sci. 126, 145–152.
Marhaba, T., Borgaonkar, A., Punburananon, K., 2009. Principal component regression model applied to dimensionally reduced spectral fluorescent signature for the determination of organic character and THM formation potential of source water. J. Hazard. Mater. 169, 998–1004.
Marhaba, T.F., Van, D., Lippincott, R.L., 2000. Changes in NOM fractions through treatment: a comparison of ozonation and chlorination. Ozone Sci. Technol. 22, 249–266.
Marhaba, T., Pu, Y., Bengraine, K., 2003. Modified dissolved organic matter fractionation technique for natural water. J. Hazard. Mater. 101, 43–53.
Markechova, D., Tomokova, M., Sadecka, J., 2013. Fluorescence excitation–emission matrix spectroscopy and parallel factor analysis in drinking water treatment: a review. Pol. J. Environ. Stud. 22, 1289–1295.
Marsh, H., Rodríguez-Reinoso, F., 2006. Activated Carbon. Elsevier, Oxford.
Martins, A.F., Wilde, M.L., Vasconcelos, T.G., Henriques, D.M., 2006. Nonylphenol polyethoxylate degradation by means of electrocoagulation and electrochemical Fenton. Sep. Purif. Technol. 50, 249–255.
Martucci, A., Pasti, L., Marchetti, N., Cavazzini, A., Dondi, F., Alberti, A., 2012. Adsorption of pharmaceuticals from aqueous solutions on synthetic zeolites. Microporous Mesoporous Mater. 148, 174–183.
Matilainen, A., Sillanpää, M., 2010. Removal of natural organic matter from drinking water by advanced oxidation processes: a review. Chemosphere 80, 351–365.
Matilainen, A., Gjessing, E.T., Lahtinen, T., Hed, L., Bhatnagar, A., Sillanpää, M., 2011. An overview of the methods used in the characterization of natural organic matter (NOM) in relation to drinking water treatment. Chemosphere 83, 1431–1442.
Matilainen, A., Iivari, P., Sallanko, J., Heiska, E., Tuhkanen, T., 2006a. The role of ozonation and activated carbon filtration in the natural organic matter removal from drinking water. Environ. Technol. 27, 1171–1180.
Matilainen, A., Vepsäläinen, M., Sillanpää, M., 2010. Natural organic matter removal by coagulation during drinking water treatment: a review. Adv. Colloid Interface Sci. 159, 189–197.
Matilainen, A., Vieno, N., Tuhkanen, T., 2006b. Efficiency of the activated carbon filtration in the natural organic matter removal. Environ. Int. 32, 324–331.
Matsui, Y., Murase, R., Sanogawa, T., Aoki, N., Mima, S., Inoue, T., Matsushita, T., 2004. Micro-ground powdered activated carbon for effective removal of natural organic matter during water treatment. Water Sci. Technol. Water Supply 4, 155–163.

Matsui, Y., Yuasa, A., Li, F., 1998. Overall adsorption isotherm of natural organic matter. J. Environ. Eng. 124, 1099–1107.
Matsukawa, S., Itoho, S., Habutnu, S., Aizawa, T., 2006. Water Sci. Technol. Water Supply 6, 67.
Mattsson, T., Kortelainen, P., Räike, A., 2005. Export of DOM from boreal catchments: impacts of land use cover and climate. Biogeochemistry 76, 373–394.
Maurice, P., Cabaniss, S., Drummond, J., Ito, E., 2002. Hydrogeochemical controls on the variations in chemical characteristics of natural organic matter at a small freshwater wetland. Chem. Geol. 187, 59–77.
Mavrov, V., Chmiel, H., Kluth, J., Meier, J., Heinrich, F., Ames, P., Backes, K., Usner, P., 1998. Comparative study of different MF and UF membranes for drinking water production. Desalination 117, 189–196.
Mawhinney, D., Rosario-Ortiz, F., Baik, S., Vanderford, B., Snyder, S., 2009. Characterization of fulvic acids by liquid chromatography-quadrupole time-of-flight mass spectrometry. J. Chromatogr. A 1216, 1319–1324.
McBain, J.W., 1909. The mechanism of the adsorption ("sorption") of hydrogen by carbon. Philos. Mag. Ser. 6, 916–935.
McGlue, M., Ellis, G., Cohen, A., Swarenzki, P., 2012. Playa Lake sedimentation and organic matter accumulation in an Andean piggyback basin: the recent record from the Cuenca de Pozuelos, north-west Argentina. Sedimentology 59, 1237–1256.
McMeen, C.R., Benjamin, M.M., 1997. NOM removal by slow sand filtration through iron oxide-coated olivine. J. Am. Water Works Assoc. 89, 57–71.
Meas, Y., Ramirez, J.A., Villalon, M.A., Chapman, T.W., 2010. Industrial wastewaters treated by electrocoagulation. Electrochim. Acta 55, 8165–8171.
Mergen, M., Adams, B., Vero, G., Price, T., Parsons, S., Jefferson, B., Jarvis, P., 2009. Characterization of natural organic matter (NOM) removed by magnetic ion exchange resin (MIEX® Resin). Water Sci. Technol. Water Supply 9, 199–205.
Mergen, M.R.D., Jefferson, B., Parsons, S.A., Jarvis, P., 2008. Magnetic ion-exchange resin treatment: impact of water type and resin use. Water Res. 42, 1977–1988.
Messaud, F., Sanderson, R., Runyon, J., Otte, T., Pasch, H., Williams, S., 2009. An overview on field-flow fractionation techniques and their applications in the separation and characterization of polymers. Prog. Polym. Sci. 34, 351–368.
Metsämuuronen, S., Sillanpää, M., Bhatnagar, A., Mänttäri, M., 2014. Natural organic matter removal from drinking water by membrane technology. Sep. Purif. Rev. 43, 1–61.
Metz, D., Meyer, M., Dotson, A., Beerendonk, E., Dionysiou, D., 2011. The effect of UV/H_2O_2 treatment on disinfection by-product formation potential under simulated distribution system conditions. Water Res. 45, 3969–3980.
Meylan, S., Hammes, F., Traber, J., Salhi, E., von Gunten, U., Pronk, W., 2007. Permeability of low molecular weight organics through nanofiltration membranes. Water Res. 41, 3968–3976.
Meyn, T., Bahn, A., Leiknes, T., 2008. Significance of flocculation for NOM removal by coagulation–ceramic membrane microfiltration. Water Sci. Technol. Water Supply 8, 691–700.
Mierzwa, J.C., da Silva, M.C.C., Veras, L.R.V., Subtil, E.L., Rodrigues, R., Li, T., Landenberger, K.R., 2012. Enhancing spiral-wound ultrafiltration performance for direct drinking water treatment through operational procedures improvement: a feasible option for the Sao Paulo Metropolitan Region. Desalination 307, 68–75.
Mierzwa, J.C., Hespanhol, I., da Silva, M.C.C., Rodrigues, L.D.B., Giorgi, C.F., 2008. Direct drinking water treatment by spiral-wound ultrafiltration membranes. Desalination 230, 41–50.
Migliorini, F.L., Braga, N.A., Alves, S.A., Lanza, M.R.V., Baldan, M.R., Ferreira, N.G., 2011. Anodic oxidation of wastewater containing the Reactive Orange 16 Dye using heavily boron-doped diamond electrodes. J. Hazard. Mater. 192, 1683–1689.

Mikola, M., Rämö, J., Sarpola, A., Tanskanen, J., 2013. Removal of different NOM fractions from surface water with aluminium formate. Sep. Purif. Technol. 118, 842–846.

Miwa, D.W., Malpass, G.R.P., Machado, S.A.S., Motheo, A.J., 2006. Electrochemical degradation of carbaryl on oxide electrodes. Water Res. 40, 3281–3289.

Mohora, E., Roncevic, S., Dalmacija, B., Agbaba, J., Watson, M., Karlovic, E., Dalmacija, M., 2012. Removal of natural organic matter and arsenic from water by electrocoagulation/flotation continuous flow reactor. J. Hazard. Mater. 235–236, 257–264.

Mollah, M.Y.A., Morkovsky, P., Gomez, J.A.G., Kesmez, M., Parga, J., Cocke, D.L., 2004. Fundamentals, present and future perspectives of electrocoagulation. J. Hazard. Mater. B114, 199–210.

Mollah, M.Y.A., Schennach, R., Parga, J.R., Cocke, D.L., 2001. Electrocoagulation (EC) – science and applications. J. Hazard. Mater. B84, 29–41.

Molnar, J., Agbaba, J., Dalmacija, B., Klašnja, M., Dalmacija, M., Kragulj, M., 2012a. A comparative study of the effects of ozonation and TiO_2-catalyzed ozonation on the selected chlorine disinfection by-product precursor content and structure. Sci. Total Environ. 425, 169–175.

Molnar, J., Agbaba, J., Dalmacija, B., Rončević, S., Prica, M., Tubić, A., 2012b. Influence of pH and ozone dose on the content and structure of haloacetic acid precursors in groundwater. Environ. Sci. Pollut. Res. 19, 3079–3086.

Molnar, J., Agbaba, J., Dalmacija, B., Tubić, A., Krčmar, D., Maletić, S., Tomašević, D., 2013. The effects of matrices and ozone dose on changes in the characteristics of natural organic matter. Chem. Eng. J. 222, 435–443.

Moncayo-Lasso, A., Pulgarin, C., Benítez, N., 2008. Degradation of DBPs' precursors in river water before and after slow sand filtration by photo-Fenton process at pH 5 in solar CPC reactor. Water Res. 42, 4125–4132.

Moncayo-Lasso, A., Rincon, A.-G., Pulgarin, C., Benitez, N., 2012. Significant decrease of THMs generated during chlorination of river water by previous photo-Fenton treatment at near neutral pH. J. Photochem. Photobiol. A Chem. 229, 46–52.

Moncayo-Lasso, A., Sanabria, J., Pulgarin, C., Benítez, N., 2009. Simultaneous *E. coli* inactivation and NOM degradation in river water via photo-Fenton process at natural pH in solar CPC reactor: a new way for enhancing solar disinfection of natural water. Chemosphere 77, 296–300.

Monteith, D.T., Stoddard, J.L., Evans, C.D., de Wit, H.A., Forsius, M., Høgåsen, T., Wilander, A., Skjelkvåle, B.L., Jeffries, D.S., Vuorenmaa, J., Keller, B., Kopácek, J., Vesely, J., 2007. Dissolved organic carbon trends resulting from changes in atmospheric deposition chemistry. Nature 450, 537–540.

Montilla, F., Michaud, P.A., Morallon, E., Vazquez, J.L., Comninellis, C., 2002. Electrochemical oxidation of benzoic acid at boron-doped diamond electrodes. Electrochim. Acta 47 (2002), 3509–3513.

Moon, J., Kang, M., Lim, J., Kim, C., Park, H., 2009. Evaluation of a low-pressure membrane filtration for drinking water treatment: pretreatment by coagulation/sedimentation for the MF membrane. Desalination 247, 271–284.

Moon, J., Kim, S., Cho, J., 2006. Characterizations of natural organic matter as nano particle using flow field-flow fractionation. Colloids Surf. A Physicochem. Eng. Aspects 287, 232–236.

Moreno, H., Cocke, D., Gomes, J., Morkovsky, P., Parga, J., Peterson, E., Garcia, C., 2009. Electrochemical reactions for electrocoagulation using iron electrodes. Ind. Eng. Chem. Res. 48, 2275–2282.

Morgan, B., Lahav, O., 2007. The effect of pH on the kinetics of spontaneous Fe(II) oxidation by O_2 in aqueous solution – basic principles and a simple heuristic description. Chemosphere 68, 2080–2084.

Morran, J., Drikas, M., Cook, D., Bursill, D., 2004. Comparison of MIEX® treatment and coagulation on NOM character. Water Sci. Technol. Water Supply 4, 129–137.

Mosteo, R., Miguel, N., Martin-Muniesa, S., Ormad, M., Ovelleiro, J., 2009. Evaluation of trihalomethane formation potential in function of oxidation processes used during the drinking water production process. J. Hazard. Mater. 172, 661–666.

Motheo, A.J., Pinhedo, L., 2000. Electrochemical degradation of humic acid. Sci. Total Environ. 256, 67–76.

Mouedhen, G., Feki, M., Wery, M.D.P., Ayedi, H.F., 2008. Behavior of aluminum electrodes in electrocoagulation process. J. Hazard. Mater. 150, 124–135.

Moussas, P., Zouboulis, A., 2008. A study on the properties and coagulation behaviour of modified inorganic polymeric coagulant—Polyferric silicate sulphate (PFSiS). Sep. Purif. Technol. 63, 475–483.

Moussas, P., Zouboulis, A., 2009. A new inorganic–organic composite coagulant, consisting of polyferric sulphate (PFS) and polyacrylamide (PAA). Water Res. 43, 3511–3524.

Moussavi, S.P., Ehrampoush, M.H., Mahvi, A.H., Ahmadian, M., Rahimi, S., 2013. Adsorption of humic acid from aqueous solution on single-walled carbon nanotubes. Asian J. Chem. 25, 5319–5324.

Murphy, K., Stedmon, C., Waite, T., Ruiz, G., 2008. Distinguishing between terrestrial and autochthonous organic matter sources in marine environments using fluorescence spectroscopy. Mar. Chem. 108, 40–58.

Murray, C., Parsons, S., 2004. Removal of NOM from drinking water: Fenton's and photo-Fenton's processes. Chemosphere 54, 1017–1023.

Murray, C., Parsons, S., 2006. Preliminary laboratory investigation of disinfection by-product precursor removal using an advanced oxidation process. Water Environ. J. 20, 123–129.

Naddeo, V., Belgiorno, V., Napoli, R., 2007. Behaviour of natural organic matter during ultrasonic irradiation. Desalination 210, 175–182.

Nagata, Y., Hirai, K., Bandow, H., Maeda, Y., 1996. Decomposition of hydrobencoic and humic acids in water by ultrasonic irradiation. Environ. Sci. Technol. 30, 1133–1138.

Naghizadeh, A., Nasseri, S., Rashidi, A.M., Kalantary, R.R., Nabizadeh, R., Mahvi, A.H., 2013. Adsorption kinetics and thermodynamics of hydrophobic natural organic matter (NOM) removal from aqueous solution by multi-wall carbon nanotubes. Water Sci. Technol. Water Supply 13, 273–285.

Najafi, M., Rostamian, R., Rafati, A.A., 2011. Chemically modified silica gel with thiol group as an adsorbent for retention of some toxic soft metal ions from water and industrial effluent. Chem. Eng. J. 168, 426–432.

Navalon, S., Alvaro, M., Garcia, H., Escrig, D., Costa, V., 2009. Photocatalytic water disinfection of *Cryptosporidium parvum* and *Giardia lamblia* using a fibrous ceramic TiO_2 photocatalysis. Water Sci. Technol. 59, 639–645.

Nederlof, M.M., Kruithof, J.C., Taylor, J.S., van der Kooij, D., Schippers, J.C., 2000. Comparison of NF/RO membrane performance in integrated membrane systems. Desalination 131, 257–269.

Ng, M., Kho, E., Liu, S., Lim, M., Amal, R., 2014. Highly adsorptive and regenerative magnetic TiO_2 for natural organic matter (NOM) removal in water. Chem. Eng. J. 246, 196–203.

Nie, Y., Hu, C., Zhou, L., Qu, J., Wei, Q., Wang, D., 2010. Degradation characteristics of humic acid over iron oxides/Fe0 core-shell nanoparticles with UVA/H_2O_2. J. Hazard. Mater. 173, 474–479.

Nimmagadda, R., McRae, C., 2007. Characterization of the backbone structures of several fulvic acids using a novel selective chemical reduction method. Org. Geochem. 38, 1061–1072.

Nkambule, T., Krause, R., Haarhoff, J., Mamba, B., 2012. A three step approach for removing organic matter from South African water sources and treatment plants. Phys. Chem. Earth 50–52 (2012), 132–139.

Nkambule, T.I., Krause, R.W., Mamba, B.B., Haarhoff, J., 2009. Removal of natural organic matter from water using ion-exchange resins and cyclodextrin polyurethanes. Phys. Chem. Earth Parts A/B/C 34, 812–818.

Nyström, M., Ruohomäki, K., Kaipia, L., 1996. Humic acid as a fouling agent in filtration. Desalination 106, 79–87.
O'Loughlin, E., Chin, Y.-P., 2001. Effect of detector wavelength on the determination of the molecular weight of humic substances by high-pressure size exclusion chromatography. Water Res. 35, 333–338.
Ødegaard, H., Østerhus, S., Melin, E., Eikebrokk, B., 2010. NOM removal technologies – Norwegian experiences. Drink. Water Eng. Sci. 3, 1–9.
Odén, S., 1919. Die Huminsaure. Kolloidchem. Beih. 11, 75.
Oh, H., Yu, M., Takizawa, S., Ohgaki, S., 2006. Evaluation of PAC behavior and fouling formation in an integrated PAC–UF membrane for surface water treatment. Desalination 192, 54–62.
Omri, A., Benzina, M., Trabelsi, W., Ammar, N., 2013. Adsorptive removal of humic acid on activated carbon prepared from almond shell: approach for the treatment of industrial phosphoric acid solution. Desalin. Water Treat., 1–12.
Orecki, A., Tomaszewska, M., Karakulski, K., Morawski, A.W., 2004. Surface water treatment by the nanofiltration method. Desalination 162, 47–54.
Owen, D.M., Amy, G.L., Chowdhury, Z.K., Paode, R., McCoy, G., Viscosil, K., 1985. NOM characterization and treatability. J. Am. Water Works Assoc. 87, 46–63.
Oyonarte, C., Mingorance, M., Durante, P., Piñero, G., Barahona, E., 2007. Indicators of change in the organic matter in arid soils. Sci. Total Environ. 378, 133–137.
Pallier, V., Feuillade-Cathalifaud, G., Serpaud, B., 2011. Influence of organic matter on arsenic removal by continuous flow electrocoagulation treatment of weakly mineralized waters. Chemosphere 83, 21–28.
Panizza, M., Cerisola, G., 2005. Application of diamond electrodes to electrochemical processes. Electrochim. Acta 51, 191–199.
Parga, J., Cocke, D., Valverde, V., Gomes, J., Kesmez, M., Moreno, H., Weir, M., Mencer, D., 2005. Characterization of electrocoagulation for removal of chromium and arsenic. Chem. Eng. Technol. 28, 605–612.
Park, J.-H., 2009. Spectroscopic characterization of dissolved organic matter and its interactions with metals in surface waters using size exclusion chromatography. Chemosphere 77, 485–494.
Park, J.-H., Lee, J.-H., Kang, S.-Y., Kim, S.-Y., 2007. Hydroclimatic controls on dissolved organic matter (DOM) characteristics and implications for trace metal transport in Hwangryong River Watershed, Korea, during a summer monsoon period. Hydrol. Processes 21, 3025–3034.
Park, N., Kwon, B., Sun, M., Ahn, H., Kim, C., Kwoak, C., Lee, D., Chae, S., Hyung, H., Cho, J., 2005. Application of various membranes to remove NOM typically occurring in Korea with respect to DBP, AOC and transport parameters. Desalination 178, 161–169.
Park, S., Yoon, T., 2007. The effects of iron species and mineral particles on advanced oxidation processes for the removal of humic acids. Desalination 208, 181–191.
Park, S., Yoon, T., 2009. Effects of iron species and inert minerals on coagulation and direct filtration for humic acid removal. Desalination 239, 146–158.
Parsi, Z., Hartog, N., Górecki, T., Poerschmann, J., 2007. Analytical pyrolysis as a tool for the characterization of natural organic matter – a comparison of different approaches. J. Anal. Appl. Pyrolysis 79, 9–15.
Parsons, S., Byrne, A., 2004. Water treatment applications. In: Parsons, S. (Ed.), Advanced Oxidation Processes for Water and Wastewater Treatment. IWA Publishing, London, pp. 329–346.
Patel, U.D., Suresh, S., 2008. Electrochemical treatment of pentachlorophenol in water and pulp bleaching effluent. Sep. Purif. Technol. 61, 115–122.
Patermarakis, G., Fountoukidis, E., 1990. Disinfection of water by electrochemical treatment. Water Res. 24, 1491–1496.

Pautler, B., Woods, G., Dubnick, A., Simpson, A., Sharp, M., Fitzimons, S., Simpson, M., 2012. Molecular characterization of dissolved organic matter in glacial ice: coupling natural abundance ^{1}H NMR and fluorescence spectroscopy. Environ. Sci. Technol. 46, 3753–3761.

Pei, X., Hu, L., Liu, W., Hao, J., 2008. Synthesis of water-soluble carbon nanotubes via surface initiated redox polymerization and their tribological properties as water-based lubricant additive. Eur. Polym. J. 44, 2458–2464.

Peiris, R.H., Hallé, C., Budman, H., Moresoli, C., Peldszus, S., Huck, P.M., Legge, R.L., 2010. Identifying fouling events in a membrane-based drinking water treatment process using principal component analysis of fluorescence exitation-emission matrices. Water Res. 44, 185–194.

Peng, X., Luan, Z., Chen, F., Tian, B., Jia, Z., 2005. Adsorption of humic acid onto pillared bentonite. Desalination 174, 135–143.

Penru, Y., Simon, F., Guastalli, A., Esplugas, S., Llorens, J., Baig, Sylvie, 2013. Characterization of natural organic matter from Mediterranean coastal seawater. J. Water Supply Res. Technol. Aqua 62, 42–51.

Pera-Titus, M., Garcia-Molina, V., Baños, M., Giménes, J., Esplugas, S., 2004. Degradation of chlorophenols by means of advanced oxidation processes: a general review. Appl. Catal. B Environ. 47, 219–256.

Pereira, V., Weinberg, H., Linden, K., Singer, P., 2007. UV degradation kinetics and modeling of pharmaceutical compounds in laboratory grade and surface water via direct and indirect photolysis at 254 nm. Environ. Sci. Technol. 41, 1682–1688.

Peuravuori, J., Ingman, P., Pihlaja, K., 2003. Critical comments on accuracy of quantitative determination of natural humic matter by solid state ^{13}C NMR spectroscopy. Talanta 59, 177–189.

Peuravuori, J., Pihlaja, K., Välimäki, N., 1997. Isolation and characterization of natural organic matter from lake water: two different adsorption chromatographic methods. Environ. Int. 23, 453–464.

Pham, A., Rose, A., Feitz, A., Waite, T., 2006. Kinetics of Fe(III) precipitation in aqueous solutions at pH 6.0–9.5 and 25 °C. Geochim. Cosmochim. Acta 70, 640–650.

Philippe, K., Hans, C., MacAdam, J., Jefferson, B., Hart, J., Parsons, S., 2010. Photocatalytic oxidation of natural organic matter surrogates and the impact on trihalomethane formation potential. Chemosphere 81, 1509–1516.

Pianta, R., Boller, M., Janex, M., Chappaz, A., Birou, B., Ponce, R., Walther, J., 1998. Micro- and ultrafiltration of karstic spring water. Desalination 117, 61–71.

Picard, T., Cathalifaud-Feuillade, G., Mazet, M., Vandensteendam, C., 2000. Cathodic dissolution in the electrocoagulation process using aluminium electrodes. J. Environ. Monit. 2, 77–80.

Pifer, A.D., Fairey, J.L., 2012. Improving on $SUVA_{254}$ using fluorescence-PARAFAC analysis and asymmetric flow-field flow fractionation for assessing disinfection byproduct formation and control. Water Res. 46, 2927–2936.

Pinhedo, L., Pelegrini, R., Bertazzoli, R., Motheo, A.J., 2005. Photoelectrochemical degradation of humic acid on a $(TiO_2)_{0.7}(RuO_2)_{0.3}$ dimensionally stable anode. Appl. Catal. B Environ. 57, 75–81.

Pirkanniemi, K., Sillanpää, M., 2002. Heterogeneous water phase catalysis as an environmental application: a review. Chemosphere 48, 1047–1060.

Pirkanniemi, K., Metsärinne, S., Sillanpää, M., 2007. Degradation of EDTA and novel complexing agents in pulp and paper mill process and waste waters by Fenton's reagent. J. Hazard. Mater. 147, 556–561.

Plottu, A., Houssais, B., Democrate, C., Gatel, D., Cavard, J., 2003. Autopsies of membranes fouled on Mery-sur-Oise pilot units: Many lessons for the behaviour of the water treatment plant. Desalination 157, 367.

Polcaro, A.M., Vacca, A., Mascia, M., Palmas, S., Pompei, R., Laconi, S., 2007. Characterization of a stirred tank electrochemical cell for water disinfection processes. Electrochim. Acta 52, 2595–2602.

Porcal, P., Koprivnjak, J.-F., Molot, L., Dillon, P., 2009. Humic substances—part 7: the biogeochemistry of dissolved organic carbon and its interactions with climate change. Environ. Sci. Pollut. Res. 16, 714–726.

Posner, A., 1963. Importance of electrolyte in the determination of molecular weights by sephadex gel with specific reference to humic acids. Nature 198, 1161–1163.

Pryor, M.J., Jacobs, E.P., Botes, J.P., Pillay, V.L., 1998. A low pressure ultrafiltration membrane system for potable water supply to developing communities in South Africa. Desalination 119, 103–111.

Pürschel, M., Ender, V., 2008. Sorption of natural organic matter by adsorber and ion exchange resins – investigations with starch and phenylalanine as model substances. In: 15th International Conference on the Properties of Water and Steam (15th ICPWS), Radisson SAS Hotel Berlin.

Qi, S., Schideman, L.C., 2008. An overall isotherm for activated carbon adsorption of dissolved natural organic matter in water. Water Res. 42, 3353–3360.

Qiao, X., Zhang, Z., Wang, N., Wee, V., Low, M., Loh, C., Ng, T., 2008. Coagulation pretreatment for a large-scale ultrafiltration process treating water from the Taihu River. Desalination 230, 305–313.

Qin, J.-J., Oo, M., Kekre, K., Knops, F., Miller, P., 2006. Impact of coagulation pH on enhanced removal of natural organic matter in treatment of reservoir water. Sep. Purif. Technol. 49, 295–298.

Qu, X., Gao, W.J., Han, M.N., Chen, A., Liao, B.Q., 2012. Integrated thermophilic submerged aerobic membrane bioreactor and electrochemical oxidation for pulp and paper effluent treatment – towards system closure. Bioresour. Technol. 116, 1–8.

Rachwal, T., Judd, S., 2006. A synopsis of membrane technologies in UK municipal potable water treatment: history, status and prospects. Water Environ. J. 20, 110–113.

Radjenovic, J., Bagastyo, A., Rozendal, R.A., Mu, Y., Keller, J., Rabaey, K., 2011. Electrochemical oxidation of trace organic contaminants in reverse osmosis concentrate using RuO_2/IrO_2-coated titanium anodes. Water Res. 45, 1579–1586.

Rahman, M., Jasim, S., Yanful, E., Ndiongue, S., Borikar, D., 2010. Advanced oxidation treatment of drinking water: part II. Turbidity, particles and organics removal from Lake Huron water. Ozone Sci. Eng. 32, 295–304.

Rajkumar, D., Kim, J.G., 2006. Oxidation of various reactive dyes with in situ electro-generated active chlorine for textile dyeing industry wastewater treatment. J. Hazard. Mater. B136, 203–212.

Rajkumar, D., Song, B.J., Kim, J.G., 2007. Electrochemical degradation of Reactive Blue 19 in chloride medium for the treatment of textile dyeing wastewater with identification of intermediate compounds. Dyes Pigm. 72, 1–7.

Rämö, J., Sillanpää, M., 2001. Degradation of EDTA by hydrogen peroxide in alkaline conditions. J. Cleaner Prod. 9, 191–195.

Rastogi, A., Al-Abed, S., Dionysiou, D., 2009. Effect of inorganic, synthetic and naturally occurring chelating agents on Fe(II) mediated advanced oxidation of chlorophenols. Water Res. 43, 684–694.

Ratnaweera, H., Gjessing, E., Oug, E., 1999. Influence of physical-chemical characteristics of natural organic matter (NOM) on coagulation properties: an analysis of eight Norwegian water sources. Water Sci. Technol. 40, 89–95.

Ravindran, V., Tsai, H., Williams, M.D., Pirbazari, M., 2009. Hybrid membrane bioreactor technology for small water treatment utilities: process evaluation and primordial considerations. J. Membr. Sci. 344, 39–54.

Reemtsma, T., 2009. Determination of molecular formulas of natural organic matter molecules by (ultra-) high-resolution mass spectrometry – status and needs. J. Chromatogr. A 1216, 3687–3701.
Reemtsma, T., These, A., 2005. Comparative investigation of low-molecular-weight fulvic acids of different origin by SEC-Q-TOF-MS: new insights into structure and formation. Environ. Sci. Technol. 39, 3507–3512.
Reemtsma, T., These, A., Springer, A., Linscheid, M., 2006. Fulvic acids as transition state of organic matter: indications from high resolution mass spectrometry. Environ. Sci. Technol. 40, 5839–5845.
Reemtsma, T., These, A., Springer, A., Linscheid, M., 2008. Differences in the molecular composition of fulvic acid size fractions detected by size-exclusion chromatography-on line Fourier transform ion cyclotron resonance (FTICR) mass spectrometry. Water Res. 42, 63–72.
Remize, P.J., Guigui, C., Cabassud, C., 2010. Evaluation of backwash efficiency, definition of remaining fouling and characterization of its contribution in irreversible fouling: case of drinking water production by air-assisted ultra-filtration. J. Membr. Sci. 355, 104–111.
Ribau Teixeira, M., Rosa, M.J., 2006. The impact of the water background inorganic matrix on the natural organic matter removal by nanofiltration. J. Membr. Sci. 279, 513–520.
Ribau Teixeira, M., Sousa, V.S., 2013. Fouling of nanofiltration membrane: effects of NOM molecular weight and microcystins. Desalination 315, 149–155.
Ribau Teixeira, M., Rosa, S.M., Sousa, V., 2011. Natural organic matter and disinfection by-products formation potential in water treatment. Water Res. Manage. 25, 3005–3015.
Richardson, S., Fasano, F., Ellington, J., Crumley, F., Buettner, K., Evans, J., Blount, B., Silva, L., Waite, T., Luther, G., McKague, A., Miltner, R., Wagner, E., Plewa, M., 2008. Occurrence and mammalian cell toxicity of iodinated disinfection by-products in drinking water. Environ. Sci. Technol. 42, 8330–8338.
Richardson, S., Plewa, M., Wagner, E., Schoeny, R., DeMarini, D., 2007. Occurrence, genotoxicity, and carcinogenicity of regulated and emerging disinfection by-products in drinking water: a review and roadmap for research. Mutat. Res. 636, 178–242.
Ricordel, C., Darchen, A., Hadjiev, D., 2010. Electrocoagulation–electroflotation as a surface water treatment for industrial uses. Sep. Purif. Technol. 74, 342–347.
Rigobello, E., Dantas, A., Di Bernardo, L., Vieira, E., 2011. Influence of the apparent molecular size of aquatic humic substances on colour removal by coagulation and filtration. Environ. Technol. 32, 1767–1777.
Rincón, A.-G., Pulgarin, C., 2007. Fe^{3+} and TiO_2 solar-light-assisted inactivation of *E.coli* at field scale. Implications in solar disinfection at low temperature of large quantities of water. Catal. Today 122, 128–136.
Ritchie, J., Perdue, E., 2008. Analytical constraints on acidic functional groups in humic acids. Org. Geochem. 39, 783–799.
Ritson, J.P., Grahamb, N.J.D., Templetonb, M.R., Clark, J.M., Gough, R., Freemand, C., 2014. The impact of climate change on the treatability of dissolved organic matter (DOM) in upland water supplies: a UK perspective. Sci. Total Environ. 473–474, 714–730.
Rizzo, L., Di Gennaro, A., Gallo, M., Belgiorno, V., 2008. Coagulation/chlorination of surface water: a comparison between chitosan and metal salts. Sep. Purif. Technol. 62, 79–85.
Roccaro, P., Vagliasindi, F., 2009. Differential vs. absolute UV absorbance approaches in studying NOM reactivity in DBPs formation: comparison and applicability. Water Res. 43, 744–750.
Rodrigo, M.A., Cañizares, P., Sanchez-Carretero, A., Saez, C., 2010. Use of conductive-diamond electrochemical oxidation for wastewater treatment. Catal. Today 151, 173–177.

Rodriguez, N., Ramirez, S., Blanco Varela, M., Guillem, M., Puig, J., 2010. Re-use of drinking water treatment plant (DWTP) sludge: characterization and technological behaviour of cement mortars with atomized sludge additions. Cem. Concr. Res. 40, 778–786.

Rodriques, F., Núñez, L., 2009. Characterization of aquatic humic substances. Water Environ. J. 23, 1–8.

Rojas, J.C., Moreno, B., Garralon, G., Plaza, F., Perez, J., Gomez, M.A., 2010. Influence of velocity gradient in a hydraulic flocculator on NOM removal by aerated spiral-wound ultrafiltration membranes (ASWUF). J. Hazard. Mater. 178, 535–540.

Rosario-Ortiz, F., Snyder, S., Suffet, I., 2007. Characterization of dissolved organic matter in drinking water sources impacted by multiple tributaries. Water Res. 41, 4115–4128.

Rosenfeldt, E., Linden, K., Canonica, S., von Gunten, U., 2006. Comparison of the efficiency of OH radical formation during ozonation and advanced oxidation processes O_3/H_2O_2 and UV/H_2O_2. Water Res. 40, 3695–3704.

Rosenfeldt, E.J., Linden, K.G., 2007. The ROH, UV concept to characterize and the model UV/H_2O_2 process in natural waters. Environ. Sci. Technol. 41, 2548–2553.

Roshani, B., Leitner, N., 2011. Effect of persulfate on the oxidation of benzotriazole and humic acid by e-beam irradiation. J. Hazard. Mater. 190, 403–408.

de la Rubia, Á., Rodríguez, M., Prats, D., 2006. pH, ionic strength and flow velocity effects on the NOM filtration with TiO_2/ZrO_2 membranes. Sep. Purif. Technol. 52, 325–331.

de la Rubia, Á., Rodríguez, M., León, V.M., Prats, D., 2008. Removal of natural organic matter and THM formation potential by ultra- and nanofiltration of surface water. Water Res. 42, 714–722.

Sagbo, O., Sun, Y., Hao, A., Gu, P., 2008. Effect of PAC addition on MBR process for drinking water treatment. Sep. Purif. Technol. 58, 320–327.

Sakalis, A., Fytianos, K., Nickel, U., Voulgaropoulos, A., 2006. A comparative study of platinized titanium and niobe/synthetic diamond as anodes in the electrochemical treatment of textile wastewater. Chem. Eng. J. 119, 127–133.

Sanchez, N., Skeriotis, A., Miller, C., 2013. Assessment of dissolved organic matter fluorescence PARAFAC components before and after coagulation–filtration in a full scale water treatment plant. Water Res. 47, 1679–1690.

Sanchez, N.P., Skeriotis, A.T., Miller, C., 2014. A PARAFAC-based long-term assessment of DOM in a multi-coagulant drinking water treatment scheme. Environ. Sci. Technol. 48, 1582–1591.

Sani, B., Basile, E., Rossi, L., Lubello, C., 2008. Effects of pre-treatment with magnetic ion exchange resins on coagulation/flocculation process. Water Sci. Technol. 57, 57–64.

Sanly, Lim M., Chiang, K., Amal, R., Fabris, R., Chow, C., Drikas, M., 2007. A study on the removal of humic acid using advanced oxidation process. Sep. Sci. Technol. 42, 1391–1404.

Santosa, S.J., Kunarti, E.S., Karmanto, 2008. Synthesis and utilization of Mg/Al hydrotalcite for removing dissolved humic acid. Appl. Surf. Sci. 254, 7612–7617.

Sarathy, S., Mohseni, M., 2007. The impact of UV/H_2O_2 advanced oxidation on molecular size distribution of chromophoric natural organic matter. Environ. Sci. Technol. 41, 8315–8320.

Sarathy, S., Mohseni, M., 2010. Effects of UV/H_2O advanced oxidation on chemical characteristics and chlorine reactivity of surface water natural organic matter. Water Res. 44, 4087–4096.

Sarathy, S., Stefan, M., Royce, A., Mohseni, M., 2011. Pilot-scale UV/H_2O_2 advanced oxidation process for surface water treatment and downstream biological treatment: effects on natural organic matter characteristics and DBP formation potential. Environ. Technol. 32, 1709–1718.

Sarathy, S.R., Mohseni, M., 2009. The fate of natural organic matter during UV/H_2O_2 advanced oxidation of drinking water. Can. J. Civ. Eng. 36, 160–169.

Särkkä, H., Kuhmonen, K., Vepsäläinen, M., Pulliainen, M., Selin, J., Rantala, P., Kukkamäki, E., Sillanpää, M., 2009. Electrochemical oxidation of sulphides in paper mill wastewater by using mixed oxide anodes. Environ. Technol. 30, 885–892.

Sarpola, A., Hellman, H., Hietapelto, V., Jalonen, J., Jokela, J., Rämö, J., Saukkoriipi, J., 2007. Hydrolysis products of water treatment chemical aluminium sulfate octadecahydrate by electrospray ionization mass spectrometry. Polyhedron 26, 2851–2858.

Schäfer, A.I., Fane, A.G., Waite, T.D., 1998. Nanofiltration of natural organic matter: removal, fouling and the influence of multivalent ions. Desalination 118, 109–122.

Schäfer, A.I., Fane, A.G., Waite, T.D., 2000. Fouling effects on rejection in the membrane filtration of natural waters. Desalination 131, 215–224.

Schäfer, A.I., Fane, A.G., Waite, T.D., 2001. Cost factors and chemical pretreatment effects in the membrane filtration of waters containing natural organic matter. Water Res. 35, 1509–1517.

Schäfer, A.I., Fane, A.G., Waite, T.D. (Eds.), 2005. Nanofiltration – Principles and Applications. Elsevier Advanced Technology, Kidlington, UK.

Schrotter, J., Bozkaya-Schrotter, B., 2010. Current and emerging membrane processes for water treatment. In: Peinemann, K., Pereira Nunes, S. (Eds.), Membranes for Water Treatment. Wiley-VCH Verlag, Weinheim, pp. 53–91.

Schulten, H., 1993. Analytical pyrolysis of humic substances and soils: geochemical, agricultural and ecological consequences. J. Anal. Appl. Pyrolysis 25, 97–122.

Scialdone, O., Galia, A., Guarisco, C., Randazzo, S., Filardo, G., 2008. Electrochemical incineration of oxalic acid at boron doped diamond anodes: the role of operative parameters. Electrochim. Acta 53, 2095–2108.

Scialdone, O., Randazzo, S., Galia, A., Filardo, G., 2009a. Electrochemical oxidation of organics at metal oxide electrodes: the incineration of oxalic acid at IrO_2–TaO_2O_5. Electrochim. Acta 54, 1210–1217.

Scialdone, O., Randazzo, S., Galia, A., Silvestri, G., 2009b. Electrochemical oxidation of organics in water: role of operative parameters in the absence and in the presence of NaCl. Water Res. 43, 2260–2272.

Seidel, A., Elimelech, M., 2002. Coupling between chemical and physical interactions in natural organic matter (NOM) fouling of nanofiltration membranes: implications for fouling control. J. Membr. Sci. 203, 245–255.

Selcuk, H., Bekbolet, M., 2008. Photocatalytic and photoelectrocatalytic humic acid removal and selectivity of TiO_2 coated photoanode. Chemosphere 73, 854–858.

Selcuk, H., Sarikaya, H., Bekbolet, M., Anderson, M., 2006. Bromate formation on the non-porous TiO_2 photoanode in the photoelectrocatalytic system. Chemosphere 62, 715–721.

Selcuk, H., Sene, J.J., Anderson, M.A., 2003. Photoelectrocatalytic humic acid degradation kinetics and effect of pH, applied potential and inorganic ions. J. Chem. Technol. Biotechnol. 78, 979–984.

Seredynska-Sobecka, B., Baker, A., Lead, A., 2007. Characterization of colloidal and particulate organic carbon in freshwaters by thermal fluorescence quenching. Water Res. 41, 3069–3076.

Shaker, A.M., Komy, Z.R., Heggy, S.E.M., El-Sayed, M.E.A., 2012. Kinetic study for adsorption humic acid on soil minerals. J. Phys. Chem. A 116, 10889–10896.

Shao, J., Hou, J., Song, H., 2011. Comparison of humic acid rejection and flux decline during filtration with negatively charged and uncharged ultrafiltration membranes. Water Res. 45, 473–482.

Sharp, E., Jarvis, P., Parsons, S., Jefferson, B., 2006a. Impact of fractional character on the coagulation of NOM. Colloids Surf. A Physicochem. Eng. Aspects 286, 104–111.

Sharp, E., Jarvis, P., Parsons, S., Jefferson, B., 2006b. The impact of zeta potential on the physical properties of ferric–NOM flocs. Environ. Sci. Technol. 40, 3934–3940.

Sharp, E., Parsons, S., Jefferson, B., 2006c. The impact of seasonal variations in DOC arising from a moorland peat catchment on coagulation with iron and aluminium salts. Environ. Pollut. 140, 436–443.

Sharp, E., Parsons, S., Jefferson, B., 2006d. Seasonal variations in natural organic matter and its impact on coagulation in water treatment. Sci. Total Environ. 363, 183–194.

Sharpless, C.M., Aeschbacher, M., Page, S.E., Wenk, J., Sander, M., McNeill, K., 2014. Photooxidation-induced changes in optical, electrochemical and photochemical properties of humic substances. Environ. Sci. Technol. 48, 2688–2696.

Shestakova, M., Sillanpää, M., 2013. Removal of dichloromethane from ground and wastewater: a review. Chemosphere 93, 1258–1267.

Shi, B., Taylor, J., 2007. Potential impact of enhanced coagulation on corrosion by-product release in a distribution system. Desalination 208, 260–268.

Shi, B., Wei, Q., Wang, D., Zhu, Z., Tang, H., 2007. Coagulation of humic acid: the performance of preformed and non-preformed Al species. Colloids Surf. A Physicochem. Eng. Aspects 296, 141–148.

Shi, D., Xie, S., Wen, D., Xi, D., 2009. Removal of bromate and natural organic matter by using biologically activated carbon. Int. J. Environ. Pollut. 38, 180–192.

Shi, W., Benjamin, M.M., 2008. Membrane interactions with NOM and an adsorbent in a vibratory shear enhanced filtration process (VSEP) system. J. Membr. Sci. 312, 23–33.

Shifu, C., Wei, Z., Wei, L., Huaye, Z., Xiaoling, Y., Yinghao, C., 2009. Preparation, characterization and activity evaluation of p-n junction photocatalyst p-$CaFe_2O_4$/n-Ag_3VO_4 under visible light irradiation. J. Hazard. Mater. 172, 1415–1423.

Shin, J., Spinette, R., O'Melia, C., 2008. Stoichiometry of coagulation revisited. Environ. Sci. Technol. 42, 2582–2589.

Shorrock, K., Drage, B., 2006. A pilot plant evaluation of the Magnetic Ion EXchange® process for the removal of dissolved organic carbon at Draycote water treatment works. Water Environ. J. 20, 65–70.

Shutova, Y., Baker, A., Bridgeman, J., Henderson, R., 2014. Spectroscopic characterization of dissolved organic matter changes in drinking water treatment: from PARAFAC analysis to online monitoring wavelengths. Water Res. 54, 159–169.

Siddiqui, M., Amy, G., Ryan, J., Odem, W., 2000. Membranes for the control of natural organic matter from surface waters. Water Res. 34, 3355–3370.

Siéliéchi, J.-M., Lartiges, B., Kayem, G., Hupont, S., Frochot, C., Thieme, J., Ghanbaja, J., d'Espinose de la Caillerie, J., Barrés, O., Kamga, R., Levitz, P., Michot, L., 2008. Changes in humic acid conformation during coagulation with ferric chloride: implications for drinking water treatment. Water Res. 42, 2111–2123.

Simate, G., Iyuke, S., Ndlovu, S., Heydenrych, M., 2012a. The heterogeneous coagulation and flocculation of brewery wastewater using carbon nanotubes. Water Res. 46, 1185–1197.

Simate, G., Iyuke, S., Ndlovu, S., Heydenrych, M., Walubita, L., 2012b. Human health effects of residual carbon nanotubes and traditional water treatment chemicals in drinking water. Environ. Int. 39, 38–49.

Simon, F., Penru, Y., Guastalli, A., Esplugas, S., Llorens, J., Baig, S., 2013. NOM characterization by LC-OCD in a SWRO desalination line. Desalin. Water Treat. 51, 1776–1780.

Simpson, A., Simpson, M., Kingery, W., Lefebvre, B., Moser, A., Williams, A., Kvasha, M., Kvelleher, B., 2006. The application of ^{1}H high-resolution magic-angle spinning NMR for the study of clay-organic associations in natural and synthetic complexes. Langmuir 22, 4498–4503.

Sindelar, H., Brown, M., Boyer, T., 2014. Evaluating UV/H_2O_2, UV/percarbonate, and UV/perborate for natural organic matter reduction from alternative water sources. Chemosphere 105, 112–118.

Singer, P., 2006. DBPs in drinking water: additional scientific and policy considerations for public health protection. J. Am. Water Works Assoc. 98, 73–80.

Singer, P., Bilyk, K., 2002. Enhanced coagulation using a magnetic ion exchange resin. Water Res. 36, 4009–4022.
Slavik, I., Müller, S., Mokosch, R., Azongbilla, J., Uhl, W., 2012. Impact of shear stress and pH changes on floc size and removal of dissolved organic matter (DOM). Water Res. 46, 6543–6553.
Smith, B., Yang, J., Bitter, J.L., Ball, W.P., Fairbrother, D.H., 2012. Influence of surface oxygen on the interactions of carbon nanotubes with natural organic matter. Environ. Sci. Technol. 46, 12839–12847.
Smith, E., Kamal, Y., 2009. Optimizing treatment for reduction of disinfection by-product (DBP) formation. Water Sci. Technol. Water Supply 9, 191–198.
Smith, E.H., 1994. Bench-scale tests and modeling of adsorption of natural organic matter by activated carbon. Water Res. 28, 1693–1702.
Snoeyink, V.L., Jenkins, D., 1980. Water Chemistry. John Wiley & Sons, New York.
Soh, Y., Roddick, F., van Leeuwen, J., 2008a. The future of water in Australia: the potential effects of climate change and ozone depletion on Australian water quality, quantity and treatability. Environmentalist 28, 158–165.
Soh, Y., Roddick, F., van Leeuwen, J., 2008b. The impact of alum coagulation on the character, biodegradability and disinfection by-product formation potential of reservoir natural organic matter (NOM) fractions. Water Sci. Technol. 58, 1173–1179.
Song, B., Niu, S., Zhang, Z., Yang, H., Li, L., Wan, S., 2012. Light and heavy fractions of soil organic matter in response to climate warming and increased precipitation in a temperate steppe. PLoS One 7 (3), e33217.
Song, H., Orr, O., Hong, Y., Karanfil, T., 2009. Isolation and fractionation of natural organic matter: evaluation of reverse osmosis performance and impact of fractionation parameters. Environ. Monit. Assess. 153, 307–321.
Song, W., Ravindran, V., Pirbazari, M., 2008. Process optimization using a kinetic model for the ultraviolet radiation-hydrogen peroxide decomposition of natural and synthetic organic compounds in groundwater. Chem. Eng. Sci. 63, 3249–3270.
Song, Y., Dong, B., Gao, N., Xia, S., 2010. Huangpu river water treatment by microfiltration with ozone pretreatment. Desalination 250, 71–75.
Specht, C., Frimmel, F., 2000. Specific interactions of organic substances in size-exclusion chromatography. Environ. Sci. Technol. 34, 2361–2366.
Spencer, R., Bolton, L., Baker, A., 2007. Freeze/thaw and pH effects on freshwater dissolved organic matter fluorescence and absorbance properties from a number of UK locations. Water Res. 41, 2941–2950.
Stedmon, C., Markager, S., 2005. Resolving the variability in dissolved organic matter fluorescence in a temperate estuary and its catchment using PARAFAC analysis. Limn. Oceanogr. 50, 686–697.
Stedmon, C., Markagera, S., Bro, R., 2003. Tracing dissolved organic matter in aquatic environments using a new approach to fluorescence spectroscopy. Mar. Chem. 82, 239–254.
Stenson, A., 2008. Reversed-phase chromatography fractionation tailored to mass spectral characterization of humic substances. Environ. Sci. Technol. 42, 2060–2065.
Stepniak, L., Stańczyk-Mazanek, E., Kusiak, M., 2012. Ultrasounds and ozone in removal of humic substances from water. Inż. Ochr. Środowiska 15, 143–154.
Sulzberger, B., Durisch-Kaiser, E., 2009. Chemical characterization of dissolved organic matter (DOM): a prerequisite for understanding UV-induced changes of DOM absorption properties and bioavailability. Aquat. Sci. 71, 104–126.
Suopajärvi, T., Liimatainen, H., Hormi, O., Niinimäki, J., 2013. Coagulation–flocculation treatment of municipal wastewater based on anionized nanocelluloses. Chem. Eng. J. 231, 59–67.
Sutherland, K., 2012. Drinking and pure water: filtration improvements progress global clean water provision. Filtr. Sep. 49, 12–16.

Suty, H., De Traversay, C., Cost, M., 2004. Applications of advanced oxidation processes: present and future. Water Sci. Technol. 49 (4), 227–233.
Swaim, P., Royce, A., Smith, T., Maloney, T., Ehlen, D., Carter, B., 2008. Effectiveness of UV advanced oxidation for destruction of mirco-pollutants. Ozone Sci. Eng. 30, 34–42.
Swietlik, J., Dabrowska, A., Raczyk-Stanislawiak, U., Nawrocki, J., 2004. Reactivity of natural organic matter fractions with chlorine dioxide and ozone. Water Res. 38, 547–558.
Syafei, A., Lin, C.-F., Wu, C.-H., 2008. Removal of natural organic matter by ultrafiltration with TiO_2-coated membrane under UV irradiation. J. Colloid Interface Sci. 323, 112–119.
Szlachta, M., Adamski, W., 2009. Effects of natural organic matter removal by integrated processes: alum coagulation and PAC-adsorption. Water Sci. Technol. 59, 1951–1957.
Tang, C.Y., Chong, T.H., Fane, A.G., 2011. Colloidal interactions and fouling of NF and RO membranes: a review. Adv. Colloid Interface Sci. 164, 126–143.
Tang, W.-W., Zeng, G.-M., Gong, J.-L., Liang, J., Xu, P., Zhang, C., Huang, B.-B., 2014. Impact of humic/fulvic acid on the removal of heavy metals from aqueous solutions using nanomaterials: a review. Sci. Total Environ. 468–469, 1014–1027.
Tang, Y., Liang, S., Yu, S., Gao, N., Zhang, J., Guo, H., Wang, Y., 2012. Enhanced adsorption of humic acid on amine functionalized magnetic mesoporous composite microspheres. Colloids Surf. A Physicochem. Eng. Aspects 406, 61–67.
Tanneru, C.T., Chellam, S., 2012. Mechanisms of virus control during iron electrocoagulation – microfiltration of surface water. Water Res. 46, 2111–2120.
Tavares, M.G., da Silva, L.V.A., Sales Solano, A.M., Tonholo, J., Martinez-Huitle, C.A., Zanta, C.L.P.S., 2012. Electrochemical oxidation of Methyl Red using $Ti/Ru_{0.3}Ti_{0.7}O_2$ and Ti/Pt anodes. Chem. Eng. J. 204–206, 141–150.
Teixeira, M.R., Sousa, V.S., 2013. Fouling of nanofiltration membrane: effects of NOM molecular weight and microcystins. Desalination 315, 149–155.
Templier, J., Derenne, S., Croué, J.-P., Largeau, C., 2005. Comparative study of two fractions of riverine dissolved organic matter using various analytical pyrolytic methods ans a ^{13}C CP/MAS NMR approach. Org. Geochem. 36, 1418–1442.
Tercero Espinoza, L., Frimmel, F., 2009. A simple simulation of the degradation of natural organic matter in homogenous and heterogenous advanced oxidation processes. Water Res. 43, 3902–3909.
Tercero Espinoza, L., Rembor, M., Matesanz, C., Heidt, A., Frimmel, F., 2009a. Formation of bromoform in irradiated titanium dioxide suspensions with varying photocatalyst, dissolved organic carbon and bromide concentrations. Water Res. 43, 4143–4148.
Tercero Espinoza, L., ter Haseborg, E., Weber, M., Frimmel, F., 2009b. Investigation of the photocatalytic degradation of brown water natural organic matter by size exclusion chromatography. Appl. Catal. B Environ. 87, 56–62.
Tercero Espinoza, L., ter Haseborg, E., Weber, M., Karle, E., Peschke, R., Frimmel, F., 2011. Effect of selected metal ions on the photocatalytic degradation of bog lake water natural organic matter. Water Res. 45, 1039–1048.
Tercero Espinoza, L., Frimmel, F., 2008. Formation of brominated products in irradiated tianium dioxide suspensions containing bromide and dissolved organic carbon. Water Res. 42, 1778–1784.
The Council of the European Union, 1998. Council directive 98/83/EC. Off. J. Eur. Communities L330, 32–54.
Theis, T.L., Singer, P.C., 1974. Complexation of iron(II) by organic matter and its effect on iron(II) oxygenation. Environ. Sci. Technol. 8, 569–573.
Thomson, J., Roddick, F., Drikas, M., 2002. Natural organic matter removal by enhanced photo-oxidation using low pressure mercury vapour lamps. Water Sci. Technol. Water Supply 2, 435–443.

Thomson, J., Roddick, F.A., Drikas, M., 2004. Vacuum ultraviolet irradiation for natural organic matter removal. J. Water Supply Res. Technol. Aqua 53, 193–206.
Thorn, K., Cox, L., 2009. N-15 NMR spectra of naturally abundant nitrogen in soil and aquatic natural organic matter samples of the International Humic Substances Society. Org. Geochem. 40, 484–499.
Thorsen, T., 1998. Flux and recovery influences on treatment efficiency in ultrafiltration of coloured drinking water. Desalination 117, 131–138.
Thorsen, T., 2004. Concentration polarization by natural organic matter (NOM) in NF and UF. J. Membr. Sci. 233, 79–91.
Thorsen, T., Fløgstad, H., 2006. Nanofiltration in drinking water treatment, literature review. Research report for Techneau Project, December 11, 2006.
Thurman, E., 1985. Organic Geochemistry of Natural Waters. Martinus Nijhoff/Dr W. Junk Publ, Dordrecht, the Netherlands.
Thurman, E., Malcolm, R., 1981. Preparative isolation of aquatic humic substances. Environ. Sci. Technol. 15, 463–466.
Tian, C., Liu, R., Liu, H., Qu, J., 2013. Disinfection by-products formation and precursors transformation during chlorination and chloramination of highly-polluted source water: significance of ammonia. Water Res. 47, 5901–5910.
Tian, J., Chen, Z., Nan, J., Liang, H., Li, G., 2010. Integrative membrane coagulation adsorption bioreactor (MCABR) for enhanced organic matter removal in drinking water treatment. J. Membr. Sci. 352, 205–212.
Tian, J., Liang, H., Li, X., You, S., Tian, S., Li, G., 2008a. Membrane coagulation bioreactor (MCBR) for drinking water treatment. Water Res. 42, 3910–3920.
Tian, J., Liang, H., Nan, J., Yang, Y., You, S., Li, G., 2009. Submerged membrane bioreactor (sMBR) for the treatment of contaminated raw water. Chem. Eng. J. 148, 296–305.
Tian, J., Liang, H., Yang, Y., Tian, S., Li, G., 2008b. Membrane adsorption bioreactor (MABR) for treating slightly polluted surface water supplies: as compared to membrane bioreactor (MBR). J. Membr. Sci. 325, 262–270.
Tomaszewska, M., Mozia, S., 2002. Removal of organic matter from water by PAC/UF system. Water Res. 36, 4137–4143.
Tony, M., Zhao, Y., Fu, J., Tayeb, A., 2008. Conditioning of aluminium-based water treatment sludge with Fenton's reagent: effectiveness and optimizing study to improve dewaterability. Chemosphere 72, 673–677.
Toor, R., Mohseni, M., 2007. UV-H_2O_2 based AOP and its integration with biological activated carbon treatment for DBP reduction in drinking water. Chemosphere 66, 2087–2095.
Tran, N., Drogui, P., 2013. Electrochemical removal of microcystin-LR from aqueous solution in the presence of natural organic pollutants. J. Environ. Manage. 114, 253–260.
Tran, T., Gray, S., Bolto, B., Farmer, T., Collings, T., 2007. Ultrasound enhancement of microfiltration performance for natural organic matter removal. Org. Geochem. 38, 1091–1096.
Tran, T., Gray, S., Naughton, R., Bolto, B., 2006. Polysilicato-iron for improved NOM removal and membrane performance. J. Membr. Sci. 280, 560–571.
Trang, V., Phuong, L., Dan, N., Thanh, B., Visvanathan, C., 2012. Assessment on the trihalomethanes formation potential of Tan Hiep water treatment plant. J. Water Sustain. 2, 43–53.
Treguer, R., Tatin, R., Couvert, A., Wolbert, D., Tazi-Pain, A., 2010. Ozonation effect on natural organic matter adsorption and biodegradation – application to a membrane bioreactor containing activated carbon for drinking water production. Water Res. 44, 781–788.
Trinh, T., Kang, L., 2011. Response surface methodological approach to optimize the coagulation–flocculation process in drinking water treatment. Chem. Eng. Res. Des. 89, 1126–1135.

Trompette, J.L., Vergnes, H., 2009. On the crucial influence of some supporting electrolytes during electrocoagulation in the presence of aluminum electrodes. J. Hazard. Mater. 163, 1282–1288.

Tsujimoto, W., Kimura, H., Izu, T., Irie, T., 1998. Membrane filtration and pre-treatment by GAC. Desalination 119, 323–326.

Tubić, A., Agbaba, J., Dalmacija, B., Perović, S., Klašnja, M., Rončević, S., Ivančev-Tumbas, I., 2011. Removal of natural organic matter from groundwater using advanced oxidation processes at a pilot scale drinking water treatment plant in the Central Banat Region (Serbia). Ozone Sci. Eng. 33, 267–278.

Tuhkanen, T.A., 2004. UV/H_2O_2 processes. In: Parsons, S. (Ed.), Advanced Oxidation Processes for Water and Wastewater Treatment. IWA Publishing, London, pp. 86–110.

Tzoupanos, N., Zouboulis, A., 2011. Preparation, characterization and application of novel composite coagulants for surface water treatment. Water Res. 45, 3614–3626.

Upadhyayula, V., Deng, S., Mitchell, M., Smith, G., 2009. Application of carbon nanotube technology for removal of contaminants in drinking water: a review. Sci. Total Environ. 408, 1–13.

Uyak, V., Toroz, I., 2007. Disinfection by-product precursors reduction by various coagulation techniques in Istanbul water supplies. J. Hazard. Mater. 141, 320–328.

Uyak, V., Koyuncu, I., Oktem, I., Cakmakci, M., Toroz, I., 2008. Removal of trihalomethanes from drinking water by nanofiltration membranes. J. Hazard. Mater. 152, 789–794.

Uyak, V., Yavuz, S., Toroz, I., Ozaydin, S., Genceli, E., 2007. Disinfection by-products precursors removal by enhanced coagulation and PAC adsorption. Desalination 216, 334–344.

Uyguner, C., Bekbolet, M., Selcuk, H., 2007a. A comparative approach to the application of a physico–chemical and advanced oxidation combined system to natural water samples. Sep. Sci. Technol. 42, 1405–1419.

Uyguner, C., Suphandag, S., Kerc, A., Bekbolet, M., 2007b. Evaluation of adsorption and coagulation characteristics of humic acids preceded by alternative advanced oxidation techniques. Desalination 210, 183–193.

Uyguner-Demirel, C., Bekbolet, M., 2011. Significance of analytical parameters for the understanding of natural organic matter in relation to photocatalytic oxidation. Chemosphere 84, 1009–1031.

Uyguner-Demirel, C., Birben, C., Bekbolet, M., 2013. Key role of common anions on the photocatalytic degradation profiles of the molecular size fractions of humic acid. Catal. Today 209, 122–126.

Valencia, S., Marín, J., Restrepo, G., Frimmel, F.H., 2014. Evaluation of natural organic matter changes from Lake Hohloh by three-dimensional excitation–emission matrix fluorescence spectroscopy during TiO_2/UV process. Water Res. 51, 124–133.

Valencia, S., Marín, J., Restrepoa, G., Frimmel, F., 2013. Application of excitation–emission fluorescence matrices and UV/Vis absorption to monitoring the photocatalytic degradation of commercial humic acid. Sci. Total Environ. 442, 207–214.

Valencia, S., Marína, J., Velásqueza, J., Restrepoa, G., Frimmel, F., 2012a. Study of pH effects on the evolution of properties of brown-water natural organic matter as revealed by size-exclusion chromatography during photocatalytic degradation. Water Res. 46, 1198–1206.

Valencia, S.H., Marín, J.A., Restrepo, G.M., 2012b. Evolution of natural organic matter by size exclusion chromatography during photocatalytic degradation by solvothermal-synthesized titanium dioxide. J. Hazard. Mater. 213–214, 318–324.

Vasyukova, E., Uhl, W., Braga, F., Simõnes, C., Baylão, T., Neder, K., 2012. Drinking water production from surface water sources in the tropics: Brasília DF, Brazil. Environ. Earth Sci. 65, 1587–1599.

Velten, S., Knappe, D.R.U., Traber, J., Kaiser, H.-P., von Gunten, U., Boller, M., Meylan, S., 2011. Characterization of natural organic matter adsorption in granular activated carbon adsorbers. Water Res. 45, 3951–3959.

Ventresque, C., Gisclon, V., Bablon, G., Chagneau, G., 2000. An outstanding feat of modern technology: the Mery-sur-Oise nanofiltration treatment plant (340,000 m^3/d). Desalination 131, 1–16.

Vepsäläinen, M., Ghiasvand, M., Selin, J., Pienimaa, J., Repo, E., Pulliainen, M., Sillanpää, M., 2009. Investigations of the effects of temperature and initial sample pH on natural organic matter (NOM) removal with electrocoagulation using response surface method (RSM). Sep. Purif. Technol. 69, 255–261.

Vepsäläinen, M., Pulliainen, M., Sillanpää, M., 2012. Effect of electrochemical cell structure on natural organic matter (NOM) removal from surface through electrocoagulation (EC). Sep. Purif. Technol. 99, 20–27.

Vepsäläinen, M., Selin, J., Pulliainen, M., Sillanpää, M., 2007. Combined electrocoagulation and chemical coagulation of paper mill mechanically cleaned water. J. Pulp Pap. Sci. 33, 233–239.

Verrelli, D., Dixon, D., Scales, P., 2009. Effect of coagulation conditions on the dewatering properties of sludges produced in drinking water treatment. Colloids Surf. A Physicochem. Eng. Aspects 348, 14–23.

Verrelli, D., Dixon, D., Scales, P., 2010. Assessing dewatering performance of drinking water treatment sludges. Water Res. 44, 1542–1552.

Vickers, J.C., Thompson, M.A., Kelkar, U.G., 1995. The use of membrane filtration in conjunction with coagulation processes for improved NOM removal. Desalination 102, 57–61.

Vidic, R.D., Suidan, M.T., 1991. Role of dissolved oxygen on the adsorptive capacity of activated carbon for synthetic and natural organic matter. Environ. Sci. Technol. 25, 1612–1618.

Vik, E.A., Carlson, D.A., Eikum, A.S., Gjessing, E.T., 1984. Electrocoagulation of potable water. Water Res. 18, 1355–1360.

Vilhunen, S., Bosund, M., Kääriäinen, M.-L., Cameron, D., Sillanpää, M., 2009. Atomic layer deposited TiO_2 films in photodegradation of aqueous salicylic acid. Sep. Purif. Technol. 66, 130–134.

Visvanathan, C., Boonthanon, N., Sathasivan, A., Jegatheesan, V., 2003. Pretreatment of seawater for biodegradable organic content removal using membrane bioreactor. Desalination 153, 133–140.

Visvanathan, C., Marsono, B.D., Basu, B., 1998. Removal of THMP by nanofiltration: effects of interference parameters. Water Res. 32, 3527–3538.

Voelker, B., Sulzberger, B., 1996. Effects of fulvic acid on Fe(II) oxidation by hydrogen peroxide. Environ. Sci. Technol. 30, 1106–1114.

Voutchkov, N., 2010. Considerations for selection of seawater filtration pretreatment system. Desalination 261, 354–364.

Väänänen, M., Kupiainen, L., Rämö, J., Sarpola, A., Tanskanen, J., 2012. Speciation and coagulation performance of novel coagulant – aluminium formate. Sep. Purif. Technol. 86, 242–247.

Wachinski, A.M., Etzel, J.E., 1997. Environmental Ion Exchange: Principles and Design. Lewis Publishers, USA.

Walker, K.M., Boyer, T.H., 2011. Long-term performance of bicarbonate-form anion exchange: removal of dissolved organic matter and bromide from the St. Johns River, FL, USA. Water Res. 45, 2875–2886.

Wan Ngah, W.S., Hanafiah, M.A.K.M., Yong, S.S., 2008. Adsorption of humic acid from aqueous solutions on crosslinked chitosan–epichlorohydrin beads: kinetics and isotherm studies. Colloids Surf. B Biointerfaces 65, 18–24.

Wang, B., Gu, L., Ma, H., 2007a. Electrochemical oxidation of pulp and paper making wastewater assisted by transition metal modified kaolin. J. Hazard. Mater. 143, 198–205.

Wang, B., Kong, W., Ma, H., 2007b. Electrochemical treatment of paper mill wastewater using three-dimensional electrodes with Ti/Co/SnO_2–Sb_2O_5 anode. J. Hazard. Mater. 146, 295–301.
Wang, G.-S., Liao, C.-H., Chen, H.-W., Yang, H.-C., 2006. Characteristics of natural organic matter degradation in water by UV/H_2O_2 treatment. Environ. Technol. 27, 277–287.
Wang, H., Kang, J., Liu, H., Qu, J., 2009a. Preparation of organically functionalized silica gel as adsorbent for copper ion adsorption. J. Environ. Sci. 21, 1473–1479.
Wang, H., Keller, A., Li, F., 2010. Natural organic matter removal by adsorption onto carbonaceous nanoparticles and coagulation. J. Environ. Eng. 136, 1075–1081.
Wang, H., Keller, A.A., Clark, K.K., 2011a. Natural organic matter removal by adsorption onto magnetic permanently confined micelle arrays. J. Hazard. Mater. 194, 156–161.
Wang, J., Wang, X., 2006. Ultrafiltration with in-line coagulation for the removal of natural humic acid and membrane fouling mechanism. J. Environ. Sci. 18, 880–884.
Wang, J., Han, X., Ma, H., Ji, Y., Bi, L., 2011b. Adsorptive removal of humic acid from aqueous solution on polyaniline/attapulgite composite. Chem. Eng. J. 173, 171–177.
Wang, K., Guo, J., Yang, M., Junji, H., Deng, R., 2009b. Decomposition of two haloacetic acids in water using UV radiation, ozone and advanced oxidation processes. J. Hazard. Mater. 162, 1243–1248.
Wang, L.K., Menon, R., 2008. Treatment of industrial effluents, municipal wastes, and portable water by membrane bioreactors. In: Wang, L.K., Chen, J.P., Hung, Y., Shammas, N.K. (Eds.), Handbook of Environmental Engineering. Membrane and Desalination Technologies, vol. 13. Springer Science & Business Media, New York, pp. 201–236.
Wang, M., Liao, L., Zhang, Z., Li, Z., 2012. Adsorption of low concentration humic acid from water by palygorskite. Appl. Clay Sci. 67–68, 164–168.
Wang, S., Peng, Y., 2010. Natural zeolites as effective adsorbents in water and wastewater treatment. Chem. Eng. J. 156, 11–24.
Wang, X., Waite, T.D., 2008. Gel layer formation and hollow fiber membrane filterability of polysaccharide dispersions. J. Membr. Sci. 322, 204–213.
Wang, X., Waite, T.D., 2009. Role of gelling soluble and colloidal microbial products in membrane fouling. Environ. Sci. Technol. 43, 9341–9347.
Wang, Y., Combe, C., Clark, M.M., 2001. The effects of pH and calcium on the diffusion coefficient of humic acid. J. Membr. Sci. 183, 49–60.
Wang, Y., Gao, B., Yue, Q., Wei, J., Li, Q., 2008. The characterization and flocculation efficiency of composite flocculant iron salts–polydimethyldiallylammonium chloride. Chem. Eng. J. 142, 175–181.
Wang, Y., Gao, B.-Y., Xu, X.-M., Xu, W.-Y., Xu, G.-Y., 2009c. Characterization of floc size, strength and structure in various aluminum coagulants treatment. J. Colloid Interface Sci. 332, 354–359.
Wang, Y.F., Shu, L., Jegatheesan, V., Gao, B.Y., 2009d. Coagulation and nano-filtration: a hybrid system for the removal of lower molecular weight organic compounds (LMWOC). Desalin. Water Treat. 11, 23–31.
Warton, B., Heitz, A., Allpike, B., Kagi, R., 2008. Size-exclusion chromatography with organic carbon detection using a mass spectrometer. J. Chromatogr. A 1207, 186–189.
Wei, J., Gao, B., Yue, Q., Wang, Y., 2009c. Strength and regrowth properties of polyferric-polymer dual-coagulant flocs in surface water treatment. J. Hazard. Mater. 175, 949–954.
Wei, J., Gao, B., Yue, Q., Wang, Y., Lu, L., 2009a. Performance and mechanism of polyferric-quaternary ammonium salt composite flocculants in treating high organic matter and high alkalinity surface water. J. Hazard. Mater. 165, 789–795.
Wei, J., Gao, B., Yue, Q., Wang, Y., Li, W., Zhu, X., 2009b. Comparison of coagulation behavior and floc structure characteristic of different polyferric-cationic polymer dual-coagulants in humic acid solution. Water Res. 43, 724–732.

Wei, Q., Feng, C., Wang, D., Shi, B., Zhang, L., Wei, Q., Tang, H., 2008. Seasonal variations of chemical and physical characteristics of dissolved organic matter and trihalomethane precursors in a reservoir: a case study. J. Hazard. Mater. 150, 257–264.

Wesley, W., Eckenfelder, J., 2000. Industrial Water Pollution Control, third ed. McGraw-Hill Series in Water Resources and Environmental Engineering.

Williams, M.D., Pirbazari, M., 2007. Membrane bioreactor process for removing biodegradable organic matter from water. Water Res. 41, 3880–3893.

Winterdahl, M., 2013. Intra-Annual Variability of Natural Organic Matter in Boreal Streams Patterns and Controls, Faculty of Natural Resources and Agricultural Sciences, Department of Aquatic Sciences and Assessment, Uppsala (Doctoral thesis). Swedish University of Agricultural Sciences, Uppsala, Sweden.

Wiszniowski, J., Robert, D., Surmacz-Gorska, J., Miksch, K., Weber, J.-V., 2002. Photocatalytic decomposition of humic acids on TiO_2 part I: discussion of adsorption and mechanism. J. Photochem. Photobiol. A Chem. 152, 267–273.

de Wit, H.A., Mulder, J., Hindar, A., Hole, L., 2007. Long-term increase in dissolved organic carbon in streamwaters in Norway is response to reduced acid deposition. Environ. Sci. Technol. 41, 7706–7713.

Wong, S., Hanna, J., King, S., Carroll, T., Eldridge, R., Dixon, D., Bolto, B., Hesse, S., Abbt-Braun, G., Frimmel, F.H., 2002. Fractionation of natural organic matter in drinking water and characterization by ^{13}C cross-polarization magic-angle spinning NMR spectroscopy and size exclusion chromatography. Environ. Sci. Technol. 36, 3497–3503.

Worrall, F., Burt, T., 2007. Trends in DOC concentration in Great Britain. J. Hydrol. 346, 81–92.

Worrall, F., Burt, T.P., 2009. Changes in DOC treatability: indications of compositional changes in DOC trends. J. Hydrol. 366, 1–8.

Wu, C.N., Wang, L., Ling, Q., Tang, Y.C., Huang, J., Pan, F.K., 2013. Removal of natural organic matter of artificial lake by modified fly ash. Adv. Mater. Res. 690–693, 1020–1023.

Wu, F., Evans, R., Dillon, P., 2003. Separation and characterization of NOM by high-performance liquid chromatography and on-line three-dimensional excitation emission matrix fluorescence detection. Environ. Sci. Technol. 37, 3687–3693.

Wu, F., Evans, R., Dillon, P., Cai, Y., 2007a. Rapid quantification of humic and fulvic acids by HPLC in natural waters. Appl. Geochem. 22, 1598–1605.

Wu, F., Kothawala, D., Evans, R., Dillon, P., Cai, Y., 2007b. Relationships between DOC concentration, molecular size and fluorescence properties of DOM in a stream. Appl. Geochem. 22, 1659–1667.

Xia, S., Liu, Y., Li, X., Yao, J., 2007. Drinking water production by ultrafiltration of Songhuajiang River with PAC adsorption. J. Environ. Sci. 19, 536–539.

Xie, Q., Xie, J., Wang, Z., Wu, D., Zhang, Z., Kong, H., 2013. Adsorption of organic pollutants by surfactant modified zeolite as controlled by surfactant chain length. Microporous Mesoporous Mater. 179, 144–150.

Xing, L., Fabris, R., Chow, C., van Leeuwen, J., Drikas, M., Wang, D., 2012. Prediction of DOM removal of low specific UV absorbance surface waters using HPSEC combined with peak fitting. J. Environ. Sci. 24, 1174–1180.

Xing, Y., 2010. Characterization of Dissolved Organic Carbon in Prairie Surface Waters Using Fourier Transform Infrared Spectroscopy (M.Sc. thesis). University of Saskatchewan, Saskatoon, Saskatchewan, Canada.

Xu, S., Zhang, X., Ng, J., Sun, D., 2009b. Preparation and application of TiO_2/Al_2O_3 microsperical photocatalyst for water treatment. Water Sci. Technol. Water Supply 9, 39–44.

Xu, X., Yu, S.-I., Shi, W., Jiang, Z., Wu, C., 2009a. Effect of acid medium on the coagulation efficiency of polysilicate-ferric (PSF)—a new kind of inorganic polymer coagulant. Sep. Purif. Technol. 66, 486–491.

Yamashita, Y., Jaffé, R., 2008. Characterizing the interactions between trace metals and dissolved organic matter using excitation–emission matrix and parallel factor analysis. Environ. Sci. Technol. 42, 7374–7379.

Yan, M., Liu, H., Wang, D., Ni, J., Qu, J., 2009a. Natural organic matter removal by coagulation: effect of kinetics and hydraulic power. Water Sci. Technol. Water Supply 9, 21–30.

Yan, M., Wang, D., Ma, X., Ni, J., Zhang, H., 2010. THMs precursor removal by an integrated process of ozonation and biological granular activated carbon for typical Northern China water. Sep. Purif. Technol. 72, 263–268.

Yan, M., Wang, D., Ni, J., Qu, J., Chow, C.W.K., Liu, H., 2008c. Mechanism of natural organic matter removal by polyaluminum chloride: effect of coagulant particle size and hydrolysis kinetics. Water Res. 42, 3361–3370.

Yan, M., Wang, D., Ni, J., Qu, J., Ni, W., Van Leeuwen, J., 2009b. Natural organic matter (NOM) removal in a typical North-China water plant by enhanced coagulation: targets and techniques. Sep. Purif. Technol. 68, 320–327.

Yan, M., Wang, D., Qu, J., Ni, J., Chow, C.W.K., 2008a. Enhanced coagulation for high alkalinity and micro-polluted water: the third way through coagulant optimization. Water Res. 42, 2278–2286.

Yan, M., Wang, D., Shi, B., Wang, M., Yan, Y., 2007. Effect of pre-ozonation on optimized coagulation of a typical North-China source water. Chemosphere 69, 1695–1702.

Yan, M., Wang, D., You, S., Qu, J., Tang, H., 2006. Enhanced coagulation in a typical North-China water treatment plant. Water Res. 40, 3621–3627.

Yan, M., Wang, D., Yu, J., Ni, J., Edwards, M., Qu, J., 2008b. Enhanced coagulation with polyaluminum chlorides: role of pH/alkalinity and speciation. Chemosphere 71, 1665–1673.

Yang, J., Yuan, D., Weng, T., 2010a. Pilot study of drinking water treatment with GAC, O_3/BAC and membrane processes in Kinmen Island, Taiwan. Desalination 263, 271–278.

Yang, S., Sun, C., Li, X., Gong, Z., Quan, X., 2010b. Enhanced photocatalytic activity for titanium dioxide by co-modifing with silica and fluorine. J. Hazard. Mater. 175, 258–266.

Yang, W., Chen, J., Li, X., Liang, H., He, W., Li, G., 2007. Using chloramine as a coagulant aid in enhancing coagulation of Yellow River water in China. J. Zhejiang Univ. Sci. A 8, 1475–1481.

Yang, X., Guo, W., Lee, W., 2013. Formation of disinfection byproducts upon chlorine dioxide preoxidation followed by chlorination or chloramination of natural organic matter. Chemosphere 91, 1477–1485.

Yang, Z.-H., Huang, J., Zeng, G.-M., Ruan, M., Zhou, C.-S., Li, L., Rong, Z.-G., 2009. Optimization of flocculation conditions for kaolin suspension using the composite flocculant of MBFGA1 and PAC by response surface methodology. Biores. Technol. 100, 4233–4239.

Yao, P., Choo, K., Kim, M., 2009. A hybridized photocatalysis–microfiltration system with iron oxide-coated membranes for the removal of natural organic matter in water treatment: effects of iron oxide layers and colloids. Water Res. 43, 4238–4248.

Yapsakli, K., Cecen, F., 2009. Use of an enriched inoculum for determination of biodegradable dissolved organic carbon (BDOC) in drinking water. Water Sci. Technol. 9, 149–157.

Yee, L.F., Abdullah, P.M., Abdullah, A., Ishak, B., Abidin, K.N.Z., 2009. Hydrophobicity characteristics of natural organic matter and the formation of THM. Malays. J. Anal. Sci. 13, 94–99.

Yildiz, Y.Ş., Koparal, A.S., Keskinler, B., 2008. Effect of initial pH and supporting electrolyte on the treatment of water containing high concentration of humic substances by electrocoagulation. Chem. Eng. J. 138, 63–72.

Yildiz, Y.Ş., Koparal, A.S., İrdemez, Ş., Keskinler, B., 2007. Electrocoagulation of synthetically prepared waters containing high concentration of NOM using iron cast electrodes. J. Hazard. Mater. B139, 373–380.

Yoon, Y., Amy, G., Cho, J., Her, N., 2005. Effects of retained natural organic matter (NOM) on NOM rejection and membrane flux decline with nanofiltration and ultrafiltration. Desalination 173, 209–221.

Yu, J., Wang, D., Yan, M., Ye, C., Yang, M., Ge, X., 2007. Optimized coagulation of high alkalinity, low temperature and particle water: pH adjustment and polyelectrolytes as coagulant aids. Environ. Monit. Assess. 131, 377–386.

Zaitseva, N., Zaitsev, V., Walcarius, A., 2013. Chromium(VI) removal via reduction–sorption on bi-functional silica adsorbents. J. Hazard. Mater. 250–251, 454–461.

Zappia, L., Alessadrino, M., Garbin, S., Pringle, P., Heitz, A., Joll, C., Masters, D., Hiller, B., Capewell, S., Franzmann, P., Plumb, J., Von Eckstaedt, S., Cadee, K., 2008. Comparison of methods for the determination of biodegradable dissolved organic carbon in potable water supply: use of a novel biofilm. Water Sci. Technol. 8, 633–641.

Zazouli, M.A., Nasseri, S., Ulbricht, M., 2010. Fouling effects of humic and alginic acids in nanofiltration and influence of solution composition. Desalination 250, 688–692.

Zepp, R., Faust, B., Hoigné, J., 1992. Hydroxyl radical formation in aqueous reactions (pH 3–8) of iron(II) with hydrogen peroxide: the photo-Fenton reaction. Environ. Sci. Technol. 26, 313–319.

Zhan, X., Gao, B., Wang, Y., Yue, Q., 2011a. Influence of velocity gradient on aluminum and iron floc property for NOM removal from low organic matter surfacewater by coagulation. Chem. Eng. J. 166, 116–121.

Zhan, X., Gao, B., Yue, Q., Wang, Y., Gao, B., 2010a. Coagulation behavior of polyferric chloride for removing NOM from surface water with low concentration of organic matter and its effect on chlorine decay model. Sep. Purif. Technol. 75, 61–68.

Zhan, Y., Lin, J., Qiu, Y., Gao, N., Zhu, Z., 2011b. Adsorption of humic acid from aqueous solution on bilayer hexadecyltrimethyl ammonium bromide-modified zeolite. Front. Environ. Sci. Eng. China 5, 65–75.

Zhan, Y., Zhu, Z., Lin, J., Qiu, Y., Zhao, J., 2010b. Removal of humic acid from aqueous solution by cetylpyridinium bromide modified zeolite. J. Environ. Sci. 22, 1327–1334.

Zhang, G., Wu, T., Li, Y., Huang, X., Wang, Y., Wang, G., 2012c. Sorption of humic acid to organo layered double hydroxides in aqueous solution. Chem. Eng. J. 191, 306–313.

Zhang, H., Zhang, Y., Shi, Q., Hu, J., Chu, M., Yu, J., Yang, M., 2012a. Study on transformation of natural organic matter in source water during chlorination and its chlorinated products using ultrahigh resolution mass spectrometry. Environ. Sci. Technol. 46, 4396–4402.

Zhang, H., Zhang, Y., Shi, Q., Ren, S., Yu, J., Ji, F., Luo, W., Yang, M., 2012b. Characterization of low molecular weight dissolved natural organic matter along the treatment trait of a waterworks using Fourier transform ion cyclotron resonance mass spectrometry. Water Res. 46, 5197–5204.

Zhang, J., Liu, F., 2010. Adsorption of natural organic matter onto a composite adsorbent prepared with chitosan and powdered activated carbon. Desalin. Water Treat. 20, 291–296.

Zhang, L., Ni, Q.-Q., Fu, Y., Natsuki, T., 2009a. One-step preparation of water-soluble single-walled carbon nanotubes. Appl. Surf. Sci. 255, 7095–7099.

Zhang, M., Li, C., Benjamin, M.M., Chang, Y., 2003. Fouling and natural organic matter removal in adsorbent/membrane systems for drinking water treatment. Environ. Sci. Technol. 37, 1663–1669.

Zhang, N., Xie, J., Varadan, V., 2002. Functionalization of carbon nanotubes by potassium permanganate assisted with phase transfer catalyst. Smart Mater. Struct. 11, 962–965.

Zhang, P., Jian, L., 2006. Ozone-enhanced photocatalytic degradation of natural organic matter in water. Water Sci. Technol. Water Supply 6, 53–61.

Zhang, T., Lu, J., Ma, J., Qiang, Z., 2008. Fluorescence spectroscopic characterization of DOM fractions isolated from a filtered river water after ozonation and catalytic ozonation. Chemosphere 71, 911–921.

Zhang, W., Zhang, X., Li, Y., Wang, J., Chen, C., 2011a. Membrane flux dynamics in the submerged ultrafiltration hybrid treatment process during particle and natural organic matter removal. J. Environ. Sci. 23, 1970–1976.

Zhang, X., Bai, R., 2003. Mechanisms and kinetics of humic acid adsorption onto chitosan-coated granules. J. Colloid Interface Sci. 264, 30–38.

Zhang, X., Pan, J.H., Fu, W., Du, A.J., Sun, D.D., 2009b. TiO_2 nanotube photocatalytic oxidation for water treatment. Water Sci. Technol. Water Supply 9, 45–49.

Zhang, Y., Tian, J., Nan, J., Gao, S., Liang, H., Wang, M., Li, G., 2011b. Effect of PAC addition on immersed ultrafiltration for the treatment of algal-rich water. J. Hazard. Mater. 186, 1415–1424.

Zhao, G., Lu, X., Zhou, Y., Gu, Q., 2013. Simultaneous humic acid removal and bromate control by O_3 and UV/O_3 processes. Chem. Eng. J. 232, 74–80.

Zhao, H., Hu, C., Liu, H., Zhao, X., Qu, J., 2008a. Role of aluminum speciation in the removal of disinfection byproduct precursors by a coagulation process. Environ. Sci. Technol. 42, 5752–5758.

Zhao, L., Luo, F., Wasikiewicz, J.M., Mitomo, H., Nagasawa, N., Yagi, T., Tamada, M., Yoshii, F., 2008b. Adsorption of humic acid from aqueous solution onto irradiation-crosslinked carboxymethylchitosan. Bioresour. Technol. 99, 1911–1917.

Zhao, X., Zhang, B., Liu, H., Chen, F., Li, A., Qu, J., 2012. Transformation characteristics of refractory pollutants in plugboard wastewater by an optimal electrocoagulation and electro-Fenton process. Chemosphere 87, 631–636.

Zhao, Y., Gao, B., Shon, H., Cao, B., Kim, J.-H., 2011. Coagulation characteristics of titanium (Ti) salt coagulant compared with aluminum (Al) and iron (Fe) salts. J. Hazard. Mater. 185, 1536–1542.

Zhao, Y.-Y., Boyd, J., Woodbeck, M., Andrews, R., Qin, F., Hrudey, S., Li, X.-F., 2008c. Formation of *N*-nitrosamines from eleven disinfection treatments of seven different surface waters. Environ. Sci. Technol. 42, 4857–4862.

Zhao, Z.-Y., Gu, D.-J., Li, H.-B., Li, X.-Y., Leung, K., 2009. Disinfection characteristics of the dissolved organic fractions at several stages of a conventional drinking water treatment plant in South China. J. Hazard. Mater. 172, 1093–1099.

Zhou, J., Li, W., Gu, J., Yu, H., Tang, Z., Wei, X., 2010. Development of a novel RAFT-UV grafting technique to modify polypropylene membrane used for NOM removal. Sep. Purif. Technol. 71, 233–240.

Zhou, Q., Cabaniss, S., Maurice, P., 2000. Considerations in the use of high-pressure size exclusion chromatography (HPSEC) for determining molecular weights of aquatic humic substances. Water Res. 34, 3505–3514.

Zhou, Q., Zhong, Y.-H., Chen, X., Liu, J.-H., Huang, X.-J., Wu, Y.-C., 2014. Adsorption and photocatalysis removal of fulvic acid by TiO_2–graphene composites. J. Mater. Sci. 49, 1066–1075.

Zhu, B., Clifford, D.A., Chellam, S., 2005. Comparison of electrocoagulation and chemical coagulation pretreatment for enhanced virus removal using microfiltration membranes. Water Res. 39, 3098–3108.

Zhu, G., Zheng, H., Chen, W., Fan, W., Zhang, P., Tshukudu, T., 2012. Preparation of a composite coagulant: polymeric aluminum ferric sulfate (PAFS) for wastewater treatment. Desalination 285, 315–323.

Zhu, X., Ni, J., Wei, J., Chen, P., 2011. Scale-up of B-doped diamond anode system for electrochemical oxidation of phenol simulated wastewater in batch mode. Electrochim. Acta 56, 9439–9447.

Ziylan, A., Ince, N., 2013. Ozonation-based advanced oxidation for pre-treatment of water with residuals of anti-inflammatory medication. Chem. Eng. J. 220, 151–160.

Zouboulis, A., Moussas, P., 2008. Polyferric silicate sulphate (PFSiS): preparation, characterization and coagulation behaviour. Desalination 224, 307–316.

Zouboulis, A., Tzoupanos, N., 2009. Polyaluminium silicate chloride—a systematic study for the preparation and application of an efficient coagulant for water or wastewater treatment. J. Hazard. Mater. 162, 1379–1389.

Zouboulis, A., Moussas, P., Vasilakou, F., 2008. Polyferric sulphate: preparation, characterization and application in coagulation experiments. J. Hazard. Mater. 155, 459–468.

Zularisam, A., Ismail, A., Salim, R., 2006. Behaviours of natural organic matter in membrane filtration for surface water treatment – a review. Desalination 194, 211–231.

Zularisam, A., Ismail, A., Salim, M., Sakinah, M., Matsuura, T., 2009. Application of coagulation–ultrafiltration hybrid process for drinking water treatment: optimization of operating conditions using experimental design. Sep. Purif. Technol. 65, 193–210.

Zularisam, A.W., Ismail, A.F., Salim, M.R., Sakinah, M., Hiroaki, O., 2007a. Fabrication, fouling and foulant analyses of asymmetric polysulfone (PSF) ultrafiltration membrane fouled with natural organic matter (NOM) source waters. J. Membr. Sci. 299, 97–113.

Zularisam, A.W., Ismail, A.F., Salim, M.R., Sakinah, M., Ozaki, H., 2007b. The effects of natural organic matter (NOM) fractions on fouling characteristics and flux recovery of ultrafiltration membranes. Desalination 212, 191–208.

LIST OF JOURNAL TITLES WITH ABBREVIATIONS

Adv. Coll. Interf. Sci.	Advances in Colloid and Interface Science
Adv. Mat. Res.	Advanced Materials Research
Agricult. Syst.	Agricultural Systems
Anal. Bioanal. Chem.	Analytical and Bioanalytical Chemistry
Anal. Chem.	Analytical Chemistry
Appl. Catal. B: Environ.	Applied Catalysis B: Environmental
Appl. Clay Sci.	Applied Clay Science
Appl. Geochem.	Applied Geochemistry
Appl. Microbiol. Biotechnol.	Applied Microbiology and Biotechnology
Appl. Surf. Sci.	Applied Surface Science
Aquatic Sci.	Aquatic Sciences
Asian J. Chem.	Asian Journal of Chemistry
Biogeochem.	Biogeochemistry
Biores. Technol.	Bioresource Technology
Brain Res. Bull.	Brain Research Bulletin
Can. J. Civil Eng.	Canadian Journal of Civil Engineering
Carbon	
Catal. Today	Catalysis Today
Cement Concrete Res.	Cement and Concrete Research
Chem. Eng. J.	Chemical Engineering Journal
Chem. Eng. Res. Design	Chemical Engineering Research and Design
Chem. Eng. Sci.	Chemical Engineering Science
Chem. Geol.	Chemical Geology
Chem. Rev.	Chemical Reviews
Chemosphere	
Chinese Sci. Bull.	Chinese Science Bulletin
Coll. Surf.	Colloids and Surfaces
A: Physicochem. Eng. Asp.	A: Physicochemical and Engineering Aspects
B: Biointerf.	B: Biointerfaces
Continental Shelf Res.	Continental Shelf Research
Desalination	
Desal. Water Treat.	Desalination and Water Treatment
Drink. Water Eng. Sci.	Drinking Water Engineering and Science
Drying Technol.	Drying Technology
Dyes and Pigments	
Electrochim. Acta	Electrochimica Acta
Energy	
Environ. Int.	Environment International
Environ. Chem. Lett.	Environmental Chemistry Letters
Environ. Earth Sci.	Environmental Earth Sciences
Environ. Mon. Assess.	Environmental Monitoring and Assessment
Environ. Pollut.	Environmental Pollution

Environ. Res.	Environmental Research
Environ. Sci Pollut. Res.	Environmental Science and Pollution Research
Environ. Sci. Technol.	Environmental Science and Technology
Environ. Technol.	Environmental Technology
Environ. Toxicol. Chem.	Environmental Toxicology and Chemistry
Environmentalist	
Eur. Polym. J.	European Polymer Journal
Filtr. Sep.	Filtration and Separation
Front. Environ. Sci. Eng. China	Frontiers of Environmental Science & Engineering in China
Geochim. Cosmochim. Acta	Geochimica et Cosmochimica Acta
Global Change Biol.	Global Change Biology
Hydrol. Process.	Hydrological Processes
Ind. Eng. Chem. Res.	Industrial and Engineering Chemistry Research
Int. J. Environ. Anal. Chem.	International Journal of Environmental Analytical Chemistry
Int. J. Environ. Health Eng.	International Journal of Environmental Health Engineering
Int. J. Environ. Pollut.	International Journal of Environment and Pollution
Inżynieria i Ochrona Środowiska	
Jordan J. Civil Eng.	Jordan Journal of Civil Engineering
J. Amer. Chem. Soc.	Journal of the American Chemical Society
J. Amer. Water Works Assoc.	Journal of the American Water Works Association
J. Anal. Appl. Pyr.	Journal of Analytical and Applied Pyrolysis
J. Anal. Chem.	Journal of Analytical Chemistry
J. Appl. Electrochem.	Journal of Applied Electrochemistry
J. Chem. Technol. Biotechnol.	Journal of Chemical Technology and Biotechnology
J. China Univ. Mining Technol.	Journal of China University of Mining and Technology
J. Chrom.	Journal of Chromatography
J. Cleaner Prod.	Journal of Cleaner Production
J. Coll. Interf. Sci.	Journal of Colloid and Interface Science
J. Environ. Eng.	Journal of Environmental Engineering
J. Environ. Manag.	Journal of Environmental Management
J. Environ. Mon.	Journal of Environmental Monitoring
J. Environ. Sci.	Journal of Environmental Sciences
J. Environ. Sci. Health Part A:	Journal of Environmental Science and Health Part A:
Environ. Sci. Eng. Toxic Hazard. Subst. Contr. Toxic/Hazardous Substances and Environmental Engineering	
J. Hazard. Mat.	Journal of Hazardous Materials
J. Hydrol.	Journal of Hydrology
J. Ind. Chem. Soc.	Journal of the Indian Chemical Society
J. Ind. Eng. Chem.	Journal of Industrial and Engineering Chemistry
J. Mass Spectrom.	Journal of Mass Spectrometry
J. Membr. Sci.	Journal of Membrane Science

J. Molec. Catal. A: Chem.	Journal of Molecular Catalysis A: Chemical
J. Non-Crystalline Solids	Journal of Non-Crystalline Solids
Journal of Ocean University of China (English Edition)	
J. Photochem. Photobiol.:	Journal of Photochemistry and Photobiology C:
Photochem. Rev.	Photochemistry Reviews
J. Phys. Chem.	Journal of Physical Chemistry
J. Pulp and Paper Sci.	Journal of Pulp and Paper Science
J. Taiwan Inst. Chem. Eng.	Journal of the Taiwan Institute of Chemical Engineers
J. Water Chem. Technol.	Journal of Water Chemistry and Technology
J. Water Supply Res. Technol. – Aqua	Journal of Water Supply: Research & Technology – Aqua
J. Water Sustain.	Journal of Water Sustainability
J. Zhejiang University: Sci. A	Journal of Zhejiang University Science A
Kolloidchemie	
Kor. J. Chem. Eng.	Korean Journal of Chemical Engineering
Langmuir	
Limn. Oceanography	Limnology and Oceanography
Malay. J. Anal. Sci.	Malaysian Journal of Analytical Sciences
Mar. Chem.	Marine Chemistry
Mat. Sci. Forum	Materials Science Forum
Microp. Mesop. Mat.	Microporous and Mesoporous Materials
Microsc. Microanal.	Microscopy and Microanalysis
Mut. Res./Rev. Mut. Res.	Mutation Research/Reviews in Mutation Research
Nature	
Org. Geochem.	Organic Geochemistry
Ozone Sci. Eng.	Ozone: Science and Engineering
Philosophical Magazine Series	
Photochem. Photobiol.	Photochemistry and Photobiology
Phys. Chem. Earth	Physics and Chemistry of the Earth
PLoS ONE	Public Library of Science ONE
Pol. J. Environ. Stud.	Polish Journal of Environmental Studies
Polyhedron	
Powder Technol.	Powder Technology
Progr. Polymer Sci.	Progress in Polymer Science
Progr. Solid State Chem.	Progress in Solid State Chemistry
React. Funct. Polym.	Reactive and Functional Polymers
RSC Advances	Royal Society of Chemistry Advances
Schweizerische Zeitschrift fur Hydrologie	
Sci. Tot. Environ.	Science of the Total Environment
Sedimentol.	Sedimentology
Sep. Pur. Rev.	Separation and Purification Reviews
Sep. Pur. Technol.	Separation and Purification Technology
Sep. Sci. Technol.	Separation Science and Technology
Smart Mat. Struct.	Smart Materials and Structures
Soil Biol. Biochem.	Soil Biology and Biochemistry
Soil Sci.	Soil Science
Solar Energy	
Spectrochim. Acta	Spectrochimica Acta

Talanta	
Turk. J. Eng. Environ. Sci.	Turkish Journal of Engineering and Environmental Sciences
Verhandlungen des Internationalen Verein Limnologie	
Vesitalous	
Water Environ. J.	Water and Environment Journal
Water Res.	Water Research
Water Res. Manag.	Water Resources Management
Water Sci. Technol.	Water Science and Technology

INDEX

Note: Page numbers with "*f*" and "*t*" denote figures and tables, respectively

A

Absorption, 8, 214–215
Acetamide, 31t
Acetic acid, 31t
Acetonitrile, 31t
Acetylfuran, 31t
ACF. *See* Activated carbon fiber
Activated alumina, 217
Activated carbon. *See also* Granular activated carbon; Powdered activated carbon
  as adsorbent, 218
  almond shell, 231t–232t
  ammonia-treated, 231t–232t
  BAC, 276–277
  filtration, 9, 277–278
  iron-impregnated, 231t–232t
Activated carbon fiber (ACF), 220–221
Activated silica, 65
Adsorbents, 216–218, 230f
  activated alumina as, 217
  activated carbon as, 218
  carbon, 218–221
  characteristics of, 231t–232t
  composite, 225–233
  hydroxide based, 223–225
  metal oxide based, 223–225
  modified, 225–233
  nanoadsorbents as, 221–223
  regeneration of, 286
  silica gel as, 217
  zeolites as, 217
Adsorption, 58f, 214–216
  advantages and disadvantages of, 9t–11t
  humic substances and, 227–228
  isotherm tests, 218–219
  membrane filtration with, 286–288
  membrane fouling and, 145, 145f
  NOM removal with, 57–58, 214, 218–233, 231t–232t
Advanced oxidation process (AOP), 9
  advantages and disadvantages of, 9t–11t
  coagulation with, 276–277
  combined treatments with, 202–203
  Fenton based, 181t–185t
  NOM removal with, 9t–11t, 160–163, 171t–179t, 181t–185t
  nonselectivity of, 203
  ozone based, 162–163, 164t–167t
  as pretreatment, 169
  UV light based, 163–170, 171t–179t
AFM. *See* Atomic force microscopy
Agricultural drain, 118t–127t
Alcoholic groups, 36t–37t
Algal toxins, 277–278
Alginate, 148
Alginic acid, 241
Aliphatic carbon chains, 40t
Almond shell activated carbon, 231t–232t
Alumina, 217, 241
Aluminum
  coagulation with, 60–63
  dissolving of, 84
  electrodes, 93f, 96t–97t
Aluminum chloride, 62t
Aluminum formate, 73
Aluminum salt, 56–57, 62t
Aluminum sulfate
  as coagulant, 60
  dose of, 61–63
  features of, 62t
  GAC with, 220
  PAC with, 220
Amberlite, 250t–269t
Aminated polyacrylonitrile fiber (APANF), 228, 231t–232t
Amine carbon, 40t
Amine functionalized magnetic mesoporous composite microspheres, 231t–232t
Amino sugars, 31t
Ammonia treatment, 220–221

Ammonia-treated activated carbon, 231t–232t
Amphoteric ion exchanger, 240
Anion, 85, 240
Anion resin, 242–243
Anionic polyacrylamides, 70t
Anode electrode. *See* Dimensional stable anode electrode
AOC. *See* Assimilable organic carbon
AOP. *See* Advanced oxidation process
APANF. *See* Aminated polyacrylonitrile fiber
Aquatic environments, 2–3
Aqueous hydrogen peroxide, 188
Aromatic carbon, 36t–37t, 40t
Aromatic rings, 36t–37t, 161
Assimilable organic carbon (AOC), 40–41
Atomic force microscopy (AFM), 41
ATP-PANI. *See* Polyaniline/attapulgite composite
Attenuated total reflectance (ATR), 38

B

BAC. *See* Biological activated carbon
Backwashing, 280
Bacteria, 149
Bacteriophage, 96t–97t
BDD electrode. *See* Boron-doped diamond electrode
BDOC. *See* Biodegradable dissolved organic carbon
Bentonite, 229–233, 231t–232t
Benzoate, 181t–185t
Benzonitrile, 31t
BET. *See* Brunauer-Emmett-Teller isotherm
Bicarbonate, 246–247
Biodegradable dissolved organic carbon (BDOC), 40–41
Biofiltration, 291
Bioflocculant, 71
Biofouling, 149
Biological activated carbon (BAC), 276–277
Biological removal methods, 9t–11t
Biological testing, 40–41
Boron-doped diamond electrode (BDD electrode), 89, 91t, 101
Bridging flocculation, 57–58
Brine, 243–244
Bromate, 164t–167t
Bromide, 190t–199t, 246–247
Brominated DBPs, 161–162
Brunauer-Emmett-Teller isotherm (BET isotherm), 214–215
Butenoic acid, 31t

C

Calcium ions, 130
Carbohydrate functional groups, 40t
Carbon. *See also* Activated carbon; Assimilable organic carbon; Dissolved organic carbon; Total organic carbon
  adsorbents, 218–221
  aliphatic, 40t
  amine, 40t
  aromatic, 36t–37t, 40t
  carbonyl, 40t
  carboxylic, 40t
  cycle, 3–4
  ester, 40t
  methyl and methylene, 39t
  modified, 220–221
  nanoparticles, 231t–232t, 288
  paraffinic, 40t
  unsaturated, 39t
Carbon nanotube (CNT), 73, 222–223
Carbon usage rate (CUR), 218–219
Carbonyl carbon, 40t
Carboxylic carbon, 40t
Carboxylic compounds, 36t–37t, 248
Carcinogens, 150
Cartridge filter, 280
Catalyst, 187
Catalytic oxidation, 180–189, 190t–199t
Catalytic wet peroxide oxidation, 189
Catechol, 31t
Cathodic dissolution, 84
Cation, 240
Cation resin, 242
Cationic polyacrylamides, 70t
Caustic brine, 243–244
Cement, 73–74
Centrifugation, 23
Characterization, 8, 41–42

Charge neutralization, 57–58
Charge repulsion, 133
Chelating agents, 200
Chelating resin, 242
Chemical oxygen demand (COD), 20
Chitosan, 225–226, 231t–232t
Chitosan/zeolite composite (CSZ), 228–229
Chloramine, 6–7
Chloride, 246–247
Chlorination, 276–277
Chlorine, 6–7
Chlorine dioxide, 6–7
Chromatography, 26–31. *See also* High performance size exclusion chromatography
  exclusion, 8
  gas, 30–31, 43t
  LC-MS, 29–30
  liquid, 26–30
  Py-GC-MS, 30
  RPHPLC, 23, 29
  SEC, 23, 26
Chromophore, 161
Clarified water, 250t–269t
Climate change, 3–4
CNT. *See* Carbon nanotube
Coagulant
  aluminum salt as, 56–57, 62t
  aluminum sulfate as, 60
  characteristics of, 284
  composite, 63–66, 72t
  dose of, 59
  ferric-based, 60–61, 63, 64t
  iron salts as, 56–57
  novel, 71–73
  pH and, 56–57
  two-stage, 285
Coagulation, 8. *See also* Electrocoagulation
  activated carbon filtration with, 277–278
  advantages and disadvantages of, 9t–11t
  aggregation mechanisms of, 57–58
  aluminum for, 60–63
  AOPs with, 276–277
  composite coagulants for, 63–66
  conventional, 74
  DOC and SUVA relationship in, 60t
  EC compared with, 99–100, 103
  efficiency of, 61–63
  enhanced, 73–74
  in-line, 282–283
  IPFs for, 63–66
  membrane filtration with, 278–279, 282–286
  MF with, 283f
  MIEX® with, 245–246, 250t–269t, 275–276
  NOM removal with, 9t–11t, 56–61, 58f, 73–74
  with organic polyelectrolytes, 66–69
  pH, 61–63
  process of, 57
  UF for, 283f
Coal, 241
COD. *See* Chemical oxygen demand
Collodine, 241
Complexation, 57–58, 58f
Composite adsorbents, 225–233
Composite coagulants, 63–66, 72t
Composite polyaluminum chloride (HPAC), 276–277
Concentrate
  membrane, 144
  RO, 102t
  treatment, 144
Concentration, 23
  divalent ion, 147–148
  polarization, 145
  salt, 141
Copper, 186–187
Coprecipitation, 57–58
Counter-ions, 241
p-Cresole, 31t
Cryptosporidium, 144
CSZ. *See* Chitosan/zeolite composite
CUR. *See* Carbon usage rate
Current density, 85
Cyclopentanone, 31t

D

DBP. *See* Disinfection by-product
DCAA. *See* Dichloroacetic acid
Decomposition, 189
Deep-well injection, 144
Desalination, 143–144
Destabilization, 58f

Diamond electrode. *See* Boron-doped diamond electrode
Diamond film, 90–91
Dichloroacetic acid (DCAA), 164t–167t
Differential absorption spectroscopy, 32–33
Diffusional back-transport coefficient, 130–131
Dimensional stable anode electrode (DSA electrode), 89, 91t
Dimethylphenol, 31t
Direct filtration, 282
Disinfection
  methods of, 6–7, 18–19
  ozone for, 162
  UV photolysis for, 163–168
Disinfection by-product (DBP)
  brominated, 161–162
  catalytic oxidation and, 190t–199t
  Fenton based AOPs and, 181t–185t
  formation of, 6, 18–20
  health effects of, 6–7
  NF and UF for removal of, 143
  NOM fractions and, 131–132
  ozone based AOPs and, 164t–167t
  UV light based AOPs and, 171t–179t
  in water treatment, 6–7
Dissolved organic carbon (DOC), 19–21
  biodegradable, 40–41
  coagulation and, 60t
  detection features of, 22t
  EC and, 96t–97t
  on-line analyzers of, 27
  SUVA and, 60t
Divalent ion concentration, 147–148
DNA, 31t
DOC. *See* Dissolved organic carbon
Dodecyl sulfonate, 231t–232t
Drinking water, 2–3
Drought, 3, 200
DSA electrode. *See* Dimensional stable anode electrode
Dual-media filter, 280–281

E

EBCT. *See* Empty bed contact time
E-beam irradiation. *See* Electron beam irradiation
EC. *See* Electrocoagulation
EF. *See* Electroflotation
Electrochemical mineralization, 82–83
Electrochemical removal methods, 9t–11t, 82–83
Electrocoagulation (EC), 82
  advantages of, 103–104
  cathodic dissolution in, 84
  chemical coagulation compared with, 99–100, 103
  DOC and, 96t–97t
  EF with, 95
  electrode arrangements in, 86f
  humic substances and, 96t–97t
  iron electrodes in, 83–84
  membrane filtration with, 95–98
  NOM removal with, 92–99, 96t–97t
  novel, 285–286
  pH and, 94
  principles of, 83–92
  reactor design in, 86
  removal rate of, 94
  theory of, 83–85
  TOC and, 96t–97t
  treatment parameters in, 85
Electrode
  aluminum, 93f, 96t–97t
  arrangements of, 86f
  BDD, 89, 91t, 101
  distance between, 85
  DSA, 89, 91t
  for EC, 83–84, 86f
  for EO, 89–92, 91t, 105
  iron, 83–84, 96t–97t
  lead oxygen, 91t
  materials for, 85
  platinum, 87, 91t
  potential, 85
  steel, 96t–97t
  titanium, 91t
  zinc, 96t–97t
Electro-Fenton process, 170–180
Electroflocculation, 82, 95–98
Electroflotation (EF), 82, 95
Electron beam irradiation (E-beam irradiation), 189–200
Electrooxidation (EO), 82
  efficiency of, 104–105
  electrodes for, 89–92, 91t, 105

NOM removal with, 100–101, 102t
pH and, 88
principles of, 83–92
reactors for, 89–92
set-up of, 92f
theory of, 87–89
Electrophoresis, 23
Electrospray ionization (ESI), 41–42
Electrostatic repulsion, 117
Elemental analyzer, 41
Eluent, 26–27
Empty bed contact time (EBCT), 218–219
Endocrine-disrupting chemicals, 162
Energy
renewable, 188
UV lamp demand for, 202
Entrapment, 57–58, 58f
Enzyme-mediated oxidative coupling, 202
EO. *See* Electrooxidation
Escherichia coli, 181t–185t
ESI. *See* Electrospray ionization
Ester carbon, 40t
Excitation emission fluorescence detection, 27
Exclusion chromatography, 8. *See also* High performance size exclusion chromatography

F

FA. *See* Fulvic acids
Faraday's law, 84
Fenton based AOPs, 181t–185t
Fenton process, 170–180, 200
Fenton reaction, 88–89
Fenton's reagent, 73–74, 170–180, 181t–185t
Ferric chloride, 63, 64t
Ferric salt. *See* Iron salt
Ferric sulfate, 63, 64t
Ferric-based coagulant, 60–61, 63, 64t
Ferrihydrite, 288
Field-flow fractionation (FFF), 23
detection features of, 43t
flow, 31
Filter
cartridge, 280
dual-media, 280–281
packed bed, 280
sand, 280–281
Filtration, 8–9. *See also* Membrane filtration; Microfiltration; Nanofiltration; Ultrafiltration
activated carbon, 9
biofiltration, 291
direct, 282
membrane, 9, 25–26
membrane filtration with, 279–282
FIX. *See* Fluidized ion exchange
Floc
formation of, 56–57
micro-, 285
size of, 60–61
Flocculant, 71. *See also* Inorganic polymer flocculant
Flocculation, 8–9, 56–57
bridging, 57–58
electro-, 82, 95–98
sweep, 57–58
Flood, 3
Flotation
EF, 82, 95
sedimentation and, 8–9
Flow FFF, 31
Fluid phase, 214–215
Fluidized ion exchange (FIX), 248
Fluorescence spectroscopy, 8, 33–35
detection features of, 36t–37t
water sample peaks with, 34t
Fluorophores, 36t–37t
Flux
decline of, 278
level of, 284–285
permeate, 130–131
Fouling. *See* Membrane fouling
Fourier transform infrared (FTIR), 27, 35–38
detection features of, 36t–37t
on humic substances, 35t
Fourier transform ion cyclotron resonance mass spectrometry (FTICR-MS), 41–42, 43t
Fractionation, 23–26. *See also* Field-flow fractionation
resin, 23–24
with SEC, 26
Free radical reaction, 160–161

Freeze drying, 23
Freshwater polishing, 116
Freundlich affinity, 226–227
Freundlich isotherm, 214–215
FTICR-MS. *See* Fourier transform ion cyclotron resonance mass spectrometry
FTIR. *See* Fourier transform infrared
Fulvic acids (FA), 4
  catalytic oxidation and, 190t–199t
  elemental composition of, 5t
  functional group distribution of, 5t
  solubility of, 20
  synthetic solution with, 190t–199t
Fulvic-like fluorophores, 36t–37t
Furfural, 31t
Furfuryl alcohol, 31t

G

GAC. *See* Granular activated carbon
Gas chromatography (GC), 30–31, 43t
Gel/cake layer formation, 145, 145f
Giardia, 144
Glass, 188
Granular activated carbon (GAC), 218–219
  with aluminum sulfate, 220
  microbiological communities in, 287
Green technology, 82
Groundwater
  EC based NOM removal in, 96t–97t
  IE and, 250t–269t
  MF and UF and, 118t–127t
  ozone based AOPs and, 164t–167t
  UV light based AOPs and, 171t–179t

H

HA. *See* Humic acids
Haloacetic acid (HAA), 6–7
  NF for, 143
  ozone based AOPs and, 164t–167t
HDTMA. *See* Hexadecyl trimethyl ammonium
Health
  DBPs and, 6–7
  NOM and, 149–150
Heavy metal ions, 4
Helium treatment, 220–221
Heterogenous photocatalysis, 180–189
Hexadecyl trimethyl ammonium (HDTMA), 223–224
HFCVD. *See* Hot filament chemical vapor deposition
High molar mass (HMM)
  NOM, 6–7
  polymers, 56–58
High performance size exclusion chromatography (HPSEC), 20
  detection features of, 43t
  detectors for, 27–28
  eluents and columns for, 26–27
  MW determination by, 28
  in water treatment monitoring, 29
HMM. *See* High molar mass
Hot filament chemical vapor deposition (HFCVD), 90–91
HPAC. *See* Composite polyaluminum chloride
HPSEC. *See* High performance size exclusion chromatography
Humic acids (HA), 4. *See also* Synthetic HA solution
  catalytic oxidation and, 190t–199t
  elemental composition of, 5t
  Fenton based AOPs and, 181t–185t
  functional group distribution of, 5t
  MF and UF and, 118t–127t
  molecular structure of, 5f
  organic compound class of, 25t
  solubility of, 20
  US irradiation and, 201t
  UV light based AOPs and, 171t–179t
Humic substances, 2–3
  adsorption and, 227–228
  classification of, 20
  composition of, 4
  EC and, 96t–97t
  FTIR on, 35t
  MF and, 118t–127t
  nonpolar contaminant interaction with, 243
  ozone and, 289–290
  UF and, 118t–127t
Humic-like fluorophores, 36t–37t
Humic-rich lake water, 190t–199t
Humin, 20
Hydrodynamic conditions

membranes and, 130–131, 142–143
NF and, 142–143
Hydrogen peroxide, 88–89, 163–168, 188
Hydrologic cycle, 2–3
Hydrophilic acid, 25t
Hydrophilic base, 25t
Hydrophilic neutral, 25t
Hydrophilic NOM, 20, 243
Hydrophobic acid, 4, 25t
Hydrophobic base, 25t
Hydrophobic interactions, 117
Hydrophobic neutral, 25t
Hydrophobic NOM, 4, 20, 243
Hydrophobicity, 20
Hydrous alumina, 217
Hydroxide
adsorbents based on, 223–225
layered double, 231t–232t
metal, 58f, 85
Hydroxypropanone, 31t

I

IE. *See* Ion exchange
IHSS. *See* International Humic Substances Society
Indole, 31t
Industrial chemicals, 277–278
Infrared. *See* Fourier transform infrared
In-line coagulation, 282–283
Inorganic micropollutants, 162
Inorganic polymer flocculant (IPF), 61
advantages and disadvantages of, 67t–68t
coagulation with, 63–66
efficiency of, 63–66
Inorganic salt, 70t
Integrated methods, 275, 294
International Humic Substances Society (IHSS), 23–24
IOCS. *See* Iron oxide-coated sand
Ion
calcium, 130
counter-, 241
divalent, 147–148
heavy metal, 4
trivalent aluminum, 62t
Ion exchange (IE). *See also* Magnetic ion exchange resin
advantages and disadvantages of, 9t–11t
capacity, 241
FIX, 248
membrane filtration with, 288–289
NOM removal with, 9t–11t, 240–270, 250t–269t
process of, 240–242, 240f
resin, 240, 242, 250t–269t
in various water types, 250t–269t
Ion exchanger, 240
amphoteric, 240
natural, 241
requirements of, 241
Ionic strength, 130, 147–148
IPF. *See* Inorganic polymer flocculant
Iron
dissolving of, 83
electrode, 83–84, 96t–97t
Iron oxide
in Fenton based AOPs, 181t–185t
on natural pumice, 188
particles, 231t–232t
Iron oxide-coated sand (IOCS), 223–224, 231t–232t
Iron salt, 56–57, 63
Iron-impregnated activated carbon, 231t–232t
Irradiation
E-beam, 189–200
solar, 180
US, 189–200, 201t
Isolation, 23
Isotherm, 214–215, 218–219

K

Keratin, 241

L

Lake water
catalytic oxidation and, 190t–199t
EC based NOM removal in, 96t–97t
humic-rich, 190t–199t
IE and, 250t–269t
Landfill leachate, 102t
Langmuir isotherm, 214–215
Layered double hydroxide, 231t–232t
LC-MS. *See* Liquid chromatography-mass spectrometry
Lead oxygen electrode, 91t

Levoglucosenone, 31t
Lignins, 31t
Liquid chromatography, 26–30. *See also* Reversed-phase high-performance liquid chromatography
Liquid chromatography-mass spectrometry (LC-MS), 29–30
Low molar mass (LMM), 6–7

M

Magnetic ion exchange resin (MIEX®), 9
  characteristics of, 244
  coagulation with, 245–246, 250t–269t, 275–276
  combined treatments with, 248–249, 250t–269t
  efficiency of, 244–245, 247
  membrane filtration with, 288–289
  process of, 244, 245f
  regeneration of, 275–276
Magnetic permanently confined micelle arrays, 231t–232t
MBR. *See* Membrane bioreactor
Membrane, 114–116
  bacteria in, 149
  capabilities of, 150–151
  concentrate disposal, 144
  hydrodynamic conditions and, 130–131, 142–143
  low-pressure, 282
  for MF, 116–117
  pressure system, 114f
  process of, 114
  rejection by, 115–116
  size exclusion by, 114–115, 115f
  solution composition and, 129–130, 140–142
  vacuum system, 114f
Membrane bioreactor (MBR)
  external, 292, 292f
  membrane filtration with, 291–294
  membrane fouling in, 95
  removal rates of, 293
  submerged, 292, 292f
Membrane filtration, 9, 25–26, 279–294
  adsorption with, 286–288
  advantages and disadvantages of, 9t–11t
  coagulation with, 278–279, 282–286
  EC with, 95–98
  filtration with, 279–282
  IE with, 288–289
  MBR with, 291–294
  MIEX® with, 288–289
  NOM removal with, 9t–11t
  oxidation with, 289–291
  pH and, 129
Membrane fouling, 145–149, 278
  adsorption and, 145, 145f
  with biofouling, 149
  divalent ion concentration and, 147–148
  fouling material, 146
  ionic strength and, 147–148
  in MBR, 95
  mechanisms of, 145f
  microbial, 149
  with NOM, 146
  pH and, 147–148
Metal hydrolysis species, 58f
Metal hydroxide, 58f, 85
Metal oxide base adsorbents, 223–225
Methanol, 181t–185t
Methoxyl groups, 40t
Methoxyphenol, 31t
Methyl and methylene carbons, 39t
N-Methylacetamide, 31t
Methylcyclopentenone, 31t
Methylfuran, 31t
Methylfurfural, 31t
Methylindol, 31t
Methylphenol, 31t
Methylpyridine, 31t
Methylpyrrole, 31t
MF. *See* Microfiltration
Micelle arrays, 231t–232t
Microbial fouling, 149
Microbiological communities, 287
Microcystins, 149
Microfiltration (MF), 116–117
  advantages of, 116
  coagulation with, 283f
  electroflocculation with, 95–98
  humic substances and, 118t–127t
  membranes for, 116–117
  NOM removal with, 118t–127t

plant capacity for, 114
size exclusion by, 115f
of various water types, 118t–127t
Microfloc, 285
Microorganisms, 162
Micropollutants, 162, 277–278
Microporous styrene resin, 244
Microspheres, 231t–232t
MIEX®. *See* Magnetic ion exchange resin
Mineralization, 82–83
Minerals, 144
MM. *See* Molar mass
Model compounds solution, 171t–179t
Molar mass (MM), 20, 114–115. *See also* High molar mass; Low molar mass
Molecular weight (MW)
HPSEC determination of, 28
oxidation and, 161–162
Moorland water, 118t–127t
Multiwalled CNT, 222–223
MW. *See* Molecular weight

N

Nanoadsorbents, 221–223
Nanofiltration (NF), 98, 132–143
charge repulsion by, 133
DBP removal with, 143
HAA removal with, 143
hydrodynamic conditions and, 142–143
NOM removal with, 132f, 134t–139t
pH and, 140–141
plant, 280, 280f
pretreatments for, 281f
size exclusion by, 115f, 133
solution composition and, 140–142
THM removal with, 143
Nanomaterials, 4
Nanoparticles, 279
bioaccumulation of, 4
carbon black, 231t–232t, 288
Natural lake water. *See* Lake water
Natural organic matter (NOM), 2, 18, 21, 128–129
in aquatic environments, 2–3
aromaticity of, 187–188
in carbon cycle, 3–4
catalytic oxidation and, 190t–199t
characterization methods for, 8, 41–42
climate change and, 3–4
composition of, 3
compounds, 2–3
DBPs and, 131–132
in drinking water, 2–3
elemental composition of, 41
fractions of, 25t
health risks from, 149–150
HMM, 6–7
hydrophilic, 20, 243
hydrophobic, 4, 20, 243
LMM, 6–7
as membrane foulant, 146
MM of, 114–115
removal of, 8, 19
adsorption for, 57–58, 214, 218–233, 231t–232t
AOP for, 9t–11t, 160–163, 171t–179t, 181t–185t
biological methods for, 9t–11t
coagulation for, 9t–11t, 56–61, 58f, 73–74
EC for, 92–99, 96t–97t
electrochemical methods for, 9t–11t, 82–83
EO for, 100–101, 102t
IE for, 9t–11t, 240–270, 250t–269t
membrane filtration for, 9t–11t
MF for, 118t–127t
NF for, 132f, 134t–139t
RO for, 134t–139t
UF for, 118t–127t, 132f
research on, 6f
seasonal variability of, 200
SUVA of, 3–4
US irradiation and, 201t
water treatment and, 6
Natural organic polyelectrolytes, 227
Natural pumice, 188, 226–227
Natural river water. *See* River water
Natural water
catalytic oxidation and, 190t–199t
EC based NOM removal in, 96t–97t
MF and UF and, 118t–127t
ozone based AOPs and, 164t–167t
US irradiation and, 201t

NF. *See* Nanofiltration
Nitrite, 170
NMR. *See* Nuclear magnetic resonance
NOM. *See* Natural organic matter
Novel coagulants, 71–73
Nuclear magnetic resonance (NMR), 8, 38–40
  ^{13}C, 40, 40t
  detection features of, 36t–37t
  ^{1}H, 39, 39t

O

Odor compounds, 171t–179t
•OH radical
  Fenton reagent and, 170–180
  as oxidant, 160–161
  ozone and, 162
  UV light based applications and, 163–170
Organic polyelectrolytes
  advantages and disadvantages of, 70t
  coagulation with, 66–69
  natural, 227
Organic probes, 181t–185t
Oxidant, 87–88
  mixed, 224–225
  •OH radical as, 160–161
Oxidation, 276–277. *See also* Advanced oxidation process; Electrooxidation.
  biofiltration with, 291
  catalytic, 180–189
  catalytic wet peroxide, 189
  membrane filtration with, 289–291
  MW shift in, 161–162
  preoxidation, 276–277, 290–291
  reaction, 87
Oxidative coupling, 202
Oxygen
  COD and, 20
  in lead oxygen electrode, 91t
  reactive oxygen species of, 88–89
Ozonation, 163, 276–277
Ozone, 6–7, 88–89
  AOP based on, 162–163, 164t–167t
  disinfection with, 162
  dosage of, 290
  humic substances and, 289–290
  •OH radical and, 162
  UV with, 163

P

PAC. *See* Powdered activated carbon
Packed bed filter, 280
PACl. *See* Polyaluminum chloride
PANF. *See* Polyacrylonitrile fiber
Paper mill wastewater, 96t–97t
Parachlorobenzoic acid
  ozone based AOPs and, 164t–167t
  UV light based AOPs and, 171t–179t
Paraffinic carbon chains, 40t
PAS. *See* Polyaluminum sulfate
PASiC. *See* Polyaluminum silicate chloride
Permanganate index, 96t–97t
Permeate flux, 130–131
Peroxyl radicals, 160–161
Persulfate, 189–200
Pesticides, 277–278
PFC. *See* Polyferric chloride
PFS. *See* Polyferric sulfate
pH
  coagulant solubility and, 56–57
  coagulation, 61–63
  EC and, 94
  EO and, 88
  membrane filtration and, 129
  membrane fouling and, 147–148
  NF and, 140–141
Pharmaceuticals, 162
Phase interface, 214–215
Phenol, 31t
Phenolic groups, 40t
Phenylacetonitrile, 31t
Photocatalysis, 169, 180–189, 200–202
Photoelectrocatalysis, 202
Photoelectrocatalytic degradation, 100
Photoelectro-Fenton process, 170–180
Photolysis, 163–169
PICl. *See* Polymeric iron chloride
Pillared bentonite, 229–233, 231t–232t
Platinum electrode, 87, 91t
Polyacrylamides, 70t
Polyacrylonitrile fiber (PANF), 228

Polyaluminum chloride (PACl), 63–65, 67t–68t
Polyaluminum silicate chloride (PASiC), 65, 67t–68t
Polyaluminum sulfate (PAS), 63–65
Polyaniline/attapulgite composite (ATP-PANI), 228–229, 231t–232t
Polydiallyldimethyl ammonium chloride, 70t
Polyelectrolytes. *See* Organic polyelectrolytes
Polyferric chloride (PFC), 66, 67t–68t
Polyferric sulfate (PFS), 66, 67t–68t
Polyhydroxybutyrates, 31t
Polymer, 56–58
Polymer flocculant. *See* Inorganic polymer flocculant
Polymeric iron chloride (PICl), 66, 67t–68t
Polyphenolic compounds, 31t
Polysaccharides
  common fragments of, 31t
  as foulant, 147–148
Polysilic acid, 66
Polystyrene resin, 246–247
Pond water, 102t
Pore blocking and constriction, 145, 145f
Potable water, 6
Powdered activated carbon (PAC), 218–219
  with aluminum sulfate, 220
  characteristics of, 231t–232t
  IE resin and, 250t–269t
Prechlorination, 276–277
Precipitative softening, 144
Preoxidation, 276–277, 290–291
Preozonation, 276–277
Pressure system, 114f
Probe
  compound, 164t–167t
  organic, 181t–185t
Process integration, 275
Propionamide, 31t
Proteins, 31t
Pumice, 188, 226–227
Py-GC-MS. *See* Pyrolysis gas chromatography-mass spectrometry
Pyridine, 31t
Pyrolysis by-products, 31t
Pyrolysis gas chromatography-mass spectrometry (Py-GC-MS), 30
Pyrrole, 31t

Q

Quaternary ammonium resin, 243

R

Raw water, 19
  catalytic oxidation and, 190t–199t
  EC based NOM removal in, 96t–97t
  Fenton based AOPs and, 181t–185t
  IE and, 250t–269t
  ozone based AOPs and, 164t–167t
  salt concentration of, 141
  UV light based AOPs and, 171t–179t
Reactive oxygen species, 88–89
Reactor
  in EC, 86
  for EO, 89–92
Renewable energy, 188
Research, 6f
Reservoir water, 118t–127t
Residual organic matter (ROM), 223
Resin. *See also* Magnetic ion exchange resin
  absorption, 8
  anion, 242–243
  cation, 242
  chelating, 242
  fractionation, 23–24
  IE, 240, 242, 250t–269t
  microporous styrene, 244
  polystyrene, 246–247
  quaternary ammonium, 243
  XAD, 23–24
Response surface method (RSM), 65–66
Reverse osmosis (RO), 23, 143–144
  concentrate, 102t
  NOM removal with, 134t–139t
  pretreatment with, 281–282
  size exclusion by, 115f
Reversed-phase high-performance liquid chromatography (RPHPLC), 23, 29
Rice husk ash (RHA), 229–233, 231t–232t

River water
- catalytic oxidation and, 190t–199t
- Fenton based AOPs and, 181t–185t
- IE and, 250t–269t
- ozone based AOPs and, 164t–167t

RO. *See* Reverse osmosis
ROM. *See* Residual organic matter
RPHPLC. *See* Reversed-phase high-performance liquid chromatography
RSM. *See* Response surface method

S

Salt
- aluminum, 56–57, 62t
- concentration of, 141
- inorganic, 70t
- iron, 56–57, 63
- in raw water, 141

SAM. *See* Self-assembled monolayer
Sand
- filter, 280–281
- IOCS, 223–224, 231t–232t

Scale-forming minerals, 144
Seawater, 143–144
SEC. *See* Size exclusion chromatography
Sedimentation/flotation, 8–9
Self-assembled monolayer (SAM), 224
Semiconductors, 180–186
Silica
- activated, 65
- gel, 217
- particles, 231t–232t

Simulated groundwater. *See* Groundwater
Simulated natural water. *See* Natural water
Size exclusion, 114–115, 115f, 133
Size exclusion chromatography (SEC), 23, 26. *See also* High performance size exclusion chromatography
Sludge, 73–74, 275–276
Slurry, 187
Softening, 144
Solar light irradiation, 180
Solar photo-Fenton, 180
Solubility, 20, 56–57
Solution. *See also* Synthetic HA solution
- model compounds, 171t–179t
- NF and, 129–130, 140–142
- synthetic FA, 190t–199t

Sonochemical decomposition, 189
Sorption, 214–215
Source water, 181t–185t
Specific UV absorbance (SUVA), 20–21
- coagulation and, 60t
- detection features of, 22t
- DOC and, 60t
- of NOM, 3–4

Spectroscopy, 8, 32–40. *See also* Fluorescence spectroscopy
- differential absorption, 32–33
- ultraviolet and visible, 32, 36t–37t

Steel electrode, 96t–97t
Styrene, 31t
Styrene resin, 244, 246–247
Sunlight
- in Fenton based AOPs, 181t–185t
- as renewable energy, 188

Support matrix, 187
Surface water, 42–44
- catalytic oxidation and, 190t–199t
- EC based NOM removal in, 96t–97t
- Fenton based AOPs and, 181t–185t
- IE and, 250t–269t
- MF and UF and, 118t–127t
- US irradiation and, 201t
- UV light based AOPs and, 171t–179t

Surfactant-modified bentonite, 231t–232t
Suspended matter, 224
SUVA. *See* Specific UV absorbance
Suwannee River, 190t–199t
Swamp water, 250t–269t
Sweep flocculation, 57–58
Synthetic FA solution, 190t–199t
Synthetic HA solution
- catalytic oxidation and, 190t–199t
- Fenton based AOPs and, 181t–185t
- ozone based AOPs and, 164t–167t
- US irradiation and, 201t
- UV light based AOPs and, 171t–179t

Synthetic water, 96t–97t, 102t

T

Tannins, 31t
Taste compounds, 171t–179t
TCAA. *See* Trichloroacetic acid
Temperature, 85
THM. *See* Trihalomethane

Titanium dioxide, 186–188, 202
Titanium electrode, 91t
Tobermorite, 231t–232t
TOC. *See* Total organic carbon
Toluene, 31t
Total organic carbon (TOC), 21
  detection features of, 22t
  EC and, 96t–97t
  of water, 4, 20
Trichloroacetic acid (TCAA), 164t–167t
Trihalomethane (THM), 6–7
  European regulation limit of, 131
  formation of, 131, 224–225
  NF of, 143
  ozone based AOPs and, 164t–167t
Trivalent aluminum ions, 62t
Tryptophan-like fluorophores, 36t–37t

U

UF. *See* Ultrafiltration
Ultracentrifugation, 23
Ultrafiltration (UF), 23, 117–132
  coagulation with, 283f
  DBP removal with, 143
  humic substances and, 118t–127t
  NOM removal with, 118t–127t, 132f
  overall plant capacity for, 114
  size exclusion by, 115f
  of various water types, 118t–127t
Ultrapure water, 171t–179t
Ultrasound irradiation (US irradiation), 189–200, 201t
Ultraviolet (UV). *See also* Specific UV absorbance
  AOP based on, 163–170, 171t–179t
  lamps, 202
  •OH radical and, 163–170
  ozone with, 163
  photolysis, 163–169
  vacuum, 169–170
  various water types and substances and, 171t–179t
Ultraviolet and visible spectroscopy, 32, 36t–37t
Unsaturated carbons, 39t
US irradiation. *See* Ultrasound irradiation
UV. *See* Ultraviolet

V

Vacuum distillation, 23
Vacuum system, 114f
Vacuum UV (VUV), 169–170

W

Wastewater
  IE and, 250t–269t
  paper mill, 96t–97t
Water. *See also* Groundwater; Lake water; Natural water; Raw water; River water; Surface water
  clarified, 250t–269t
  drinking, 2–3
  fluorescence peaks for, 34t
  fresh, 116
  moorland, 118t–127t
  pond, 102t
  potable, 6
  purification of, 44
  quality parameters of, 20
  reservoir, 118t–127t
  sea, 143–144
  source, 181t–185t
  swamp, 250t–269t
  synthetic, 96t–97t, 102t
  TOC of, 4, 20
  ultrapure, 171t–179t
Water treatment
  DBPs in, 6–7
  HPSEC in, 29
  NOM and, 6
  optimization of, 7

X

XAD resin, 23–24

Z

Zeolites
  as adsorbent, 217
  characteristics of, 231t–232t
  in CSZ, 228–229
  as natural ion exchanger, 241
Zeta potential window, 59
Zinc electrodes, 96t–97t

www.ingramcontent.com/pod-product-compliance
Ingram Content Group UK Ltd.
Pitfield, Milton Keynes, MK11 3LW, UK
UKHW021238300926
15471UKWH00013B/339

9 780128 015032